Ronald Christensen

Linear Models for Multivariate, Time Series, and Spatial Data

With 40 Illustrations

Springer

Ronald Christensen
Department of Mathematics and Statistics
University of New Mexico
Albuquerque, NM 87131
USA

Mathematics Subject Classification: 62H17

Library of Congress Cataloging-in-Publication Data
Christensen, Ronald, 1951-
    Linear models for multivariate, time series, and spatial data/
Ronald Christensen.
        p. cm.
    Includes bibliographical references and index.
    ISBN 0-387-97413-X
        1. Linear models. (Statistics) I. Title.
QA279.C477 1990   90-10377
519.5–dc20   CIP

Printed on acid-free paper.

Photocomposed copy prepared from the author's LaTeX file.
Printed and bound by Edwards Brothers Inc., Ann Arbor, Michigan
Printed in the United States of America.

9 8 7 6 5 4 3 2 (Corrected second printing, 1997)

ISBN 0-387-97413-X Springer-Verlag New York Berlin Heidelberg
ISBN 3-540-97413-X Springer-Verlag Berlin Heidelberg New York SPIN 10629034

Springer Texts in Statistics

Advisors:
George Casella   Stephen Fienberg   Ingram Olkin

Springer
New York
Berlin
Heidelberg
Barcelona
Budapest
Hong Kong
London
Milan
Paris
Santa Clara
Singapore
Tokyo

Springer Texts in Statistics

Alfred: Elements of Statistics for the Life and Social Sciences
Berger: An Introduction to Probability and Stochastic Processes
Blom: Probability and Statistics: Theory and Applications
Brockwell and Davis: An Introduction to Times Series and Forecasting
Chow and Teicher: Probability Theory: Independence, Interchangeability, Martingales, Third Edition
Christensen: Plane Answers to Complex Questions: The Theory of Linear Models, Second Edition
Christensen: Linear Models for Multivariate, Time Series, and Spatial Data
Christensen: Log-Linear Models
Creighton: A First Course in Probability Models and Statistical Inference
du Toit, Steyn and Stumpf: Graphical Exploratory Data Analysis
Edwards: Introduction to Graphical Modelling
Finkelstein and Levin: Statistics for Lawyers
Flury: A First Course in Multivariate Statistics
Jobson: Applied Multivariate Data Analysis, Volume I: Regression and Experimental Design
Jobson: Applied Multivariate Data Analysis, Volume II: Categorical and Multivariate Methods
Kalbfleisch: Probability and Statistical Inference, Volume I: Probability, Second Edition
Kalbfleisch: Probability and Statistical Inference, Volume II: Statistical Inference, Second Edition
Karr: Probability
Keyfitz: Applied Mathematical Demography, Second Edition
Kiefer: Introduction to Statistical Inference
Kokoska and Nevison: Statistical Tables and Formulae
Lehmann: Testing Statistical Hypotheses, Second Edition
Lindman: Analysis of Variance in Experimental Design
Lindsey: Applying Generalized Linear Models
Madansky: Prescriptions for Working Statisticians
McPherson: Statistics in Scientific Investigation: Its Basis, Application, and Interpretation
Mueller: Basic Principles of Structural Equation Modeling
Nguyen and Rogers: Fundamentals of Mathematical Statistics: Volume I: Probability for Statistics
Nguyen and Rogers: Fundamentals of Mathematical Statistics: Volume II: Statistical Inference
Noether: Introduction to Statistics: The Nonparametric Way
Peters: Counting for Something: Statistical Principles and Personalities
Pfeiffer: Probability for Applications
Pitman: Probability

Continued at end of book

To Wes, Russ and Ed

Preface

This is a companion volume to *Plane Answers to Complex Questions: The Theory of Linear Models*. It consists of six additional chapters written in the same spirit as the last six chapters of the earlier book. Brief introductions are given to topics related to linear model theory. No attempt is made to give a comprehensive treatment of the topics. Such an effort would be futile. Each chapter is on a topic so broad that an in depth discussion would require a book-length treatment.

People need to impose structure on the world in order to understand it. There is a limit to the number of unrelated facts that anyone can remember. If ideas can be put within a broad, sophisticatedly simple structure, not only are they easier to remember but often new insights become available. In fact, sophisticatedly simple models of the world may be the only ones that work. I have often heard Arnold Zellner say that, to the best of his knowledge, this is true in econometrics. The process of modeling is fundamental to understanding the world.

In statistics, the most widely used models revolve around linear structures. Often the linear structure is exploited in ways that are peculiar to the subject matter. Certainly this is true of frequency domain time series and geostatistics. The purpose of this volume is to take three fundamental ideas from standard linear model theory and exploit their properties in examining multivariate, time series, and spatial data. In decreasing order of importance to the presentation, the three ideas are: best linear prediction, projections, and Mahalanobis's distance. (Actually, Mahalanobis's distance is a fundamentally multivariate idea that has been appropriated for use in linear models.) Numerous references to results in *Plane Answers* are made. Nevertheless, I have tried to make this book as independent as possible. Typically, when a result from *Plane Answers* is needed not only is the reference given but also the result itself. Of course, for proofs of these results the reader will have to refer to the original source.

I want to reemphasize that this is a book about linear models. It is not traditional multivariate analysis, time series, or geostatistics. Multivariate linear models are viewed as linear models with a nondiagonal covariance matrix. Discriminant analysis is related to the Mahalanobis distance and multivariate analysis of variance. Principal components are best linear predictors. Frequency domain time series involves linear models with a peculiar design matrix. Time domain analysis involves models that are linear in the parameters but have random design matrices. Best linear predictors are used for forecasting time series; they are also fundamental to the estimation techniques used in time domain analysis. Spatial data analysis involves linear models in which the covariance matrix is modeled from the data; a primary objective in analyzing spatial data is making best linear unbiased predictions of future observables. While other approaches to these problems may yield different insights, there is value in having a unified approach to looking at these problems. Developing such a unified approach is the purpose of this book.

There are two well-known models with linear structure that are conspicuous by their absence in my two volumes on linear models. One is Cox's (1972) proportional hazards model. The other is the generalized linear model of Nelder and Wedderburn (1972). The proportional hazards methodology is a fundamentally nonparametric technique for dealing with censored data having linear structure. The emphasis on nonparametrics and censored data would make its inclusion here awkward. The interested reader can see Kalbfleisch and Prentice (1980). Generalized linear models allow the extension of linear model ideas to many situations that involve independent nonnormally distributed observations. Beyond the presentation of basic linear model theory, these volumes focus on methods for analyzing correlated observations. While it is true that generalized linear models can be used for some types of correlated data, such applications do not flow from the essential theory. McCullagh and Nelder (1989) give a detailed exposition of generalized linear models and Christensen (1990) contains a short introduction.

ACKNOWLEDGMENTS

I would like to thank MINITAB for providing me with a copy of release 6.1.1, BMDP for providing me with copies of their programs 4M, 1T, 2T, and 4V, and Dick Lund for providing me with a copy of MSUSTAT. Nearly all of the computations were performed with one of these programs. Many were performed with more than one.

I would not have tackled this project but for Larry Blackwood and Bob Shumway. Together Larry and I reconfirmed, in my mind anyway, that multivariate analysis is just the same old stuff. Bob's book put an end to a specter that has long haunted me: a career full of half-hearted attempts at figuring out basic time series analysis.

At my request, Ed Bedrick, Bert Koopmans, Wes Johnson, Bob Shumway, and Dale Zimmerman tried to turn me from the errors of my ways. I sincerely thank them for their valuable efforts. The reader must judge how successful they were with a recalcitrant subject. As always, I must thank my editors Steve Fienberg and Ingram Olkin for their suggestions. Jackie Damrau did an exceptional job in typing the first draft of the manuscript.

Finally, I have to recognize the contribution of Magic Johnson. I was so upset when the 1987-88 Lakers won a second consecutive NBA title that I began writing this book in order to block the mental anguish. I am reminded of Woody Allen's dilemma: is the importance of life more accurately reflected in watching *The Sorrow and the Pity* or in watching the Knicks? (In my case, the Jazz and the Celtics.) It's a tough call. Perhaps life is about actually making movies and doing statistics.

Ronald Christensen
Albuquerque, New Mexico
April 19, 1990

BMDP Statistical Software is located at 1440 Sepulveda Boulevard, Los Angeles, CA 90025, telephone: (213) 479-7799.

MINITAB is a registered trademark of Minitab, Inc., 3081 Enterprise Drive, State College, PA 16801, telephone: (814) 238-3280, telex: 881612.

MSUSTAT is marketed by the Research and Development Institute Inc., Montana State University, Bozeman, MT 59717-0002, Attn: R.E. Lund.

Contents

Chapter I

Multivariate Linear Models

Chapters I, II, and III examine topics in multivariate analysis. Specifically, they discuss multivariate linear models, discriminant analysis, principal components, and factor analysis. The basic ideas behind these subjects are closely related to linear model theory. Multivariate linear models are simply linear models with more than one dependent variable. Discriminant analysis is closely related to both Mahalanobis's distance (cf. Christensen, 1987, Section XIII.1) and multivariate one-way analysis of variance. Principal components are user-constructed variables which are best linear predictors (cf. Christensen, 1987, Section VI.3) of the original data. Factor analysis has ties to both multivariate linear models and principal components.

These three chapters are introductory in nature. The discussions benefit from the advantage of being based on linear model theory. They suffer from the disadvantage of being relatively brief. More detailed discussions are available in numerous other sources, e.g., Anderson (1984), Arnold (1981), Dillon and Goldstein (1984), Eaton (1983), Gnanadesikan (1977), Johnson and Wichern (1988), Mardia, Kent, and Bibby (1979), Morrison (1976), Muirhead (1982), Press (1982), and Seber (1984).

As mentioned above, the distinction between multivariate linear models and standard (univariate) linear models is simply that multivariate linear models involve more than one dependent variable. Let the dependent variables be $y_1, \ldots, y_q$. If n observations are taken on each dependent variable, we have $y_{i1}, \ldots, y_{iq}$, $i = 1, \ldots, n$. Let $Y_1 = [y_{11}, \ldots, y_{n1}]'$ and, in general, $Y_h = [y_{1h}, \ldots, y_{nh}]'$, $h = 1, \ldots, q$. For each h, the vector Y_h is the vector of n responses on the variable y_h and can be used as the response vector for a linear model. For $h = 1, \ldots, q$, write the linear model

$$Y_h = X\beta_h + e_h \ , \ \mathrm{E}(e_h) = 0 \ , \ \mathrm{Cov}(e_h) = \sigma_{hh}I \tag{1}$$

where X is a known $n \times p$ matrix that is the same for all dependent variables, but β_h and the error vector $e_h = [e_{1h}, \ldots, e_{nh}]'$ are peculiar to the dependent variable.

The multivariate linear model consists of fitting the q linear models simultaneously. Write the matrices

$$\begin{aligned} Y_{n \times q} &= [Y_1, \ldots, Y_q] \ , \\ B_{p \times q} &= [\beta_1, \ldots, \beta_q] \end{aligned}$$

and

$$e_{n \times q} = [e_1, \ldots, e_q] \ .$$

The multivariate linear model is

$$Y = XB + e. \tag{2}$$

The key to the analysis of the multivariate linear model is the random nature of the $n \times q$ error matrix $e = [e_{ih}]$. At a minimum, we assume that $E(e) = 0$ and

$$\text{Cov}(e_{ih}, e_{i'h'}) = \begin{cases} \sigma_{hh'} & \text{if } i = i' \\ 0 & \text{if } i \neq i' \end{cases}.$$

Let

$$\delta_{ii'} = \begin{cases} 1 & \text{if } i = i' \\ 0 & \text{if } i \neq i', \end{cases}$$

then the covariances can be written simply as

$$\text{Cov}(e_{ih}, e_{i'h'}) = \sigma_{hh'}\delta_{ii'}.$$

To construct tests and confidence regions we assume that the e_{ij}'s have a multivariate normal distribution with the previously indicated mean and covariances. Note that this covariance structure implies that the error vector in model (1) has $\text{Cov}(e_h) = \sigma_{hh}I$ as indicated previously.

An alternative but equivalent way to state the multivariate linear model is by examining the rows of model (2). Write

$$Y = \begin{bmatrix} y_1' \\ \vdots \\ y_n' \end{bmatrix},$$

$$X = \begin{bmatrix} x_1' \\ \vdots \\ x_n' \end{bmatrix},$$

and

$$e = \begin{bmatrix} \varepsilon_1' \\ \vdots \\ \varepsilon_n' \end{bmatrix}.$$

The multivariate linear model is

$$y_i' = x_i'B + \varepsilon_i'$$

$i = 1, \ldots, n$. The error vector ε_i has the properties

$$E(\varepsilon_i) = 0,$$

$$\text{Cov}(\varepsilon_i) = \Sigma_{q \times q} = [\sigma_{hh'}],$$

and for $i \neq j$

$$\mathrm{Cov}(\varepsilon_i, \varepsilon_j) = 0.$$

To construct tests and confidence regions, the vectors ε_i are assumed to have independent multivariate normal distributions.

EXERCISE 1.1.   For any two columns of Y say Y_r and Y_s, show that $\mathrm{Cov}(Y_r, Y_s) = \sigma_{rs} I$.

I.1   Estimation

The key to estimation in the multivariate linear model is rewriting the model as a univariate linear model. The model

$$Y = XB + e \ , \ \mathrm{E}(e) = 0 \ , \ \mathrm{Cov}(e_{ih}, e_{i'h'}) = \sigma_{hh'}\delta_{ii'} \tag{1}$$

can be rewritten as

$$\begin{bmatrix} Y_1 \\ Y_2 \\ \vdots \\ Y_q \end{bmatrix} = \begin{bmatrix} X & 0 & \cdots & 0 \\ 0 & X & & \vdots \\ \vdots & & \ddots & 0 \\ 0 & 0 & & X \end{bmatrix} \begin{bmatrix} \beta_1 \\ \beta_2 \\ \vdots \\ \beta_q \end{bmatrix} + \begin{bmatrix} e_1 \\ e_2 \\ \vdots \\ e_q \end{bmatrix} \tag{2}$$

where the error vector has mean zero and covariance matrix

$$\begin{bmatrix} \sigma_{11}I_n & \sigma_{12}I_n & \cdots & \sigma_{1q}I_n \\ \sigma_{12}I_n & \sigma_{22}I_n & \cdots & \sigma_{2q}I_n \\ \vdots & \vdots & \ddots & \vdots \\ \sigma_{1q}I_n & \sigma_{2q}I_n & \cdots & \sigma_{qq}I_n \end{bmatrix}. \tag{3}$$

Recalling that the Vec operator (cf. Christensen, 1987, Definition B.6) stacks the columns of a matrix, the dependent variable in model (2) is precisely $\mathrm{Vec}(Y)$. Similarly, the parameter vector and the error vector are $\mathrm{Vec}(B)$ and $\mathrm{Vec}(e)$. The design matrix in (2) can be rewritten using Kronecker products (cf. Christensen, 1987, Definition B.5). The design matrix is $I_q \otimes X$ where I_q is a $q \times q$ identity matrix. Model (2) can now be rewritten as

$$\mathrm{Vec}(Y) = [I_q \otimes X]\mathrm{Vec}(B) + \mathrm{Vec}(e) . \tag{4}$$

The first two moments of $\mathrm{Vec}(e)$ are

$$\mathrm{E}[\mathrm{Vec}(e)] = 0$$

and, rewriting (3),

$$\mathrm{Cov}[\mathrm{Vec}(e)] = \Sigma \otimes I_n . \tag{5}$$

EXERCISE 1.2.   Show that $[A \otimes B][C \otimes D] = [AC \otimes BD]$ where the matrices are of conformable sizes.

For estimation, the nice thing about model (1) is that least squares estimates are optimal. In particular, it will be shown that optimal estimation is based on

$$\hat{Y} = X\hat{B} = MY$$

where $M = X(X'X)^- X'$ is, as always, the perpendicular projection operator onto the column space of X, $C(X)$. This is a simple generalization of the univariate linear model results of Christensen (1987, Chapter II). To show that least squares estimates are best linear unbiased estimates, (BLUE's), apply Christensen's (1987) Theorem 10.4.5 to model (2). Theorem 10.4.5 states that for a univariate linear model $Y_{n \times 1} = X\beta + e$, $\mathrm{E}(e) = 0$, $\mathrm{Cov}(e) = \sigma^2 V$, least squares estimates are BLUE's if and only if $C(VX) \subset C(X)$.

The design matrix in (2) is $[I_q \otimes X]$. The covariance matrix is $[\Sigma \otimes I_n]$. We need to show that $C\left([\Sigma \otimes I_n][I_q \otimes X]\right) \subset C\left([I_q \otimes X]\right)$. Using either Exercise 1.2 or simply using the forms given in (2) and (3)

$$
\begin{aligned}
[\Sigma \otimes I_n][I_q \otimes X] &= [\Sigma \otimes X] \\
&= \begin{bmatrix} \sigma_{11}X & \cdots & \sigma_{1q}X \\ \vdots & & \vdots \\ \sigma_{1q}X & \cdots & \sigma_{qq}X \end{bmatrix} \\
&= [I_q \otimes X][\Sigma \otimes I_p] .
\end{aligned}
$$

Recalling that $C(RS) \subset C(R)$ for any conformable matrices R and S, it is clear that

$$C([\Sigma \otimes I_n][I_q \otimes X]) = C([I_q \otimes X][\Sigma \otimes I_p]) \subset C([I_q \otimes X]).$$

Applying Christensen's Theorem 10.4.5 establishes that least squares estimates are best linear unbiased estimates.

To find least squares estimates, we need the perpendicular projection operator onto $C\left([I_q \otimes X]\right)$. The projection operator is

$$P = [I_q \otimes X]\left([I_q \otimes X]'[I_q \otimes X]\right)^- [I_q \otimes X].$$

Because $[A \otimes B]' = [A' \otimes B']$, we have

$$
\begin{aligned}
[I_q \otimes X]'[I_q \otimes X] &= [I_q \otimes X'][I_q \otimes X] \\
&= [I_q \otimes X'X] .
\end{aligned}
$$

It is easily seen from the definition of a generalized inverse that

$$([I_q \otimes X'X])^- = [I_q \otimes (X'X)^-].$$

It follows that

$$P \;=\; [I_q \otimes X][I_q \otimes (X'X)^-][I_q \otimes X]'$$
$$=\; [I_q \otimes X(X'X)^- X']$$
$$=\; [I_q \otimes M].$$

By Christensen (1987, Theorem 2.2.1), in a univariate linear model $Y_{n\times1} = X\beta + e$ least squares estimates $\hat\beta$ satisfy $X\hat\beta = MY_{n\times1}$; thus for the univariate linear model (4), least squares estimates of $\mathrm{Vec}(B)$, say $\mathrm{Vec}(\hat B)$, satisfy

$$[I_q \otimes X]\mathrm{Vec}(\hat B) = [I_q \otimes M]\mathrm{Vec}(Y) \, ,$$

i.e.,

$$\begin{bmatrix} X\hat\beta_1 \\ \vdots \\ X\hat\beta_q \end{bmatrix} = \begin{bmatrix} MY_1 \\ \vdots \\ MY_q \end{bmatrix}.$$

In terms of the multivariate linear model (1), this is equivalent to

$$X\hat B = MY.$$

Maximum Likelihood Estimates

Write the matrices Y and X using their component rows,

$$Y = \begin{bmatrix} y_1' \\ \vdots \\ y_n' \end{bmatrix} \qquad \text{and} \qquad X = \begin{bmatrix} x_1' \\ \vdots \\ x_n' \end{bmatrix}.$$

To find maximum likelihood estimates (MLE's), we assume that Σ is nonsingular. We also assume that the rows of Y are independent and $y_i \sim N(B'x_i, \Sigma)$. The likelihood function for Y is

$$L(XB, \Sigma) = \prod_{i=1}^{n} (2\pi)^{-q/2} |\Sigma|^{-1/2} \exp\left[-(y_i - B'x_i)'\Sigma^{-1}(y_i - B'x_i)/2\right]$$

and the log of the likelihood function is

$$\ell(XB, \Sigma) = -\frac{nq}{2}\log(2\pi) - \frac{n}{2}\log(|\Sigma|) - \frac{1}{2}\sum_{i=1}^{n}(y_i - B'x_i)'\Sigma^{-1}(y_i - B'x_i).$$

Consider model (2). As for any other univariate linear model, if the nonsingular covariance matrix is fixed, then the MLE of $[I_q \otimes X]\mathrm{Vec}(B)$ is the same as the BLUE. As we have just seen, least squares estimates are BLUE's. The least squares estimate of XB does not depend on the covariance matrix; hence, for any value of Σ, $X\hat{B} = MY$ maximizes the likelihood function. It remains only to find the MLE of Σ.

The log-likelihood, and thus the likelihood, are maximized for any Σ by substituting a least squares estimate for B. Write $\hat{B} = (X'X)^{-}X'Y$. We need to maximize

$$\ell(X\hat{B}, \Sigma) = -\frac{nq}{2}\log(2\pi) - \frac{n}{2}\log(|\Sigma|)$$
$$-\frac{1}{2}\sum_{i=1}^{n}(y_i - Y'X(X'X)^{-}x_i)'\Sigma^{-1}(y_i - Y'X(X'X)^{-}x_i)$$

subject to the constraint that Σ is positive definite. The last term on the right-hand side can be simplified. Define the $n \times 1$ vector

$$\rho_i = (0, \ldots, 0, 1, 0, \ldots, 0)'$$

with the 1 in the ith place.

$$\sum_{i=1}^{n}(y_i - Y'X(X'X)^{-}x_i)'\Sigma^{-1}(y_i - Y'X(X'X)^{-}x_i)$$
$$= \sum_{i=1}^{n}\rho_i'(Y - X(X'X)^{-}X'Y)\Sigma^{-1}(Y' - Y'X(X'X)^{-}X')\rho_i$$
$$= \sum_{i=1}^{n}\rho_i'(I - M)Y\Sigma^{-1}Y'(I - M)\rho_i$$
$$= \operatorname{tr}[(I - M)Y\Sigma^{-1}Y'(I - M)]$$
$$= \operatorname{tr}[\Sigma^{-1}Y'(I - M)Y].$$

Thus our problem is to maximize

$$\ell(X\hat{B}, \Sigma) = -\frac{nq}{2}\log(2\pi) - \frac{n}{2}\log(|\Sigma|) - \frac{1}{2}\operatorname{tr}[\Sigma^{-1}Y'(I - M)Y]. \qquad (6)$$

We will find the maximizing value by setting all the partial derivatives (with respect to the σ_{ij}'s) equal to zero. To find the partial derivatives, we need part (3) of Proposition 12.4.1 in Christensen (1987) and a variation on part (4) of the proposition, cf. Exercise 1.8.14. The variation on part (4) is that

$$\frac{\partial}{\partial\sigma_{ij}}\log|\Sigma| = \operatorname{tr}\left[\Sigma^{-1}\frac{\partial\Sigma}{\partial\sigma_{ij}}\right] \qquad (7)$$
$$= \operatorname{tr}\left[\Sigma^{-1}T_{ij}\right].$$

where the symmetric $q \times q$ matrix T_{ij} has ones in row i column j and row j column i and zeros elsewhere. Part (3) of Christensen's Proposition 12.4.1

gives

$$\frac{\partial}{\partial \sigma_{ij}} \Sigma^{-1} = -\Sigma^{-1} \frac{\partial \Sigma}{\partial \sigma_{ij}} \Sigma^{-1} \qquad (8)$$

$$= -\Sigma^{-1} T_{ij} \Sigma^{-1}.$$

We need one final result involving the derivative of a trace. Let $\dot{A}(s) = [a_{ij}(s)]$ be an $r \times r$ matrix function of the scalar s.

$$\frac{d}{ds} \text{tr}[A(s)] = \frac{d}{ds}[a_{11}(s) + \cdots + a_{rr}(s)]$$

$$= \sum_{i=1}^{r} \frac{da_{ii}(s)}{ds} \qquad (9)$$

$$= \text{tr}\left[\frac{dA(s)}{ds}\right].$$

From (8), (9), and the chain rule

$$\frac{\partial}{\partial \sigma_{ij}} \text{tr}[\Sigma^{-1} Y'(I - M)Y] = \text{tr}\left[\frac{\partial}{\partial \sigma_{ij}}\{\Sigma^{-1} Y'(I - M)Y\}\right]$$

$$= \text{tr}\left[\left\{\frac{\partial \Sigma^{-1}}{\partial \sigma_{ij}}\right\} Y'(I - M)Y\right] \qquad (10)$$

$$= \text{tr}[-\Sigma^{-1} T_{ij} \Sigma^{-1} Y'(I - M)Y].$$

Applying (7) and (10) to (6), we get

$$\frac{\partial}{\partial \sigma_{ij}} \ell(X\hat{B}, \Sigma) = -\frac{n}{2} \text{tr}[\Sigma^{-1} T_{ij}] + \frac{1}{2} \text{tr}[\Sigma^{-1} T_{ij} \Sigma^{-1} Y'(I - M)Y].$$

Setting the partial derivatives equal to zero leads to finding a positive definite matrix Σ that solves

$$\text{tr}[\Sigma^{-1} T_{ij}] = \text{tr}[\Sigma^{-1} T_{ij} \Sigma^{-1} Y'(I - M)Y/n] \qquad (11)$$

for all i and j.

Let $\hat{\Sigma} = \frac{1}{n} Y'(I - M)Y$; this is clearly nonnegative definite (positive semidefinite). If $\hat{\Sigma}$ is positive definite, then $\hat{\Sigma}$ is our solution. Substituting $\hat{\Sigma}$ for Σ in (11) gives

$$\text{tr}[\hat{\Sigma}^{-1} T_{ij}] = \text{tr}[\hat{\Sigma}^{-1} T_{ij} \hat{\Sigma}^{-1} Y'(I - M)Y/n]$$

$$= \text{tr}[\hat{\Sigma}^{-1} T_{ij}].$$

Obviously this holds for all i and j. Moreover, under weak conditions $\hat{\Sigma}$ is positive definite with probability one. (See the discussion following Theorem 1.2.2.)

UNBIASED ESTIMATION OF Σ

The MLE $\hat{\Sigma}$ is a biased estimate just as the MLE of the variance in a standard univariate linear model is biased. (Note that the univariate linear model is just the special case where $q = 1$.) The usual unbiased estimate of Σ does not depend on the assumption of normality and is generalized from the univariate result. An unbiased estimate of Σ is

$$S = Y'(I - M)Y/[n - r(X)].$$

To see this, consider the i, j element of $Y'(I - M)Y$.

$$
\begin{aligned}
\mathrm{E}[Y_i'(I - M)Y_j] &= \mathrm{E}[(Y_i - X\beta_i)'(I - M)(Y_j - X\beta_j)] \\
&= \mathrm{E}\{\mathrm{tr}[(Y_i - X\beta_i)'(I - M)(Y_j - X\beta_j)]\} \\
&= \mathrm{E}\{\mathrm{tr}[(I - M)(Y_j - X\beta_j)(Y_i - X\beta_i)']\} \\
&= \mathrm{tr}\{\mathrm{E}[(I - M)(Y_j - X\beta_j)(Y_i - X\beta_i)']\} \\
&= \mathrm{tr}\{(I - M)\mathrm{E}[(Y_j - X\beta_j)(Y_i - X\beta_i)']\} \\
&= \mathrm{tr}\{(I - M)\mathrm{Cov}(Y_j, Y_i)\} \\
&= \mathrm{tr}\{(I - M)\sigma_{ji}I\} \\
&= \sigma_{ij}(n - r(X)).
\end{aligned}
$$

Thus, each element of S in an unbiased estimate of the corresponding element of Σ.

EXAMPLE 1.1.1.   *Partial Correlation Coefficients*
   Partial correlations were discussed in Christensen (1987, Section VI.5). Suppose we have n observations on two dependent variables y_1, y_2 and $p-1$ independent variables $x_1, \ldots, x_{p-1}$. Write

$$
Y = \begin{bmatrix} y_{11} & y_{12} \\ \vdots & \vdots \\ y_{n1} & y_{n2} \end{bmatrix} = [Y_1, Y_2]
$$

and

$$
Z = \begin{bmatrix} x_{11} & \cdots & x_{1\,p-1} \\ \vdots & & \vdots \\ x_{n1} & \cdots & x_{n\,p-1} \end{bmatrix}.
$$

Write a multivariate linear model as

$$Y = [J, Z]B + e$$

where J is an $n \times 1$ vector of 1's. As discussed above, the unbiased estimate of Σ is $S = [s_{ij}]$ where

$$
\begin{aligned}
S &= Y'(I - M)Y/[n - r(X)] \\
&= \frac{1}{n - r(X)} \begin{bmatrix} Y_1'(I - M)Y_1 & Y_1'(I - M)Y_2 \\ Y_2'(I - M)Y_1 & Y_2'(I - M)Y_2 \end{bmatrix}.
\end{aligned}
$$

From Christensen (1987, Section VI.5), the sample partial correlation coefficient is

$$r_{y \cdot x} = \frac{Y_1'(I - M)Y_2}{[Y_1'(I - M)Y_1 \ Y_2'(I - M)Y_2]^{1/2}}$$

$$= \frac{s_{12}}{\sqrt{s_{11}s_{22}}} \, .$$

The sample partial correlation coefficient is just the sample correlation coefficient as estimated in a multivariate linear model in which the effects of the x variables have been eliminated.

I.2   Testing Hypotheses

Consider testing the multivariate model

$$Y = XB + e \tag{1}$$

against a reduced model

$$Y = X_0\Gamma + e \tag{2}$$

where $C(X_0) \subset C(X)$ and the elements of e are multivariate normal. The covariance matrix $[\Sigma \otimes I_n]$ from model (1.1.2) is unknown, so standard univariate methods of testing do not apply. Let $M_0 = X_0(X_0'X_0)^- X_0'$ be the perpendicular projection operator onto $C(X_0)$. Multivariate tests of model (2) versus model (1) are based on the hypothesis statistic

$$H \equiv Y'(M - M_0)Y$$

and the error statistic

$$E \equiv Y'(I - M)Y.$$

These statistics look identical to the sums of squares used in univariate linear models. The difference is that the univariate sums of squares are scalars, while in multivariate models these statistics are matrices. The matrices have diagonals that consist of sums of squares for the various dependent variables and off-diagonals that are sums of cross-products of the different dependent variables.

For univariate models, the test statistic is proportional to the scalar $Y'(M - M_0)Y[Y'(I - M)Y]^{-1}$. For multivariate models, the test statistic is often taken as a function of the matrix $Y'(M - M_0)Y[Y'(I - M)Y]^{-1}$ or some closely related matrix. For multivariate models, there is no one test statistic that is the universal standard for performing tests. Various test statistics are discussed in the next subsection.

The procedure for testing an hypothesis about an estimable parametric function, say

$$H_0 : \Lambda' B = 0 \quad \text{versus} \quad H_A : \Lambda' B \neq 0 \tag{3}$$

where $\Lambda' = P'X$, follows from model testing exactly as in Christensen (1987, Section III.3). The test is based on the hypothesis statistic

$$\begin{aligned} H &\equiv Y' M_{MP} Y \\ &= (\Lambda'\hat{B})' [\Lambda'(X'X)^- \Lambda]^- (\Lambda'\hat{B}) \end{aligned}$$

and the error statistic

$$E \equiv Y'(I - M)Y .$$

The projection operator in H is $M_{MP} = MP(P'MP)^- P'M$. Its use follows from the fact that $\Lambda' B = 0$ puts a constraint on the model that requires $E(Y_h) \in C[(I - M_{MP})X] = C(M - M_{MP})$ for each h. Thus, the reduced model can be written

$$Y = (M - M_{MP})\Gamma + e$$

and the hypothesis statistic is

$$Y'[M - (M - M_{MP})]Y = Y' M_{MP} Y .$$

Just as in Christensen (1987, Sections III.2, III.3), both the reduced model hypothesis and the parametric function hypothesis can be generalized. The reduced model can be generalized to

$$Y = X_0 \Gamma + Z + e \tag{4}$$

where Z is a known $n \times q$ matrix with $C(Z) \subset C(X)$. As in Christensen (1987, Section III.2), model (1) is rewritten as

$$(Y - Z) = X B_* + e \tag{5}$$

for some appropriate reparameterization B_*. Model (4) is rewritten as

$$(Y - Z) = X_0 \Gamma + e . \tag{6}$$

The test of (4) versus (1) is performed by testing (6) against (5). The hypothesis statistic for the test is

$$H \equiv (Y - Z)'(M - M_0)(Y - Z) .$$

The error statistic is

$$\begin{aligned} E &\equiv (Y - Z)'(I - M)(Y - Z) \\ &= Y'(I - M)Y \end{aligned}$$

Similarly for a known matrix W, a test can be performed of

$$H_0\colon \Lambda'B = W \quad \text{versus} \quad H_A\colon \Lambda'B \neq W$$

where $\Lambda' = P'X$ and the equation $\Lambda'B = W$ has at least one solution. Let G be a known solution $\Lambda'G = W$, then the hypothesis statistic is

$$\begin{aligned} H &\equiv (Y - XG)'M_{MP}(Y - XG) \\ &= (\Lambda'\hat{B} - W)'[\Lambda'(X'X)^-\Lambda]^-(\Lambda'\hat{B} - W) \end{aligned}$$

and the error statistic is

$$\begin{aligned} E &\equiv (Y - XG)'(I - M)(Y - XG) \\ &= Y'(I - M)Y. \end{aligned}$$

An interesting variation on the hypothesis $H_0\colon \Lambda'B = 0$ is

$$H_0\colon \Lambda'B\xi = 0 \quad \text{versus} \quad H_A\colon \Lambda'B\xi \neq 0 \tag{7}$$

where ξ can be any $q \times 1$ vector and again $\Lambda' = P'X$. To test (7), transform model (1) into

$$Y\xi = XB\xi + e\xi.$$

This is a standard univariate linear model with dependent variable vector $Y\xi$, parameter vector $B\xi$, and error vector $e\xi \sim N(0, \xi'\Sigma\xi\, I_n)$. It is easily seen that the least squares estimate of $B\xi$ in the univariate model is $\hat{B}\xi$. From univariate theory, a test for (7) is based on the noncentral F distribution, in particular

$$\frac{(\Lambda'\hat{B}\xi)'[\Lambda'(X'X)^-\Lambda]^-(\Lambda'\hat{B}\xi)/r(\Lambda)}{\xi'Y'(I - M)Y\xi/(n - r(X))} \sim F(r(\Lambda), n - r(\Lambda), \pi)$$

where

$$\pi = \xi'B'\Lambda[\Lambda'(X'X)^-\Lambda]^-\Lambda'B\xi/2\xi'\Sigma\xi.$$

This test can also be generalized in several ways. For example, let Z be a known $q \times r$ matrix with $r(Z) = r < q$. A test of

$$H_0\colon \Lambda'BZ = 0 \quad \text{versus} \quad H_A\colon \Lambda'BZ \neq 0 \tag{8}$$

can be performed by examining the transformed multivariate linear model

$$YZ = XBZ + eZ.$$

Here the dependent variable matrix is YZ, the error matrix is eZ, and the parameter matrix is BZ. The test of (8) follows precisely the form of (3). The hypothesis statistic is

$$H_* \equiv Z'Y'M_{MP}YZ = (\Lambda'\hat{B}Z)'(\Lambda'(X'X)^-\Lambda)^-(\Lambda'\hat{B}Z)$$

and the error statistic is

$$E_* \equiv Z'Y'(I - M)YZ .$$

It was convenient to assume that Z has full column rank. If B can be any $p \times q$ matrix and Z has full column rank then BZ can be any $p \times r$ matrix. Thus BZ can serve as the parameter matrix for a multivariate linear model. If Z does not have full column rank then YZ has linear dependencies, $\mathrm{Cov}(Z'\varepsilon_i) = Z'\Sigma Z$ is singular and BZ is not an arbitrary $p \times r$ matrix. None of these problems is an insurmountable difficulty for conducting the analysis of the transformed model but proving that the analysis works for a non-full rank Z is more trouble than it is worth.

EXERCISE 1.3.   Show that under the multivariate linear model

$$\mathrm{E}\left[Y'(M - M_0)Y\right] = r(M - M_0)\Sigma + B'X'(M - M_0)XB.$$

TEST STATISTICS

Various functions of the hypothesis and error statistics have been proposed as test statistics. Four of the more commonly used are discussed here. A complete survey will not be attempted and an exhaustive treatment of the related distribution theory will certainly not be given.

In this subsection, we consider testing only reduced models such as (2) or parametric hypotheses such as (3). Hence

$$E = Y'(I - M)Y$$

and, according to the context, either

$$H = Y'(M - M_0)Y$$

or

$$H = Y'M_{MP}Y .$$

Adjustments for other hypotheses are easily made.

The test statistics discussed in this section are all functions of H and E. Under normality, the null distributions of these statistics depend on H and E only through the fact that they have independent central Wishart distributions.

Definition 1.2.1.   Let $w_1, w_2, \cdots, w_n$ be independent $N(\mu_i, \Sigma)$ then

$$W = \sum_{i=1}^{n} w_i w_i'$$

has a noncentral Wishart distribution with n degrees of freedom, covariance matrix Σ, and noncentrality parameter matrix Q where

$$Q = \frac{1}{2}\Sigma^{-1} \sum_{i=1}^{n} \mu_i\mu_i' \, .$$

If $Q = 0$ the distribution is a central Wishart. In general, write

$$W \sim W(n, \Sigma, Q) \, .$$

Under the full model and assuming normal distributions, H and E have independent Wishart distributions. In particular,

$$E \sim W\big(n - r(X), \Sigma, 0\big)$$

and

$$H \sim W\left(r(X) - r(X_0), \Sigma, \frac{1}{2}\Sigma^{-1}B'X'(M - M_0)XB\right) .$$

The reduced model is true if and only if

$$H \sim W\big(r(X) - r(X_0), \Sigma, 0\big).$$

EXERCISE 1.4.   a) Use Definition 1.2.1 to show that E and H have the distributions indicated above.
b) Show that H and E are independent.
c) Show that MY and E are independent.
Hint: For (b) show that $(I - M)Y$ and $(M - M_0)Y$ are independent.

A more traditional approach to the distribution theory of multivariate linear models would be to define the distributions to be used below, i.e., U, ϕ_{max}, T^2, and V, as various functions of random matrices with independent central Wishart distributions. One could then show that the corresponding functions of H and E have these distributions. For the present purposes, we are only interested in these distributions because they are interesting functions of the hypothesis and error statistics. There seems to be little point in defining the distributions as anything other than the appropriate functions of H and E.

We begin by considering the likelihood ratio test statistic. This is simply the maximum of the likelihood under H_0 divided by the overall maximum of the likelihood. The overall maximum of the likelihood function is obtained at the MLE's so substituting $\hat\Sigma$ for Σ in (1.1.6) gives the maximum value of the log-likelihood as

$$\begin{aligned}
\ell(X\hat B, \hat\Sigma) &= -\frac{nq}{2}\log(2\pi) - \frac{n}{2}\log(|\hat\Sigma|) - \frac{1}{2}\mathrm{tr}[\hat\Sigma^{-1}n\hat\Sigma] \\
&= -\frac{nq}{2}\log(2\pi) - \frac{n}{2}\log(|\hat\Sigma|) - \frac{nq}{2} \, .
\end{aligned}$$

The maximum value of the likelihood function is

$$L(X\hat{B}, \hat{\Sigma}) = 2\pi^{-nq/2}|\hat{\Sigma}|^{-n/2}e^{-nq/2}.$$

Similarly, if we assume that the reduced model (2) is true, then the MLE's are $\hat{\Gamma} = (X_0'X_0)^- X_0'Y$ and $\hat{\Sigma}_H = Y'(I - M_0)Y/n$. The maximum value of the likelihood function under the assumption that H_0 is true is

$$L(X_0\hat{\Gamma}, \hat{\Sigma}_H) = 2\pi^{-nq/2}|\hat{\Sigma}_H|^{-n/2}e^{-nq/2}.$$

The likelihood ratio test statistic is

$$\frac{L(X_0\hat{\Gamma}, \hat{\Sigma}_H)}{L(X\hat{B}, \hat{\Sigma})} = \frac{|\hat{\Sigma}_H|^{-n/2}}{|\hat{\Sigma}|^{-n/2}}$$

$$= \left[\frac{|\hat{\Sigma}|}{|\hat{\Sigma}_H|}\right]^{n/2}.$$

The null hypothesis is rejected if the maximum value of the likelihood under H_0 is too much smaller than the overall maximum, i.e., if the likelihood ratio test statistic is too small. Because the function $f(x) = x^{2/n}$ is strictly increasing, the likelihood ratio test is equivalent to rejecting H_0 when

$$U = f([|\hat{\Sigma}|/|\hat{\Sigma}_H|]^{n/2})$$

$$= \frac{|\hat{\Sigma}|}{|\hat{\Sigma}_H|}$$

is too small. Noting that $\hat{\Sigma} = E/n$ and $\hat{\Sigma}_H = (E + H)/n$, we see that

$$U = \frac{|E|}{|E + H|}.$$

Multiplying the numerator and denominator of U by $|E^{-1}|$, we get a slightly different form

$$U = \frac{|I|}{|I + HE^{-1}|} = |I + HE^{-1}|^{-1}.$$

When H_0 is true, U has some distribution say,

$$U \sim U(q, d, n - r(X))$$

where d is either $r(X) - r(X_0)$ or $r(\Lambda)$ depending on which kind of hypothesis is being tested. An α-level test is rejected if the observed value of U is smaller than the α percentile of the $U(q, d, n - r(X))$ distribution.

The likelihood ratio test statistic (LRTS) is often referred to as *Wilks's* Λ. The symbol Λ is used here for other purposes, so to minimize internal confusion the LRTS is denoted U. In reading applications and computer output, the reader needs to remember that references to Wilks's Λ are references to the LRTS.

Rao (1951) has established the following approximate distribution for U when H_0 is true. Let

$$
\begin{aligned}
r &= r(X) \\
d &= r(X) - r(X_0) = r(\Lambda) \\
s &= \frac{qd}{2} + 1 \\
f &= (n - r) + d - \frac{1}{2}(d + q + 1)
\end{aligned}
$$

and

$$
t = \begin{cases} \left[(q^2 d^2 - 4)/(q^2 + d^2 - 5)\right]^{\frac{1}{2}} & \text{if } \min(q, d) \geq 2 \\ 1 & \text{if } \min(q, d) = 1 \end{cases}
$$

then it is approximately true that

$$
\frac{1 - U^{1/t}}{U^{1/t}} \; \frac{ft - s}{qd} \sim F(qd, ft - s).
$$

The test is rejected for large values of $(1 - U^{1/t})/U^{1/t}$. If $\min(q, d)$ is 1 or 2, the distribution is exact. For properties and tables of the U distribution see a text on multivariate analysis, e.g., Seber (1984, p. 413).

In the discussion of the previous paragraphs the matrix E^{-1} was used. E is a random matrix. It is by no means clear that E is nonsingular. Fortunately, the following theorem establishes indirectly that E is nonsingular with probability one.

Theorem 1.2.2.   Let Y be an $n \times q$ random matrix and let A be a fixed symmetric $n \times n$ matrix. If the joint distribution of the nq elements of Y admits a density function with respect to Lesbesgue measure on $\mathbf{R}^{nq}$ then

$$
\Pr\left[r(Y'AY) = \min(q, r(A))\right] = 1.
$$

PROOF. See Okamoto (1973). $\qquad\qquad\square$

With

$$
Y = \begin{bmatrix} y_1' \\ \vdots \\ y_n' \end{bmatrix}
$$

and the rows of Y independent with

$$
y_i \sim N(B'x_i, \Sigma),
$$

the density condition of Theorem 1.2.2 holds. Thus, with probability one,

$$\begin{aligned}
r(E) &= r\big(Y'(I-M)Y\big) \\
&= \min\big(q, r(I-M)\big) \\
&= \min\big(q, n - r(X)\big).
\end{aligned}$$

We will henceforth assume that the number of observations n is large enough so that

$$q \le n - r(X).$$

In this case, with probability one, E is a nonnegative definite $q \times q$ matrix of rank q, hence E is almost surely positive definite. In the discussion below we will simply treat E as positive definite.

An alternative to the likelihood ratio test statistic can be generated by Roy's union-intersection principle, cf. Roy (1953). His method is based on the fact that the reduced model (2) holds if and only if, for every $q \times 1$ vector ξ, the univariate linear model

$$Y\xi = X_0\Gamma\xi + e\xi \ , \ e\xi \sim N(0, (\xi'\Sigma\xi)I) \tag{9}$$

holds. If we denote $H_{0\xi}$ as the hypothesis that model (9) holds and H_0 as the hypothesis that model (2) holds, then

$$H_0 = \bigcap_{\text{all } \xi} H_{0\xi} \, .$$

The intersection is appropriate because the equivalence requires the validity of *every* model of the form (9).

Similarly, the full model (1) holds if and only if every model

$$Y\xi = XB\xi + e\xi \tag{10}$$

holds. The alternative hypothesis H_A that (1) holds but (2) does not is simply that for some ξ, the univariate model (10) holds, but model (9) does not. Let $H_{A\xi}$ be the univariate alternative. Because H_A requires only one of the $H_{A\xi}$ to be true, we have

$$H_A = \bigcup_{\text{all } \xi} H_{A\xi} \, .$$

This reformulation of H_A and H_0 in terms of unions and intersections is the genesis of the name "union-intersection principle."

For testing model (9) against model (10), there is a standard univariate F statistic available, $F_\xi = [\xi'Y'(M - M_0)Y\xi/(r(X) - r(X_0))]/[\xi'Y'(I - M)Y\xi/(n - r(X))]$. The union-intersection principle test statistic is simply

$$\psi_{max} \equiv \sup_{\xi}\{F_\xi\} \, .$$

The test is rejected if ψ_{max} is too large. In practice, an alternative but equivalent test statistic is used. Because

$$F_\xi = \frac{n - r(X)}{r(X) - r(X_0)} \frac{\xi' H \xi}{\xi' E \xi} \; ,$$

it is equivalent to reject H_0 if

$$\phi_{max} \equiv \sup_\xi \left[\frac{\xi' H \xi}{\xi' E \xi} \right]$$

is too large.

We will show that ϕ_{max} is distributed as the largest eigenvalue of HE^{-1}. The argument is based on the following result.

Lemma 1.2.3.   If E is a positive definite matrix then there exists a nonsingular matrix $E^{1/2}$ such that

$$E = E^{1/2} E^{1/2}$$

and

$$E^{-1/2} E \, E^{-1/2} = I.$$

PROOF.   As in the proof of Christensen (1987, Theorem B.22), write

$$E = PD(\lambda_i)P'$$

where P is an orthogonal matrix, $D(\lambda_i)$ is diagonal, and the λ_i's are all positive. Pick

$$E^{1/2} = PD(\sqrt{\lambda_i})P'$$

and the results are easily shown. $\square$

Note that any vector ξ can be written as $E^{-1/2}\rho$ for some vector ρ, and any vector ρ determines a vector ξ, so

$$\phi_{max} = \sup_\rho \left[\frac{\rho' E^{-1/2} H E^{-1/2} \rho}{\rho' \rho} \right] .$$

Writing $E^{-1/2} H E^{-1/2}$ as $P_0 D(\phi_i) P_0'$ where P_0 is an orthogonal matrix and the ϕ_i's are the eigenvalues of $E^{-1/2} H E^{-1/2}$, we can replace ρ with $v = P_0 \rho$ giving

$$\phi_{max} = \sup_v \left[\frac{v' D(\phi_i) v}{v' v} \right] .$$

Because

$$v' D(\phi_i) v / v' v = \sum_{i=1}^{q} v_i^2 \phi_i \Big/ \sum_{i=1}^{q} v_i^2$$

is a weighted average of the ϕ_i's, the maximum value is attained when all of the weight is placed on the largest eigenvalue. Thus

$$\phi_{max} = \max_i \phi_i \, .$$

Finally, we need to establish that $E^{-1/2}HE^{-1/2}$ and HE^{-1} have the same eigenvalues. Let w be an eigenvector for $E^{-1/2}HE^{-1/2}$ corresponding to the eigenvalue ϕ. Because $E^{-1/2}HE^{-1/2}w = \phi w$, we have

$$E^{1/2}E^{-1/2}HE^{-1/2}E^{-1/2}E^{1/2}w = \phi E^{1/2}w$$

and

$$HE^{-1}E^{1/2}w = \phi E^{1/2}w \, .$$

Clearly, $E^{1/2}w$ is an eigenvector of HE^{-1} corresponding to ϕ. Thus we have shown that ϕ_{max} is the largest eigenvalue of HE^{-1}.

Rather than using ϕ_{max}, tests are often performed using θ_{max} where θ_{max} is the maximum eigenvalue of $H(E+H)^{-1}$. It is a simple matter to see that the eigenvalues of $H(E+H)^{-1}$ and HE^{-1} are related by the equation $\theta = \phi/(1+\phi)$. Thus θ_{max} is a one to one increasing transformation of ϕ_{max} and the test based on θ_{max} is equivalent to the one based on ϕ_{max}. Tables of the distribution of θ_{max} under H_0 were worked out by Heck (1960). Derivations and tables for the distribution of ϕ_{max} or θ_{max} under the null hypothesis can be found in many standard multivariate analysis books, e.g., Seber (1984, Section 2.5).

EXERCISE 1.5.   Show that θ is an eigenvalue of $H(E+H)^{-1}$ if and only if ϕ is an eigenvalue of HE^{-1} where $\theta = \phi/(1+\phi)$. Find U, ϕ_{max}, T^2, and V in terms of the eigenvalues of HE^{-1}.

Hints: Show that the eigenvalues of AB^{-1} and $B^{-1}A$ are the same. Use the fact that

$$I = (E+H)^{-1}H + (E+H)^{-1}E.$$

Although the likelihood ratio and union-intersection test statistics are probably the best known, several other test statistics have also been proposed. Two of these are the Lawley-Hotelling trace

$$T^2 = \big(n - r(X)\big)\mathrm{tr}[HE^{-1}] = \mathrm{tr}[HS^{-1}]$$

and Pillai's trace

$$V = \mathrm{tr}\big[H(E+H)^{-1}\big] \, .$$

Though their exact distributions require special tables, the distributions of these statistics when H_0 is true can be approximated using standard tables. Let $d = r(X) - \mathrm{r}(X_0) = r(\Lambda)$ and let $n - r(X) = n - r$. A very good approximation to T^2 has been suggested by McKeon (1974). He advocates using

$$GT^2 \sim F(qd, D)$$

where

$$D = 4 + \frac{qd + 2}{B - 1}$$

$$B = \frac{(n - r + d - q - 1)(n - r - 1)}{(n - r - q - 3)(n - r - q)}$$

and

$$G = (qd)^{-1} \left[\frac{D}{D - 2} \right] \left[\frac{n - r - q - 1}{n - r} \right] .$$

This gives the exact distribution for $\min(q, d) = 1$.

Note that for large n, the term $B - 1$ gets very small so that D, the denominator degrees of freedom, gets large. In particular, for large n, the distribution of GT^2 is approximately that of a $\chi^2(qd)$ divided by its degrees of freedom. This result is equivalent to the standard asymptotic approximation

$$T^2 \sim \chi^2(qd) .$$

The null distribution of Pillai's trace can be approximated by

$$\frac{n - r - q + s}{|q - d| + s} \frac{V}{s - V} \sim F(s[|q - d| + s], s[n - r - q + s]).$$

where

$$s = \min(q, d) .$$

Asymptotically,

$$(n - r)V \sim \chi^2(qd) .$$

Seber (1984, p. 414) compares all four test statistics. He also provides tables and other details about the distributions under H_0, (cf. Seber, 1984, Section 2.5). Kres (1983) is another good source for tables related to multivariate linear models.

Formal tests based on these statistics depend on the assumption of multivariate normality. Nonetheless, these are reasonable test statistics even without the normality assumption. To evaluate the test statistics intuitively, we need some idea of the values of the test statistics when H_0 is true. As we have seen, even without the assumption of normality

$$E(S) = E\left(E/\left[n - r(X)\right]\right) = \Sigma$$

and from Exercise 1.3, under H_0

$$E\left(H/r(M - M_0)\right) = \Sigma .$$

Using these equalities as crude approximations, it follows that if H_0 is true

$$\begin{aligned}
U &\doteq \left| I + \left(\frac{r(X) - r(X_0)}{n - r(X)}\right) I \right|^{-1} \\
&= \left[\frac{n - r(X)}{n - r(X_0)}\right]^q ,
\end{aligned}$$

$$\phi_{max} \doteq \frac{r(X) - r(X_0)}{n - r(X)} ,$$

$$\theta_{max} \doteq \frac{n - r(X)}{n - r(X_0)} ,$$

$$\begin{aligned}
T^2 &\doteq \mathrm{tr}(I)\left[r(X) - r(X_0)\right] \\
&= q\left[r(X) - r(X_0)\right]
\end{aligned}$$

and

$$\begin{aligned}
V &\doteq \mathrm{tr}(I)\left[r(X) - r(X_0)\right] / \left[n - r(X_0)\right] \\
&= q\left[r(X) - r(X_0)\right] / \left[n - r(X_0)\right] .
\end{aligned}$$

These comparison values can be very useful in exploring the data. If the observed value of U is much smaller than $([n - r(X)] / [n - r(X_0)])^q$ or if the observed values of the other test statistics are much larger than the comparison values, the null hypothesis is called in question. In a formal test, the null distribution of the test statistic is used to quantify the meaning of the word "much" in the previous sentence.

An important parameter in the distributions of the test statistics is the rank of HE^{-1}. With probability one, E is nonsingular so

$$r\left(HE^{-1}\right) = r(H) .$$

By definition,

$$H = Y'(M - M_0)Y$$

is a $q \times q$ matrix. Because $r(M - M_0) = r(X) - r(X_0)$, Theorem 1.2.2 implies that if Y has a density, $r(HE^{-1}) = r(H) = \min\left(q, r(X) - r(X_0)\right)$ with probability one.

PREDICTION REGIONS

Suppose we wish to predict the value of a new observation vector y_0' with $E(y_0') = x_0'B$ where $x_0'B$ is estimable, i.e., $x_0'B = \rho_0'XB$ for some vector ρ_0. It is natural to assume that y_0 is generated by the same process as Y, thus $\mathrm{Cov}(y_0) = \Sigma$ and y_0 is independent of Y. It follows that the best linear unbiased predictor of y_0 is $\hat{y}_0 \equiv \hat{B}'x_0$, cf. Christensen (1987, Sections XII.2), Section III.1, and Section VI.2.

A prediction region for y_0 can be based on the distribution of $y_0 - \hat{y}_0$. Clearly $E(y_0 - \hat{y}_0) = 0$ and, recalling Exercise 1.1,

$$
\begin{aligned}
\mathrm{Cov}(y_0 - \hat{y}_0) &= \Sigma + \mathrm{Cov}(\hat{y}_0) \\
&= \Sigma + \mathrm{Cov}(Y'M\,\rho_0) \\
&= \Sigma + \mathrm{Cov}\left(\begin{bmatrix} Y_1'M\,\rho_0 \\ \vdots \\ Y_q'M\,\rho_0 \end{bmatrix}\right) \\
&= \Sigma + \rho_0'M\,\rho_0\Sigma \\
&= (1 + x_0'(X'X)^- x_0)\Sigma.
\end{aligned}
$$

If y_0 and Y are multivariate normal, then $y_0 - \hat{y}_0$ is also normal, independent of E and

$$
\left(1 + x_0'(X'X)^- x_0\right)^{-1} (y_0 - \hat{y}_0)(y_0 - \hat{y}_0)' \sim W(1, \Sigma, 0).
$$

It follows that

$$
\frac{\mathrm{tr}\left[(y_0 - \hat{y}_0)(y_0 - \hat{y}_0)'S^{-1}\right]}{1 + x_0'(X'X)^- x_0}
$$

or equivalently

$$
\frac{(y_0 - \hat{y}_0)'S^{-1}(y_0 - \hat{y}_0)}{1 + x_0'(X'X)^- x_0}
$$

has the same distribution as the null distribution of the Lawley-Hotelling T^2 statistic with $d = 1$. From our discussion of McKeon's approximation with $dfE \equiv n - r(X)$, the exact distribution is

$$
\frac{(y_0 - \hat{y}_0)'S^{-1}(y_0 - \hat{y}_0)}{(dfE)\,(1 + x_0'(X'X)^- x_0)} \frac{dfE + 1 - q}{q} \sim F(q, dfE + 1 - q, 0).
$$

A $(1 - \alpha)100\%$ prediction ellipsoid consists of all y_0 vectors that satisfy

$$
(y_0 - \hat{y}_0)'S^{-1}(y_0 - \hat{y}_0) \leq
$$
$$
F(1 - \alpha, q, dfE + 1 - q)\frac{q}{dfE + 1 - q}\,(dfE)\,\left(1 + x_0'(X'X)^- x_0\right).
$$

The methods illustrated here can also be used to obtain confidence ellipsoids. Examples of such ellipsoids are given in Sections 3 and 4.

MULTIPLE COMPARISON METHODS

A number of multiple comparison methods for univariate linear models where discussed in Christensen (1987, Chapter V). Three of these are easily adapted for use with multivariate linear models. If a number of preplanned hypotheses are to be tested, a Bonferroni adjustment to the levels of the individual tests will control the experimentwise error rate, cf. Christensen (1987, Section V.3). The basic idea of the Least Significant Difference method (cf. Christensen, 1987, Section V.2) can also be applied in a limited way. An overall test for, say, treatment differences in a one-way multivariate ANOVA (MANOVA), is performed. If the overall test is not rejected no more tests are performed. If the overall test rejects the hypothesis of no treatment differences then contrasts in the differences can be tested in the usual way. This procedure does not allow for hypotheses involving relationships between the dependent variables. The third approach involves ideas from the discussion of Scheffé's method in Christensen (1987, Section V.1). These can either be combined with ideas from Roy's method of test construction or can be applied to the Lawley-Hotelling T^2. The method based on Roy's construction was developed by Roy and Bose (1953) and is the primary subject of the current subsection.

Consider a univariate linear model

$$Y_{n\times 1} = X\beta + e, \ e \sim N(0, \sigma^2 I).$$

Scheffé's method controls the error rate for testing any and all one degree of freedom hypotheses that put a constraint on some specified subspace of $C(X)$. For example, the hypotheses could be all contrasts in a one-way ANOVA or all interaction contrasts in a two-way ANOVA. One method of specifying the subspace is to identify it as the constraint space associated with

$$H_0\colon \Lambda'\beta = 0,$$

cf. Definition 3.3.2 in Christensen (1987). For testing all contrasts in a one-way ANOVA with a treatments, $\Lambda'\beta$ can be any vector of contrasts that contains $a - 1$ linearly independent elements. Recall that contrasts $\lambda_i'\beta$ are linearly independent if and only if the vectors λ_i that define the contrasts are linearly independent. In general, a one degree of freedom hypothesis $H_0\colon \lambda'\beta = 0$ puts a constraint on the subspace determined by $\Lambda'\beta = 0$ if and only if $\lambda'\beta = \zeta'\Lambda'\beta$ for some vector ζ, cf. Exercise 1.6.

The key to applying these univariate results to the multivariate linear model

$$Y_{n\times q} = XB + e, \ \varepsilon_i \text{ indep. } N(0, \Sigma)$$

is in recalling that Roy's method of test construction amounts to testing all univariate linear models

$$Y\xi = XB\xi + e\xi, \quad \xi'\varepsilon_i \text{ indep. } N(0, \xi'\Sigma\xi).$$

Combining the results on univariate models and multivariate models we see that testing the multivariate hypothesis

$$H_0\colon \Lambda' B = 0$$

is equivalent to testing all the univariate hypotheses

$$H_0\colon \zeta' \Lambda' B \xi = 0.$$

In particular, the statistics for the univariate and multivariate tests have the relationship

$$\frac{\left(\zeta'\Lambda'\hat{B}\xi\right)' \left[\zeta'\Lambda'(X'X)^{-}\Lambda\zeta\right]^{-1} \left(\zeta'\Lambda'\hat{B}\xi\right) \Big/ r(\Lambda)}{\xi' S \xi} \leq \frac{n - r(X)}{r(\Lambda)} \phi_{max} \quad (11)$$

where ϕ_{max} is Roy's maximum root statistic for testing $\Lambda' B = 0$. It follows that the multiple comparison procedure that rejects $H_0\colon \zeta'\Lambda'B\xi = 0$ if and only if

$$\frac{\left(\zeta'\Lambda'\hat{B}\xi\right)^2 \Big/ r(\Lambda)}{\left[\zeta'\Lambda'(X'X)^{-}\Lambda\zeta\right]\xi'S\xi} > \frac{n - r(X)}{r(\Lambda)} \phi_{max}\left(1 - \alpha, q, r(\Lambda), n - r(X)\right)$$

has an experimentwise error rate no greater that α when applied to testing any or all hypotheses of the form $\zeta'\Lambda'B\xi = 0$.

For testing all hypotheses of the form $H_0\colon \zeta'\Lambda'B\xi = 0$, the experimentwise error rate is precisely α. This follows from the fact that there exists a linear parametric function $\zeta'\Lambda'B\xi$ for which the test statistics in (11) are equal. From Christensen (1987, Section V.1), for any fixed ξ there exists a ζ so that the test statistic for $H_0\colon \zeta'\Lambda'B\xi = 0$ is equal to the test statistic for $H_0\colon \Lambda'B\xi = 0$. From our discussion of Roy's method of test construction, there exists a ξ so that the test statistic for $H_0\colon \Lambda'B\xi = 0$ equals the test statistic for $H_0\colon \Lambda'B = 0$. Together these imply that equality is attained in (11) for some ζ and ξ.

As with Scheffé's method, the procedure for controlling the simultaneous error rate of multiple tests can be adapted to providing simultaneous confidence intervals. With confidence coefficient $(1 - \alpha)100\%$, the intervals

$$\zeta'\Lambda'\hat{B}\xi \pm \sqrt{\xi'E\xi\left[\zeta'\Lambda'(X'X)^{-}\Lambda\zeta\right]\phi_{max}\left(1 - \alpha, q, r(\Lambda), n - r(X)\right)}$$

contain all parameters of the form $\zeta'\Lambda'B\xi$.

The Lawley-Hotelling T^2 can also be used to test all hypotheses of the form $H_0\colon \zeta'\Lambda'BZ = 0$. Let T_*^2 be the Lawley-Hotelling statistic for $H_0\colon \Lambda'BZ = 0$ and let T_1 be the statistic for $H_0\colon \zeta'\Lambda'BZ = 0$. As in Christensen (1987, Section V.1), the column space of the perpendicular projection operator for testing $H_0\colon \zeta'\Lambda'BZ = 0$ is contained in the column space of the projection operator for $H_0\colon \Lambda'BZ = 0$. In Christensen's

Section V.1, this implied that the sum of squares for the one degree of freedom hypothesis was no greater than the sum of squares for the larger hypothesis. In multivariate applications H_1, the hypothesis matrix for the one degree of freedom test $H_0: \zeta'\Lambda'BZ = 0$, is smaller than H_* the hypothesis statistic for the larger hypothesis $H_0: \Lambda'BZ = 0$. Specifically, $H_* - H_1$ is nonnegative definite. The covariance matrix estimate appropriate for testing both of these hypotheses is $S_* = Z'Y'(I - M)YZ/[n - r(X)]$. It is not difficult to show that $[H_* - H_1] S_*^{-1}$ is nonnegative definite, cf. Exercise 1.8.13. Thus for any hypothesis $\zeta'\Lambda'BZ = 0$ the Lawley-Hotelling test statistic satisfies $T_*^2 = \text{tr}[H_* S_*^{-1}] \geq \text{tr}[H_1 S_*^{-1}] = T_1^2$. It follows that if the one degree of freedom hypothesis is rejected only when T_1^2 exceeds the critical point appropriate for T_*^2, the experimentwise error rate will be controlled. The problem with this procedure is that if the $q \times r$ matrix Z has $r > 1$, there may be no vector ζ nor even a collection of vectors $\zeta_1, \ldots, \zeta_r$ such that the test of

$$H_0: \begin{bmatrix} \zeta_1' \\ \vdots \\ \zeta_r' \end{bmatrix} \Lambda'BZ = 0$$

has the same T^2 value as the hypothesis $\Lambda'BZ = 0$. This multiple comparison method is applied in Example 1.5.3.

EXERCISE 1.6.   Show that the constraint imposed by any hypothesis $H_0: \zeta'\Lambda'\beta = 0$ is contained in the constraint subspace determined by $H_0: \Lambda'\beta = 0$.   Hint: Review Christensen (1987, Section III.3).

EXERCISE 1.7.   Prove the inequality (11).

I.3   One-Sample Problems

The multivariate one-sample problem has the same linear structure as the univariate one-sample problem that was explored in Christensen (1987, Exercises 2.3 and 3.3). Let $y_1, \ldots, y_n$ be i.i.d. $N(\mu, \Sigma)$ where μ is $q \times 1$ and Σ is $q \times q$. Write $y_i' = (y_{i1}, \ldots, y_{iq})$,

$$Y = \begin{bmatrix} y_1' \\ \vdots \\ y_n' \end{bmatrix},$$

and

$$Y = J\mu' + e$$

where again J is an $n \times 1$ vector of 1's. Least squares estimates satisfy

$$J\hat{\mu}' = \frac{1}{n} J_n^n Y = J\bar{y}'.$$

where $\bar{y}' = \frac{1}{n}\sum_{i=1}^{n} y_i' = (\bar{y}_{\cdot 1}, \ldots, \bar{y}_{\cdot q})$. The sample mean $\bar{y}_{\cdot}$ is also the MLE of μ. The MLE of Σ is

$$
\begin{aligned}
\hat{\Sigma} = \frac{n-1}{n}S &= \frac{1}{n}Y'\left(I - \frac{1}{n}J_n^n\right)Y \\
&= \frac{1}{n}\left[\left(I - \frac{1}{n}J_n^n\right)Y\right]'\left[\left(I - \frac{1}{n}J_n^n\right)Y\right] \\
&= \frac{1}{n}[(y_1 - \bar{y}_{\cdot}), \ldots, (y_n - \bar{y}_{\cdot})]\begin{bmatrix}(y_1 - \bar{y}_{\cdot})' \\ \vdots \\ (y_n - \bar{y}_{\cdot})'\end{bmatrix} \\
&= \frac{1}{n}\sum_{i=1}^{n}(y_i - \bar{y}_{\cdot})(y_i - \bar{y}_{\cdot})'.
\end{aligned}
$$

In particular, writing $\hat{\Sigma} = [\hat{\sigma}_{jj'}]$, we have

$$
\begin{aligned}
\hat{\sigma}_{jj'} &= \frac{1}{n}Y_j'\left(I - \frac{1}{n}J_n^n\right)Y_{j'} \\
&= \frac{1}{n}\left[\left(I - \frac{1}{n}J_n^n\right)Y_j\right]'\left[\left(I - \frac{1}{n}J_n^n\right)Y_{j'}\right] \\
&= \frac{1}{n}\sum_{i=1}^{n}(y_{ij} - \bar{y}_{\cdot j})(y_{ij'} - \bar{y}_{\cdot j'}).
\end{aligned}
$$

To test the hypothesis,

$$
H_0 : \mu = \mu_0 \qquad \text{versus} \qquad H_A : \mu \neq \mu_0,
$$

we need to recognize that, because the one-sample model is a regression model, μ is estimable and Λ' is just the scalar 1. The hypothesis and error statistics are

$$
\begin{aligned}
H &= (\Lambda'\hat{B} - W)'[\Lambda'(X'X)^{-}\Lambda]^{-}(\Lambda'\hat{B} - W) \\
&= (\bar{y}_{\cdot}' - \mu_0')'[1(J'J)^{-1}1]^{-1}(\bar{y}_{\cdot}' - \mu_0') \\
&= n(\bar{y}_{\cdot} - \mu_0)(\bar{y}_{\cdot} - \mu_0)'
\end{aligned}
$$

and

$$
E = (n-1)S.
$$

In the one-sample problem, the Lawley-Hotelling trace is the famous Hotelling T^2 statistic.

$$
\begin{aligned}
T^2 &= (n-1)\operatorname{tr}[HE^{-1}] \\
&= (n-1)\operatorname{tr}\left[n(\bar{y}_{\cdot} - \mu_0)(\bar{y}_{\cdot} - \mu_0)'\left\{\frac{1}{n-1}S^{-1}\right\}\right] \\
&= n\operatorname{tr}[(\bar{y}_{\cdot} - \mu_0)'S^{-1}(\bar{y}_{\cdot} - \mu_0)] \\
&= n(\bar{y}_{\cdot} - \mu_0)'S^{-1}(\bar{y}_{\cdot} - \mu_0).
\end{aligned}
$$

The reason the test statistic simplifies so nicely is because H is a rank one matrix. Moreover, we will see that in this case all of the test statistics discussed in Section 2 are equivalent. First note that E is nonsingular (with probability one) so HE^{-1} has rank one. A rank one matrix has only one nonzero eigenvalue, hence the maximum eigenvalue equals the sum of the eigenvalues. In other symbols

$$\phi_{max} = \text{tr}[HE^{-1}]$$

or equivalently

$$\phi_{max} = \frac{1}{n-1}T^2 .$$

Thus, the union-intersection test statistic is equivalent to Hotelling's T^2.

The likelihood ratio test uses the statistic

$$U = |I + HE^{-1}|^{-1} .$$

It is easy to see that the eigenvalues of $I + HE^{-1}$ are just one plus the eigenvalues of HE^{-1}. The determinant is the product of the eigenvalues and HE^{-1} has only one nonzero eigenvalue so

$$|I + HE^{-1}| = 1 + \phi_{max}$$

and

$$U = (1 + \phi_{max})^{-1} .$$

The statistic U is a strictly decreasing function of ϕ_{max}; thus the likelihood ratio test is equivalent to the other tests.

The equivalence of Pillai's trace to Hotelling's T^2 is established in Exercise 1.8.

In the one-sample problem, there seems to be little question about the appropriate choice of a test statistic. Hotelling's T^2 is used almost universally. Under H_0,

$$\frac{T^2}{(n-1)}\frac{n-q}{q} \sim F(q, n-q).$$

This follows from the exact part of McKeon's approximation, see also Seber (1984, Section 2.4). The test is rejected for large values of T^2. There is no need for any unusual tables to perform the test.

Arguments similar to those given in Section I.2 for the development of the prediction region yield a $(1 - \alpha)100\%$ confidence ellipsoid for μ consisting of all μ vectors that satisfy

$$(\mu - \bar{y}.)'S^{-1}(\mu - \bar{y}.) \leq F(1 - \alpha, q, n-q)\frac{q}{n-q}\frac{n-1}{n} .$$

EXAMPLE 1.3.1.   Mosteller and Tukey (1977) consider data from *The Coleman Report* on the relationships between several variables and mean verbal test scores for sixth graders at twenty schools in the New England and Mid-Atlantic regions of the United States. The data are also given at the end of Christensen (1987, Chapter XIV) in Exercise 14.10. In this example, we consider only two variables, x_2 — the percentage of sixth-graders' fathers employed in white collar jobs and x_5 — one-half of the sixth-graders' mothers' mean number of years of schooling. The data are given in Table I.1. In particular, we will test the null hypothesis that the percentage of white collar fathers is 50% and that the mothers are on average high school graduates, i.e., have twelve years of schooling. To perform a formal test of this hypothesis we need to establish that the data have a multivariate normal distribution. A normal plot (cf. Christensen, 1987, Section XIII.2) of x_2 is given in Figure I.1. It is not very encouraging. It has a very noticeable shoulder at the low end. It also has a gap and a flat spot in the middle. The Wilk-Francia statistic for the plot is $W' = .904$ which has a P value of about .05. Figure I.2 contains a normal plot of $\sqrt{x_2}$. This is much better behaved and has a W' value of .933. The variance stabilizing transformation $\mathrm{Arcsin}(\sqrt{x_2/100})$ was also considered but its behavior was actually worse than that of $\sqrt{x_2}$. A normal plot for x_5 has a minor shoulder near the top but a W' value of .976; we leave x_5 untransformed. Figure I.3 contains a plot of $\sqrt{x_2}$ versus x_5. This should look elliptical. Except for the smallest value of x_5, it is not too bad.

TABLE I.1. Coleman Report Data

x_2	$\sqrt{x_2}$	x_5	x_2	$\sqrt{x_2}$	x_5
28.87	5.37	6.19	12.20	3.49	5.62
20.10	4.48	5.17	22.55	4.75	5.34
69.05	8.31	7.04	14.30	3.78	5.80
65.40	8.09	7.10	31.79	5.64	6.19
29.59	5.44	6.15	11.60	3.41	5.62
44.82	6.69	6.41	68.47	8.27	6.94
77.37	8.80	6.86	42.64	6.53	6.33
24.67	4.97	5.78	16.70	4.09	6.01
65.01	8.06	6.51	86.27	9.29	7.51
9.99	3.16	5.57	76.73	8.76	6.96

To test the hypothesis we need the statistics

$$S = \begin{bmatrix} 4.288 & 1.244 \\ 1.244 & 0.428 \end{bmatrix}$$

and

$$\bar{y}' = (6.069, 6.255).$$

The hypothesized mean is

$$\mu_0' = (\sqrt{50}, 12/2)$$

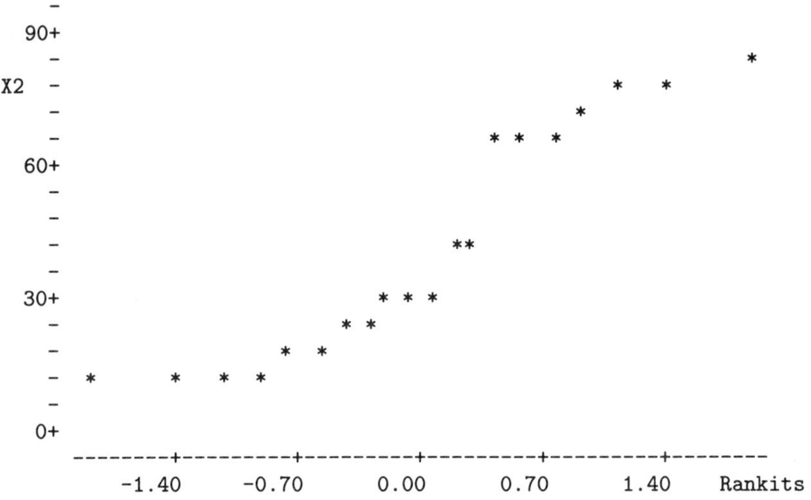

FIGURE I.1. Normal Plot of x_2

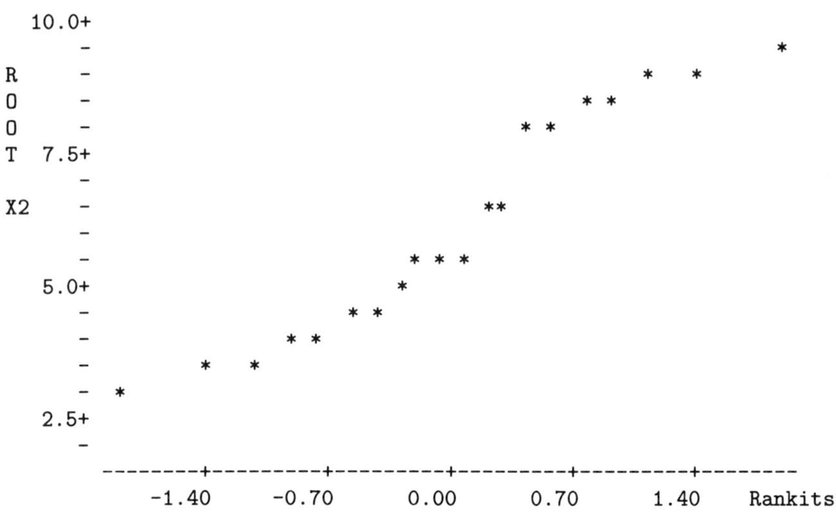

FIGURE I.2. Normal Plot of Root(x_2)

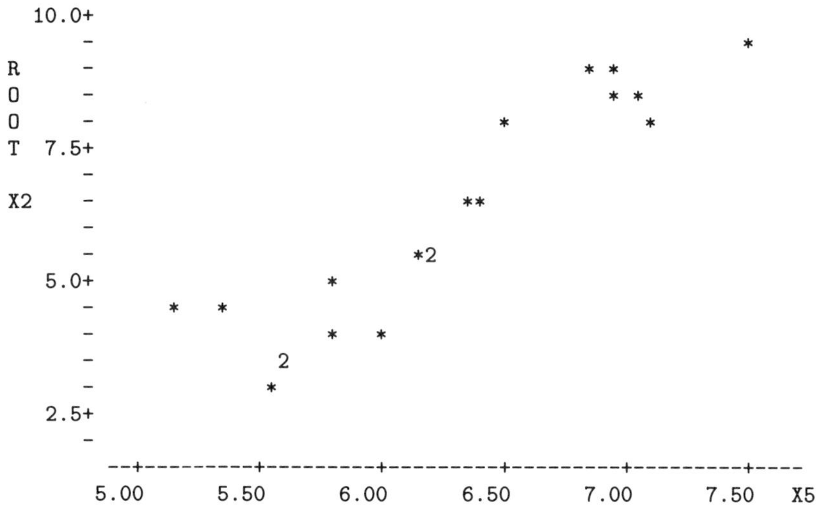

FIGURE I.3. Scatter Plot of Root(x_2) and x_5

and Hotelling's statistic is

$$T^2 = 20(\bar{y}. - \mu_0)'S^{-1}(\bar{y}. - \mu_0) = 93.45.$$

The comparison value for T^2 is 2 so T^2 seems to be highly significant. To perform a formal test based on multivariate normality, compute

$$\frac{T^2}{n-1}\frac{n-q}{q} = \frac{93.45}{19}\frac{18}{2} = 44.27$$

and compare it to an $F(2, 18)$ distribution. Again, the result is highly significant.

We have not yet sufficiently analyzed the question of multivariate normality. Not only must the marginal distributions of a multivariate normal be normal but all linear combinations of the variables must also be normal. Figure I.4 contains a normal plot of $\sqrt{x_2} + x_5$. While this has a not unrespectable W' value of .925 the plot looks horrible. The normal plot for $\sqrt{x_2} - x_5$ is also disturbing. Nevertheless, the null hypothesis is so clearly untrue that the lack of multivariate normality is probably not crucial.

EXERCISE 1.8.   Use Proposition 13.5.1 in Christensen (1987) to show that the test based on Pillai's trace is equivalent to the test based on Hotelling's T^2. Christensen's Proposition 13.5.1 states that $(A + a'b)^{-1} = A^{-1} - A^{-1}a'(I + bA^{-1}a')^{-1}bA^{-1}$.

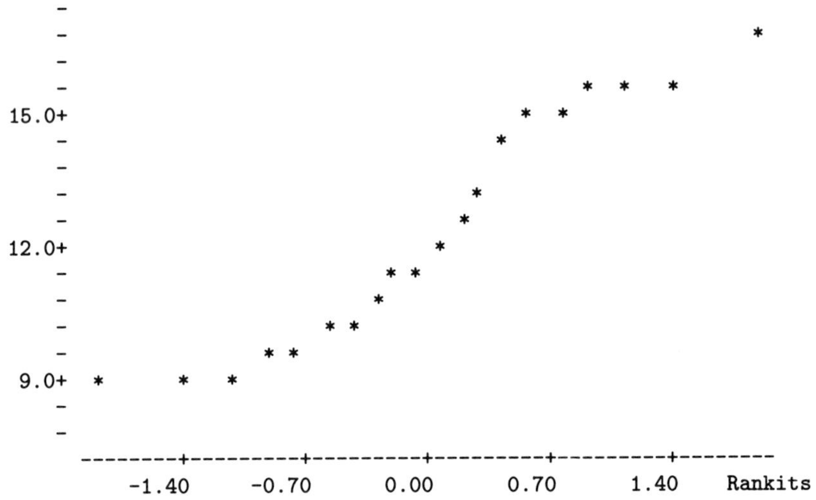

FIGURE I.4. Normal Plot of Root$(x_2) + x_5$

I.4 Two-Sample Problems

Christensen (1987, Exercises 2.2 and 3.2) examines the univariate two-sample problem. The multivariate two-sample problem has a similar linear structure. Let $y_{11}, \ldots, y_{1r}$ be i.i.d. $N_q(\mu_1, \Sigma)$, let $y_{21}, \ldots, y_{2t}$ be i.i.d. $N_q(\mu_2, \Sigma)$ and let the two samples be independent. Write

$$Z_1 = \begin{bmatrix} y'_{11} \\ \vdots \\ y'_{1r} \end{bmatrix}, \quad Z_2 = \begin{bmatrix} y'_{21} \\ \vdots \\ y'_{2t} \end{bmatrix}$$

and

$$Y = \begin{bmatrix} Z_1 \\ Z_2 \end{bmatrix}.$$

The multivariate linear model is

$$Y = \begin{bmatrix} J_r & 0 \\ 0 & J_t \end{bmatrix} \begin{bmatrix} \mu'_1 \\ \mu'_2 \end{bmatrix} + e.$$

While the analysis is quite simple, the actual computations are left to the reader. Only the final results are given.

$$\hat{\mu}_1 = \bar{y}_{1\cdot} \tag{1}$$

$$\hat{\mu}_2 = \bar{y}_{2\cdot} \tag{2}$$

$$S = \frac{(r-1)S_1 + (t-1)S_2}{r+t-2} \tag{3}$$

where

$$S_1 = \frac{1}{r-1} \sum_{j=1}^{r} (y_{1j} - \bar{y}_{1\cdot})(y_{1j} - \bar{y}_{1\cdot})' \tag{4}$$

and

$$S_2 = \frac{1}{t-1} \sum_{j=1}^{t} (y_{2j} - \bar{y}_{2\cdot})(y_{2j} - \bar{y}_{2\cdot})' . \tag{5}$$

To test $\mu_1 = \mu_2$, write the hypothesis as

$$H_0 : \mu_1 - \mu_2 = 0 \qquad \text{versus} \qquad H_A : \mu_1 - \mu_2 \neq 0 .$$

The null hypothesis is equivalent to assuming a reduced model

$$Y = J_n \mu' + e$$

where $n = r + t$ and μ is a $q \times 1$ vector. The hypothesis statistic is

$$
\begin{aligned}
H &= (\bar{y}_{1\cdot} - \bar{y}_{2\cdot}) \left[(1,-1) \begin{pmatrix} r & 0 \\ 0 & t \end{pmatrix}^{-1} \begin{pmatrix} 1 \\ -1 \end{pmatrix} \right]^{-1} (\bar{y}_{1\cdot} - \bar{y}_{2\cdot})' \\
&= \left(\frac{1}{r} + \frac{1}{t} \right)^{-1} (\bar{y}_{1\cdot} - \bar{y}_{2\cdot})(\bar{y}_{1\cdot} - \bar{y}_{2\cdot})' .
\end{aligned}
$$

The error statistic is

$$E = (r + t - 2)S .$$

H is again a rank one matrix so the Lawley-Hotelling trace is

$$
\begin{aligned}
T^2 &= (r + t - 2)\operatorname{tr}[HE^{-1}] \\
&= \left(\frac{1}{r} + \frac{1}{t} \right)^{-1} (\bar{y}_{1\cdot} - \bar{y}_{2\cdot})'S^{-1}(\bar{y}_{1\cdot} - \bar{y}_{2\cdot}) \\
&= \left(\frac{rt}{r+t} \right) (\bar{y}_{1\cdot} - \bar{y}_{2\cdot})'S^{-1}(\bar{y}_{1\cdot} - \bar{y}_{2\cdot}) .
\end{aligned}
$$

This is Hotelling's T^2 for the two-sample problem. Under H_0,

$$\frac{T^2}{r+t-2} \frac{r+t-q-1}{q} \sim F(q, r+t-q-1) .$$

Once again, because H has rank one, Hotelling's T^2 test is equivalent to the likelihood ratio test, Roy's maximum root of HE^{-1} test, and Pillai's trace test.

Arguments similar to those given in Section I.3 for the development of the prediction interval yield a $(1 - \alpha)100\%$ confidence ellipsoid consisting of all $\mu_1 - \mu_2$ vectors that satisfy

$$\big((\mu_1 - \mu_2) - (\bar{y}_{1\cdot} - \bar{y}_{2\cdot})\big)'S^{-1}\big((\mu_1 - \mu_2) - (\bar{y}_{1\cdot} - \bar{y}_{2\cdot})\big) \tag{6}$$

$$\leq F(1 - \alpha, q, r+t-q-1)\frac{q(r+t-2)}{r+t-q-1}\left(\frac{1}{r} + \frac{1}{t} \right) .$$

EXAMPLE 1.4.1.   Lubischew (1962) presents data on four characteristics
of male flea-beetles from two species within the genus Haltica. The four
characteristics are y_1 — the distance, in microns, from the posterior border
of the prothorax to the transverse groove, y_2 — the length, in .01 mm units,
of the elytra, y_3 — the length, in microns, of the second antennal joint, and
y_4 — the length, in microns, of the third antennal joint. While various plots
of the data do not look too bad, there is reason to question the assumptions
of multivariate normality and equal covariance matrices for the two groups.
For the purpose of this example, we will ignore such problems. Example
2.2.2 includes a discussion of difficulties involved with the assumptions of
multivariate normality and equality of covariance matrices.

There are $r = 19$ and $t = 20$ observations on the two species. The mean
vectors are

$$\bar{y}_{1\cdot} = \begin{bmatrix} 194.47 \\ 267.05 \\ 137.37 \\ 185.95 \end{bmatrix} \quad \text{and} \quad \bar{y}_{2\cdot} = \begin{bmatrix} 179.55 \\ 290.80 \\ 157.20 \\ 209.25 \end{bmatrix}.$$

The individual estimated covariance matrices are

$$S_1 = \begin{bmatrix} 187.596 & 176.863 & 48.371 & 113.582 \\ 176.863 & 345.386 & 75.980 & 118.781 \\ 48.371 & 75.980 & 66.357 & 16.243 \\ 113.582 & 118.781 & 16.243 & 239.941 \end{bmatrix}$$

and

$$S_2 = \begin{bmatrix} 101.839 & 128.063 & 36.989 & 32.592 \\ 128.063 & 389.010 & 165.358 & 94.368 \\ 36.989 & 165.358 & 167.537 & 66.526 \\ 32.592 & 94.368 & 66.526 & 177.882 \end{bmatrix}.$$

The pooled estimate of the covariance matrix is

$$S = \begin{bmatrix} 143.559 & 151.803 & 42.527 & 71.993 \\ 151.803 & 367.788 & 121.877 & 106.245 \\ 42.527 & 121.877 & 118.314 & 42.064 \\ 71.993 & 106.245 & 42.064 & 208.073 \end{bmatrix}.$$

To test the hypothesis $H_0 : \mu_1 - \mu_2 = 0$, we need the statistics

$$(\bar{y}_{1\cdot} - \bar{y}_{2\cdot})' = (14.92, -23.75, -19.83, -23.30)$$

and

$$S^{-1} = \begin{bmatrix} 0.0132580 & -0.0053492 & 0.0015135 & -0.0021618 \\ -0.0053492 & 0.0066679 & -0.0047338 & -0.0005969 \\ 0.0015135 & -0.0047338 & 0.0130491 & -0.0007445 \\ -0.0021618 & -0.0005969 & -0.0007445 & 0.0060093 \end{bmatrix}.$$

These yield

$$(\bar{y}_{1\cdot} - \bar{y}_{2\cdot})' S^{-1} (\bar{y}_{1\cdot} - \bar{y}_{2\cdot})' = 13.70$$

and Hotelling's test statistic

$$T^2 = \left(\frac{19 \cdot 20}{19 + 20} \right) 13.70 = 133.5 .$$

The test statistic is huge, so one can feel reasonably confident that the populations are different in spite of any doubts about the validity of the assumptions. Standardizing T^2 so that it can be compared to an F distribution gives

$$\frac{T^2}{19 + 20 - 2} \frac{19 + 20 - 4 - 1}{4} = 30.7.$$

The corresponding reference distribution is an $F(4, 34)$.

EXERCISE 1.9.   Prove formulas (1), (2), (3), (4), and (5).

EXERCISE 1.10.   Prove that the confidence ellipsoid for $\mu_1 - \mu_2$ given in (6) is correct.

I.5 One-Way Analysis of Variance and Profile Analysis

Consider a multivariate one-way analysis of variance

$$y'_{ij} = \mu'_i + \varepsilon'_{ij} \tag{1}$$

where $i = 1, \ldots, a$, $j = 1, \ldots, n_i$, the ε_{ij}'s are independent $N(0, \Sigma)$ random vectors, and $\mu_i = (\mu_{i1}, \cdots, \mu_{iq})'$. One can do the standard test of $H_0 \colon \mu_1 = \cdots = \mu_a$ by testing the MANOVA model against the reduced model

$$y'_{ij} = \mu' + \varepsilon'_{ij} . \tag{2}$$

Write the dependent variable matrix as

$$Y = [y_{ij,h}]$$

where the pair (i, j) denote a row of Y and $h = 1, \ldots, q$ denotes a column. Let X be the design matrix for the one-way MANOVA. As in Christensen (1987, Chapter IV), $X = [x_{ij,k}]$ where $x_{ij,k} = \delta_{ik}$. This is just the standard design matrix for a univariate one-way ANOVA that is parameterized without a grand mean. Let

$$B = \begin{bmatrix} \mu'_1 \\ \vdots \\ \mu'_a \end{bmatrix}$$

and, in conformance with $Y = [y_{ij,h}]$, write a matrix with ε'_{ij} as the ij row, say

$$e = [\varepsilon'_{ij}] .$$

The full model (1) is

$$Y = XB + e$$

and the reduced model (2) is

$$Y = J\mu' + e .$$

Because the linear structure of X is just that of a one–way ANOVA, the analysis is similar to that of Christensen (1987, Chapter IV). Consider the estimation of Σ. The error matrix is

$$E = Y'(I - M)Y = [Y'_h(I - M)Y_{h'}]$$

where, from Christensen (1987, Chapter IV), the elements of this matrix are

$$
\begin{aligned}
Y'_h(I - M)Y_{h'} &= [(I - M)Y_h]' \, [(I - M)Y_{h'}] \\
&= \sum_{i=1}^{a} \sum_{j=1}^{n_i} (y_{ij,h} - \bar{y}_{i\cdot,h}) (y_{ij,h'} - \bar{y}_{i\cdot,h'}) .
\end{aligned}
$$

It follows easily that,

$$E = Y'(I - M)Y = \sum_{i=1}^{a} \sum_{j=1}^{n_i} (y_{ij} - \bar{y}_{i\cdot}) (y_{ij} - \bar{y}_{i\cdot})' .$$

Thinking of the observations as separate samples and using the results of Section 3, define

$$S_i = \sum_{j=1}^{n_i} (y_{ij} - \bar{y}_{i\cdot}) (y_{ij} - \bar{y}_{i\cdot})' / (n_i - 1) .$$

From the fact that

$$S = E/(n - a)$$

we can write S as a weighted average of the S_i's,

$$S = \sum_{i=1}^{a} (n_i - 1)S_i/(n - a) .$$

This is the usual pooled estimate of Σ.

Estimation of the parameter μ_{ih} is performed exactly as in a one–way ANOVA based on the model

$$Y_h = X\beta_h + e_h$$

where $\beta_h = (\mu_{1h}, \cdots, \mu_{ah})'$. The estimate is $\hat{\mu}_{ih} = \bar{y}_{i\cdot,h}$. This implies that

$$\hat{\mu}_i = \bar{y}_{i\cdot}$$

for $i = 1, \ldots, a$ and

$$\hat{B} = \begin{bmatrix} \bar{y}'_{1\cdot} \\ \vdots \\ \bar{y}'_{a\cdot} \end{bmatrix} .$$

Finally, the hypothesis matrix for testing the reduced model of no treatment effects, i.e., $H_0\colon \mu_1 = \cdots = \mu_a$, is

$$H = Y' \left(M - \frac{1}{n} J^n_n \right) Y = \sum_{i=1}^{a} n_i \left(\bar{y}_{i\cdot} - \bar{y}_{\cdot\cdot} \right) \left(\bar{y}_{i\cdot} - \bar{y}_{\cdot\cdot} \right)' .$$

EXAMPLE 1.5.1.   *One-Way Analysis of Variance with Repeated Measures*
A study was conducted to examine the effects of two drugs on heart rates. Thirty women were randomly divided into three groups of ten. An injection was given to each person. Depending on their group, women received either a placebo, drug A, or drug B. Repeated measurements of their heart rates were taken beginning at two minutes after the injection and at five minute intervals thereafter. Four measurements were taken on each individual. The data are given in Table I.2.

TABLE I.2. Heart Rate Data

	Placebo				DRUG A				B			
TIME	1	2	3	4	1	2	3	4	1	2	3	4
SUBJECT												
1	80	77	73	69	81	81	82	82	76	83	85	79
2	64	66	68	71	82	83	80	81	75	81	85	73
3	75	73	73	69	81	77	80	80	75	82	80	77
4	72	70	74	73	84	86	85	85	68	73	72	69
5	74	74	71	67	88	90	88	86	78	87	86	77
6	71	71	72	70	83	82	86	85	81	85	81	74
7	76	78	74	71	85	83	87	86	67	73	75	66
8	73	68	64	64	81	85	86	85	68	73	73	66
9	76	73	74	76	87	89	87	82	68	75	79	69
10	77	78	77	73	77	75	73	77	73	78	80	70

Clearly, observations taken over time on the same individual are correlated. We can consider the heart rate measurements taken at the four times to be four dependent variables. This is a completely randomized design so a multivariate one–way analysis of variance is appropriate. The treatments

are the two drugs and the placebo. The multivariate model can be written
as
$$y'_{ij} = \mu'_i + e'_{ij}$$
where $i = 1, 2, 3$ and $j = 1, 2, \ldots, 10$. Because $q = 4$,
$$y'_{ij} = (y_{ij1}, y_{ij2}, y_{ij3}, y_{ij4})$$
and
$$\mu'_i = (\mu_{i1}, \mu_{i2}, \mu_{i3}, \mu_{i4}) \ .$$
The least squares estimate of B is

$$
\hat{B} \;=\; \begin{bmatrix} \hat{\mu}'_1 \\ \hat{\mu}'_2 \\ \hat{\mu}'_3 \end{bmatrix} = \begin{bmatrix} \bar{y}'_{1\cdot} \\ \bar{y}'_{2\cdot} \\ \bar{y}'_{3\cdot} \end{bmatrix} = \begin{bmatrix} \bar{y}_{1\cdot1} & \bar{y}_{1\cdot2} & \bar{y}_{1\cdot3} & \bar{y}_{1\cdot4} \\ \bar{y}_{2\cdot1} & \bar{y}_{2\cdot2} & \bar{y}_{2\cdot3} & \bar{y}_{2\cdot4} \\ \bar{y}_{3\cdot1} & \bar{y}_{3\cdot2} & \bar{y}_{3\cdot3} & \bar{y}_{3\cdot4} \end{bmatrix}
$$
$$
=\; \begin{bmatrix} 73.8 & 72.8 & 72.0 & 70.3 \\ 82.9 & 83.1 & 83.4 & 82.9 \\ 72.9 & 79.0 & 79.6 & 72.0 \end{bmatrix} \ .
$$

The unbiased estimates of the covariance matrix computed within each
treatment group are

$$
S_1 \;=\; \begin{bmatrix} 18.62 & 15.07 & 8.22 & 0.73 \\ 15.07 & 17.07 & 11.11 & 3.07 \\ 8.22 & 11.11 & 13.33 & 8.89 \\ 0.73 & 3.07 & 8.89 & 11.34 \end{bmatrix} ,
$$

$$
S_2 \;=\; \begin{bmatrix} 10.54 & 13.57 & 12.93 & 6.77 \\ 13.57 & 22.54 & 18.62 & 10.12 \\ 12.93 & 18.62 & 21.82 & 12.82 \\ 6.77 & 10.12 & 12.82 & 8.99 \end{bmatrix} ,
$$

and

$$
S_3 \;=\; \begin{bmatrix} 24.10 & 25.11 & 19.18 & 18.89 \\ 25.11 & 28.22 & 23.11 & 22.33 \\ 19.18 & 23.11 & 24.93 & 19.00 \\ 18.89 & 22.33 & 19.00 & 22.00 \end{bmatrix} \ .
$$

In general, a weighted average of these gives S. Because the sample sizes
for the drugs are all equal, the weights are equal and a simple average gives

$$
S \;=\; \begin{bmatrix} 17.76 & 17.91 & 13.44 & 8.80 \\ 17.91 & 22.61 & 17.61 & 11.84 \\ 13.44 & 17.61 & 20.03 & 13.57 \\ 8.80 & 11.84 & 13.57 & 14.11 \end{bmatrix} \ .
$$

The error matrix is

$$
E = (30 - 3)S \;=\; \begin{bmatrix} 479.4 & 483.7 & 363.0 & 237.5 \\ 483.7 & 610.5 & 475.6 & 319.7 \\ 363.0 & 475.6 & 540.8 & 366.4 \\ 237.5 & 319.7 & 366.4 & 381.0 \end{bmatrix} \ .
$$

The correlation matrix is

$$R = \begin{bmatrix} 1.000 & .894 & .713 & .556 \\ .894 & 1.000 & .828 & .663 \\ .713 & .828 & 1.000 & .807 \\ .556 & .663 & .807 & 1.000 \end{bmatrix}.$$

It consists of the correlations between all the variables, e.g., $r_{12} = .894 = 17.91/\sqrt{17.76}\sqrt{22.61}$.

The reason for treating this data as a multivariate one–way ANOVA was our initial claim that the observations made on an individual are correlated. This certainly seems to be borne out by the large off-diagonal elements of the correlation matrix. In fact, for normal data we could test whether the correlations are zero. In Christensen (1987, Section VI.5), a t test for partial correlations was presented. While these are not partial correlations in the usual sense, the correlations are pooled over three groups so these correlations represent a special case of partial correlations. A test of H_0: $\rho_{34} = 0$ can be based on comparing $\sqrt{30 - 4}\,(.807)/\sqrt{1 - .807^2} = 6.97$ to a $t(30 - 4)$ distribution. The correlation is highly significant.

Of course, there are methods available for analyzing correlated data other than the multivariate linear model. One alternative is to consider a split plot model where individual women are whole plots, drugs are whole plot treatments, and the four times are subplot treatments. (There are no blocks in the whole plots.) As discussed in Christensen (1987, Chapter XI), the split plot model assumes that observations in different whole plots are un-correlated while all observations in the same whole plot have a common positive correlation. In the multivariate one–way, the covariance matrix for observations on the same person is denoted by Σ. The split plot model assumes that

$$\Sigma = \sigma^2 \begin{bmatrix} 1 & \rho & \cdots & \rho \\ \rho & 1 & \cdots & \rho \\ \vdots & \vdots & \ddots & \vdots \\ \rho & \rho & \cdots & 1 \end{bmatrix}.$$

If this assumption is correct, the split plot model is more appropriate than the multivariate one–way ANOVA. Although the correlation matrix R does not seem very supportive of the split plot model assumption, a detailed comparison of the two analyses will be made in Example 1.5.3.

Returning to the multivariate one–way analysis of this data, we might wish to test for differences in the treatment means. If we fit the reduced model

$$y'_{ij} = \mu' + e'_{ij}$$

we obtain

$$S_0 = \begin{bmatrix} 37.64 & 31.52 & 28.02 & 33.72 \\ 31.52 & 39.60 & 37.10 & 32.29 \\ 28.02 & 37.10 & 41.89 & 35.39 \\ 33.72 & 32.29 & 35.39 & 45.37 \end{bmatrix}$$

and

$$E_0 = (30 - 1)S_0 = \begin{bmatrix} 1091.5 & 914.2 & 812.7 & 977.9 \\ 914.2 & 1148.3 & 1076.0 & 936.4 \\ 812.7 & 1075.9 & 1214.7 & 1026.3 \\ 977.9 & 936.4 & 1026.3 & 1315.9 \end{bmatrix}.$$

The hypothesis matrix can be computed as

$$\begin{aligned} H &= Y'(M - M_0)Y = Y'(I - M_0)Y - Y'(I - M)Y \\ &= E_0 - E \\ &= \begin{bmatrix} 612.1 & 430.5 & 449.7 & 740.4 \\ 430.5 & 537.8 & 600.4 & 616.7 \\ 449.7 & 600.4 & 673.9 & 659.9 \\ 740.4 & 616.7 & 659.9 & 934.9 \end{bmatrix}. \end{aligned}$$

If there are no differences in the drug means, then H divided by its degrees of freedom should estimate Σ. Computing this gives

$$\frac{1}{2}H = \begin{bmatrix} 306 & 215 & 225 & 370 \\ 215 & 269 & 300 & 308 \\ 225 & 300 & 337 & 330 \\ 370 & 308 & 330 & 467 \end{bmatrix}.$$

Even though this estimate has only two degrees of freedom, it is clear that this is not estimating the same thing that S is estimating.

If the data have a multivariate normal distribution, formal tests can be performed. The various test statistics, comparison values, and $\alpha = .01$ normal theory critical values are given below.

Statistic	Observed Value	Comparison Value	Critical Value
U	.0628	.75	.440
ϕ_{max}	5.52	.07	less than 1.19
T^2	188.0	8	26.66
V	1.44	.276	less than .725

In particular,

$$T^2 = 187.99$$

which is much larger than the intuitive comparison value $q\,[r(X) - r(X_0)] = 4(2) = 8$. Comparing T^2 to the exact small sample distribution, McKeon's approximate distribution or the asymptotic χ^2 distribution with $q\,[r(X) - r(X_0)] = 8$ degrees of freedom leads to the clear conclusion that the drugs have different multivariate means.

The validity of multivariate linear model tests depends on the data having a multivariate normal distribution. In particular, each dependent variable must be normal and any linear combinations of the variables must also be normal. As in Christensen (1987, Section XIII.2), the normality of data can be evaluated using normal plots. If the sample size for each drug were large, it would be appropriate to check for normality within the treatment groups. Since it is difficult to draw distributional conclusions using only ten observations, the residual matrix $\hat{e} = (I - M)Y$ was used. In general, the standardized residuals are more appropriate for normal plots but for this model all cases have the same leverage, so the residuals are equivalent.

Normal plots of the residuals were made for each dependent variable and also for one linear combination of the residuals. The linear combination was the sum of the residuals for the four variables. All of the normal plots looked reasonably linear. The Wilk–Francia statistic W' was also computed for each plot (cf. Christensen, 1987, Section XIII.2). The results are given below.

Variable	W'
Time 1	.986
Time 2	.974
Time 3	.949
Time 4	.980
Sum of Variables	.960

Comparing these values to those tabled in Weisberg (1974) gives no cause for concern. Of course, to do a proper check for multivariate normality, one should inspect the normal plot for every linear combination of the columns of the residual matrix. Unfortunately, that can be rather time consuming. The data analyst must usually be satisfied with evaluating some finite number of linear combinations.

Another assumption of the multivariate linear model is that the covariance matrix is the same for every observation vector y_i. As in Christensen (1987, Section XIII.4), residual plots can be performed checking for constant variance in each of the dependent variables. Residual plots for linear combinations of the columns of the residual matrix should also display constant variance.

In a one-way ANOVA, the assumption of a constant covariance matrix can also be checked by comparing the estimated covariance matrices for each of the individual treatments. The sample covariance matrices S_1, S_2, and S_3 were given earlier. Although they display some differences that seem to be fairly substantial (the estimated covariances between Time 1 and Time 4 vary from .73 to 18.89), taken as a whole the covariance matrices are reasonably consistent. In fact, Bartlett's modification to the likelihood ratio test for equality of the covariance matrices gives a P value greater than .05 (based on a $\chi^2(20)$ approximation). This in spite of the fact that the test is so notoriously sensitive to nonnormality that it is rarely used.

A visual approach to evaluating the individual covariance matrices can be based on plotting pairs of variables. The ten pairs of observations at times 1 and 2 can be plotted for each of the three drugs. Each of these three plots should be roughly elliptical and the ellipses should have the same orientation in two-dimensional space. A similar set of $a = 3$ plots can

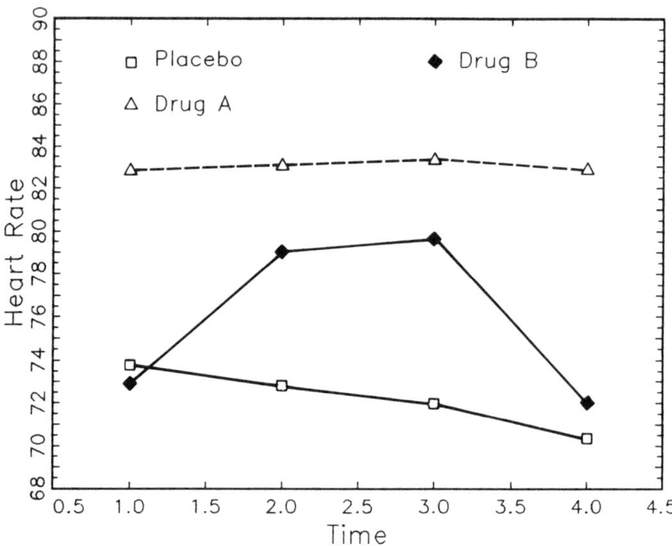

FIGURE I.5. Heart Rate Profiles

be made for each of the $\binom{q}{2} = 6$ pairs of dependent variables. Given the difficulties of evaluating plots based on only ten observations, there seems to be no reason to doubt the assumption of equal covariance matrices for this example. In particular, the orientations of the three plots in each set are consistent.

PROFILE ANALYSIS

Profile analysis seeks to examine possible similarities between the μ_i vectors. The name derives from the plots of the a theoretical curves defined by (h, μ_{ih}), $h = 1, \ldots, q$. These curves are referred to as profiles. (The q points available for each i are connected by line segments to obtain the a different profiles.) Because the μ_i's are not available, the estimated profiles $(h, \hat{\mu}_{ih})$ provide a valuable visual display.

EXAMPLE 1.5.2.   Figure I.5 contains a plot of the estimated profiles from Example 1.5.1.

Three questions are commonly asked in profile analysis. First, whether the curves are parallel. Second, whether the curves have the same average level. The average level for each curve is defined as the average over different dependent variables. The third question is whether the average curve is horizontal. The average curve is obtained by averaging over treatments. Note that the hypotheses involving averages are of particular interest when the curves are parallel. In this case, if the average levels are the same the profiles are the same and if the average profile is horizontal all profiles are horizontal.

To test these questions, it is convenient to define, for a general value r, the $(r-1) \times r$ matrix

$$\Lambda_r' = [J_{r-1}, -I_{r-1}].$$

Note that

$$\Lambda_a' B = \begin{bmatrix} \mu_1' - \mu_2' \\ \mu_1' - \mu_3' \\ \vdots \\ \mu_1' - \mu_a' \end{bmatrix}$$

so $H_0 \colon \Lambda_a' B = 0$ is precisely $H_0 \colon \mu_1 = \mu_2 = \cdots = \mu_a$.

The hypothesis that the profiles are parallel is that for every possible choice of i and i'.

$$\mu_{i1} - \mu_{i'1} = \mu_{i2} - \mu_{i'2} = \cdots = \mu_{iq} - \mu_{i'q}.$$

Equivalently, the hypothesis is that

$$\mu_{11} - \mu_{i1} = \cdots = \mu_{1q} - \mu_{iq} \quad \text{for} \quad i = 2, \ldots, a.$$

Finally, the hypothesis can be written $(\mu_{11} - \mu_{i1}) - (\mu_{12} - \mu_{i2}) = 0$, $(\mu_{11} - \mu_{i1}) - (\mu_{13} - \mu_{i3}) = 0$, $\ldots$, $(\mu_{11} - \mu_{i1}) - (\mu_{1q} - \mu_{iq}) = 0$ for $i = 2, \ldots, a$. With $\Lambda_a' B$ illustrated above, it is not difficult to see that the profiles are parallel if and only if $\Lambda_a' B \Lambda_q = 0$. The test of parallel profiles is

$$H_0 \colon \Lambda_a' B \Lambda_q = 0 \qquad \text{versus} \qquad H_A \colon \Lambda_a' B \Lambda_q \neq 0.$$

This is a standard multivariate hypothesis and yields the hypothesis statistic

$$H_* = (\Lambda_a' \hat{B} \Lambda_q)'[\Lambda_a'(X'X)^{-1}\Lambda_a]^{-1}(\Lambda_a' \hat{B} \Lambda_q).$$

Recall that $X'X$ is $Diag(n_1, \ldots, n_a)$. The hypothesis statistic is compared to

$$E_* = \Lambda_q' E \Lambda_q .$$

To test whether the average level of the curves are the same, we need the average level for each curve. For the i'th curve, the average level is $\bar{\mu}_{i\cdot} = \frac{1}{q} \sum_{h=1}^{q} \mu_{ih}$. The hypothesis $\bar{\mu}_{1\cdot} = \cdots = \bar{\mu}_{a\cdot}$ can be written as $\bar{\mu}_{1\cdot} - \bar{\mu}_{2\cdot} = 0$, $\bar{\mu}_{1\cdot} - \bar{\mu}_{3\cdot} = 0$, $\ldots$, $\bar{\mu}_{1\cdot} - \bar{\mu}_{a\cdot} = 0$. Equivalently, we can look for equality of the curve totals where the i'th curve total is $\mu_{i\cdot} = \mu_i' J_q = \sum_{h=1}^{q} \mu_{ih}$. Recalling the form of $\Lambda_a' B$, it is easily seen that the null hypothesis is

$$H_0 \colon \Lambda_a' B J_q = 0$$

and

$$H_* = (\Lambda_a' \hat{B} J_q)'[\Lambda_a'(X'X)^{-1}\Lambda_a]^{-1}(\Lambda_a' \hat{B} J_q).$$

The error statistic is

$$E_* = J_q' E J_q \, .$$

The third hypothesis is that the average curve is horizontal. The average curve is based on $\bar{\mu}'_\cdot = \frac{1}{a} J_a' B = \frac{1}{a} \sum_{i=1}^{a} \mu_i'$. Write $\bar{\mu}'_\cdot = (\bar{\mu}_{\cdot 1}, \ldots, \bar{\mu}_{\cdot q})$. Testing whether the curve is horizontal amounts to testing that $\bar{\mu}_{\cdot 1} - \bar{\mu}_{\cdot 2} = 0, \ldots, \bar{\mu}_{\cdot 1} - \bar{\mu}_{\cdot q} = 0$. Equivalently, we can test whether the total curve is horizontal by examining $\mu'_\cdot = J_a' B$. Clearly, the test that the total curve is horizontal is

$$H_0 : J_a' B \Lambda_q = 0 \qquad \text{versus} \qquad H_A : J_a' B \Lambda_q \neq 0$$

with

$$H_* = (J_a' \hat{B} \Lambda_q)' [J_a' (X'X)^{-1} J_a]^{-1} (J_a' \hat{B} \Lambda_q)$$

and

$$E_* = \Lambda_q' E \Lambda_q \, .$$

Finally, another word about combining the hypotheses. The curves are both parallel and have the same average level if and only if the curves are identical, i.e., $\mu_1 = \cdots = \mu_a$. In other notation, $\Lambda_a' B \Lambda_q = 0$ and $\Lambda_a' B J_q = 0$ if and only if $\Lambda_a' B = 0$. Also, the curves are parallel and the average curve is horizontal if and only if all of the curves are horizontal. Putting it another way, $\Lambda_a' B \Lambda_q = 0$ and $J_a' B \Lambda_q = 0$ if and only if $B \Lambda_q = 0$. The hypothesis that $B \Lambda_q = 0$ is simply that all of the curves are horizontal.

EXAMPLE 1.5.3.   We again consider the heart rate data of Example 1.5.1. It was mentioned earlier that using a split plot model is an alternative method of analyzing this data. The nature of profile analysis will be clearer if we contrast profile analysis with the more familiar split plot analysis. We begin with the split plot analysis (cf. Christensen, 1987, Chapter XI).

The split plot model for this experimental design is

$$y_{ijk} = \mu + \delta_i + \eta_{ij} + \tau_k + (\delta\tau)_{ik} + e_{ijk}$$

$i = 1, 2, 3$; $j = 1, \ldots, 10$; $k = 1, 2, 3, 4$. Here δ_i indicates the drug effect, τ_k indicates the time effect, η_{ij} indicates a random error for the ij individual, and e_{ijk} is a random error specific to the observation on the ij individual at the k'th time period. As usual, we assume $\text{Var}(\eta_{ij}) = \sigma_w^2$, $\text{Var}(e_{ijk}) = \sigma_s^2$, and that all of the random errors have zero covariance with all of the other random errors. For the construction of tests and confidence regions, it is assumed that the errors have a joint multivariate normal distribution. To construct the ANOVA table for the split plot model, it is often convenient to compute the sums of squares treating the data as a complete factorial experiment and then combine terms to get the correct table. Considering the data as a complete factorial, there are three treatments, Drugs (D), Times (T), and Individuals (I). The corresponding ANOVA table is

Source	df	SS	MS
Drugs	2	2438.5	1219.2
Individuals	9	404.3	44.93
$D \times I$	18	1221.5	67.86
Times	3	222.3	74.10
$T \times D$	6	320.1	53.36
$T \times I$	27	64.0	2.37
$T \times D \times I$	54	321.9	5.96

Since individuals do not constitute whole plot blocks, the whole plot error is found by pooling the Individuals and the $D \times I$ terms. The subplot error contains the $T \times I$ and $T \times D \times I$ terms. The correct split plot ANOVA table is

Source	df	SS	MS	F
Drugs	2	2438.5	1219.2	20.25
Whole Plot Error	27	1625.8	60.22	
Times	3	222.3	74.10	15.56
$T \times D$	6	320.1	53.36	11.20
Subplot Error	81	385.9	4.76	

All of the effects are highly significant. The next step in the split plot analysis might be to examine contrasts in the interactions. This involves looking at contrasts in the 3×4 table of means $\bar{y}_{i \cdot k}$.

The three basic tests in profile analysis are analogous to the tests for whole plot treatments, subplot treatments, and interaction. The test for parallelism is equivalent to testing for interaction. The test of whether the average levels of the curves are the same is equivalent to testing for whole plot treatments. The test of whether the average curve is horizontal is equivalent to testing for subplot treatments. Profile analysis is based on examining the structure of

$$\hat{B} = \begin{bmatrix} 73.8 & 72.8 & 72.0 & 70.3 \\ 82.9 & 83.1 & 83.4 & 82.9 \\ 72.9 & 79.0 & 79.6 & 72.0 \end{bmatrix} .$$

This is the 3×4 table of means, $\bar{y}_{i \cdot k}$. The split plot analysis examines the structure of exactly the same means table. While not all interaction contrasts have obvious multivariate tests, any interaction contrast that can be written as $\lambda' B \xi$ can also be tested in the multivariate model.

Depending on the software one has available, to test the hypothesis of parallel profiles H_0: $\Lambda_3' B \Lambda_4 = 0$, it may be convenient to do a one-way ANOVA on a new set of variables $(y_{ij1}^*, y_{ij2}^*, y_{ij3}^*) = (y_{ij1} - y_{ij2}, y_{ij1} - y_{ij3}, y_{ij1} - y_{ij4}) = y_{ij}' \Lambda_4$. The one-way ANOVA on the transformed data yields

$$S_* = \Lambda_4' S \Lambda_4 = \begin{bmatrix} 4.54 & 4.01 & 2.89 \\ 4.01 & 10.90 & 9.09 \\ 2.89 & 9.09 & 14.27 \end{bmatrix}$$

and

$$E_* = \Lambda_4' \hat{E} \Lambda_4 = \begin{bmatrix} 122.5 & 108.3 & 77.9 \\ 108.3 & 294.2 & 245.3 \\ 77.9 & 245.3 & 385.4 \end{bmatrix} . \tag{3}$$

Fitting the model of no drug effects yields

$$S_{0*} = \Lambda_4' S_0 \Lambda_4 = \begin{bmatrix} 14.19 & 15.19 & 4.68 \\ 15.19 & 23.48 & 11.28 \\ 4.68 & 11.28 & 15.57 \end{bmatrix}$$

and

$$E_{0*} = \Lambda_4' E_0 \Lambda_4 = \begin{bmatrix} 411.4 & 440.6 & 135.7 \\ 440.6 & 680.8 & 327.2 \\ 135.7 & 327.2 & 451.5 \end{bmatrix} .$$

The hypothesis statistic is

$$H_* = E_{0*} - E_* = \begin{bmatrix} 288.9 & 332.3 & 57.8 \\ 332.3 & 386.6 & 81.9 \\ 57.8 & 81.9 & 66.1 \end{bmatrix} .$$

Dividing by the degrees of freedom gives

$$\frac{1}{2} H_* = \begin{bmatrix} 144.5 & 166.1 & 28.8 \\ 166.1 & 193.3 & 40.8 \\ 28.8 & 40.8 & 33.1 \end{bmatrix} .$$

It is clear that $\frac{1}{2} H_*$ and S_* are not estimating the same thing. The test statistics, comparison values, and normal theory critical values for an $\alpha = .05$ test are given below.

Statistic	Observed Value	Comparison Value	Critical Value
U	.204	.807	.510
ϕ_{max}	3.22	.07	less than .886
T^2	91.4	6	22.08
V	.902	.207	less than.569

All of the values are far from their comparison values and are in the critical regions of the tests. The hypothesis that the profiles are parallel is rejected. In other words, the relationship between the time means depends on which drug you look at, i.e., there is interaction.

To explore further the lack of parallelism, consider the orthogonal contrasts in drugs determined by $\lambda_1' = (1, -1, 0)$ and $\lambda_2' = (-1, -1, 2)$. The test of H_0: $\lambda_1' B \Lambda_4 = 0$ examines whether the placebo curve is parallel to the drug A curve. H_0: $\lambda_2' B \Lambda_4 = 0$ hypothesizes that the drug B curve is parallel to the average of the others. Define the hypothesis matrices

$$H_i = (\lambda_i' \hat{B} \Lambda_4)' \left[\lambda_i' (X'X)^{-1} \lambda_i \right]^{-1} (\lambda_i' \hat{B} \Lambda_4) .$$

Thus

$$H_1 \;=\; \begin{bmatrix} 1.2 \\ 2.3 \\ 3.5 \end{bmatrix} (5)[1.2, 2.3, 3.5]$$

$$=\; \begin{bmatrix} 7.20 & 13.80 & 21.00 \\ 13.80 & 26.42 & 40.25 \\ 21.00 & 40.25 & 61.25 \end{bmatrix}$$

and

$$H_2 \;=\; \begin{bmatrix} -13 \\ -14.7 \\ -1.7 \end{bmatrix} (5/3)[-13, -14.7, -1.7]$$

$$=\; \begin{bmatrix} 281.67 & 318.50 & 36.83 \\ 318.50 & 360.15 & 41.65 \\ 36.83 & 41.65 & 4.82 \end{bmatrix} .$$

Because the contrasts λ_i are orthogonal

$$H_* = H_1 + H_2 \;.$$

The test statistics, comparison values, and $\alpha = .01$ critical values are given below.

Statistic	$\lambda_1' B\Lambda_4 = 0$	$\lambda_2' B\Lambda_4 = 0$	Comp. Value	Crit. Value
U	.853	.237	.900	617
ϕ_{max}	.17	3.21	.07	.56
T^2	4.65	86.75	3.00	15.16
V	.147	.763	.103	.359

Clearly, the predominant cause of nonparallelism is due to the fact that drug B is not parallel to the placebo and drug A. The placebo and drug A appear to be reasonably parallel.

A problem with doing formal tests for the orthogonal contrasts considered above is that they were chosen after looking at the sample profiles in Figure I.5. This invalidates the distributions used. One nice property of the Lawley–Hotelling T^2 is that the experimentwise error rate for the multiple tests can be controlled. We use an argument similar to that on which Scheffé's multiple comparison method is based, cf. Christensen (1987, Section V.1). Since $H_* = H_1 + H_2$, we have

$$T_*^2 = T_1^2 + T_2^2$$

where $T_*^2 = \operatorname{tr}(H_* S_*^{-1})$, $T_1^2 = \operatorname{tr}(H_1 S_*^{-1})$, and $T_2^2 = \operatorname{tr}(H_2 S_*^{-1})$. Moreover,

$$T_i^2 \leq T_*^2$$

If we reject $H_0: \lambda_i' B \Lambda_4 = 0$ only when T_i^2 is greater than the critical value appropriate for an α-level test based on T_*^2 then the experimentwise error rate is no greater than α (and probably much less). Since $T_2^2 = 86.75$ for the second contrast and the critical value appropriate for T_*^2 is 22.08, the profile for drug B and the average profile for the placebo and drug A display a significant lack of parallelism, regardless of the fact that the contrast was chosen after examining the data.

Although not every contrast in the drug–time interaction can be written as $\lambda' B \xi$ for some drug contrast vector λ and time contrast vector ξ, these are often the most interpretable contrasts. Clearly, any such contrast can be tested in the multivariate model. As just illustrated, orthogonal contrasts in the drugs are a useful tool, especially in relation to T^2. Because of the correlation between times, orthogonal contrasts are of little interest relative to the times. For example, it is natural in the split plot model to examine orthogonal polynomial contrasts in the times. Unfortunately, the relationship between orthogonal polynomial contrasts and polynomial regression depends on the validity of the least squares analysis. Using the standard polynomial contrasts is not particularly appropriate for the multivariate model. Fitting a polynomial in the times requires the use of a growth curve model, cf. Section 6.

In univariate ANOVA, when there is interaction, the tests of main effects are difficult to interpret. This is precisely because they involve averaging over the interactions. In profile analysis, when the profiles are not parallel, the hypotheses that involve averages are also difficult to interpret. For example, if the curves are not parallel, testing whether the average curve is horizontal does not seem too interesting. If this null hypothesis were true, it would tell us very little. On the other hand, if the profiles are parallel and the average profile is horizontal then every profile must be horizontal. Because there is clear evidence that the profiles are not parallel, the other two standard tests in profile analysis are of little interest. In spite of this fact, we now illustrate their computation and interpretation.

The test for equality of the average levels, $H_0: \Lambda_3' B J_4 = 0$, can be performed by doing a univariate one–way ANOVA on the dependent variable constructed by adding together the four time variables. The test corresponds to testing for drug main effects in the split plot model. The ANOVA table is given below.

Source	df	SS	MS	F
Drugs	2	9754	4877	20.25
Error	27	6503	241	
Total	29	16,257		

The F test is highly significant, thus indicating that the drugs affect the average level of the curves. In other words, averaging over times, there are differences in the drugs. Note that this is precisely the split plot model test for Drug main effects. The sums of squares for Drugs and Error are exactly four times those reported earlier for the split plot model. This is a result of the fact that $q = 4$.

The test of H_0: $J_3' B \Lambda_4 = 0$ is a test of whether the average curve is horizontal. It tests for differences in Times averaging over Drugs. Using the dependent variables formed to test for parallelism, we get

$$\hat{B}\Lambda_4 = \begin{bmatrix} 1.0 & 1.8 & 3.5 \\ -0.2 & -0.5 & 0.0 \\ -6.1 & -6.7 & 0.9 \end{bmatrix}$$

and

$$J_3'\hat{B}\Lambda_4 = (-5.3, -5.4, 4.4) .$$

From standard ANOVA theory

$$(X'X) = \begin{bmatrix} 10 & 0 & 0 \\ 0 & 10 & 0 \\ 0 & 0 & 10 \end{bmatrix}$$

and

$$\left[J_3'(X'X)^{-1}J_3 \right]^{-1} = \frac{10}{3} .$$

It follows that

$$H_* = \frac{10}{3} \begin{bmatrix} -5.3 \\ -5.4 \\ 4.4 \end{bmatrix} [-5.3, -5.4, 4.4]$$

$$- \begin{bmatrix} 93.6 & 95.4 & -77.7 \\ 95.4 & 97.2 & -79.2 \\ -77.7 & -79.2 & 64.5 \end{bmatrix} .$$

Clearly this matrix has rank one, hence one degree of freedom. The error matrix E_* is the same as used to test parallelism. It is given in equation (3). An ad hoc evaluation of the null hypothesis compares $H_*/1$ with S_*. The matrices are vastly different. Formal tests confirm the conclusion that the average curve is not horizontal.

It is interesting to examine the relationship between the split plot model test for Time main effects and the multivariate test for average time effects. With one degree of freedom for H_*, the standard test statistics are equivalent. In particular,

$$\begin{aligned} T^2 &= [n - r(X)]\,\mathrm{tr}\left[E_*^{-1}H_*\right] \\ &= \mathrm{tr}\left[(\Lambda_4'S\Lambda_4)^{-1}H_*\right] \end{aligned}$$

If we require that S have the structure of a split plot covariance matrix, write

$$S = \hat{\sigma}^2 \begin{bmatrix} 1 & \hat{\rho} & \hat{\rho} & \hat{\rho} \\ \hat{\rho} & 1 & \hat{\rho} & \hat{\rho} \\ \hat{\rho} & \hat{\rho} & 1 & \hat{\rho} \\ \hat{\rho} & \hat{\rho} & \hat{\rho} & 1 \end{bmatrix}$$
$$= \hat{\sigma}^2(1 - \hat{\rho})I_4 + \hat{\sigma}^2\hat{\rho}J_4^4$$

then it is easily seen that

$$\Lambda_4' S \Lambda_4 = \hat{\sigma}^2(1 - \hat{\rho})\Lambda_4'\Lambda_4$$

so

$$T^2 = \operatorname{tr}\left[(\Lambda_4'\Lambda_4)^{-1}H_*\right] / \left[\hat{\sigma}^2(1 - \hat{\rho})\right] .$$

Recall from Christensen (1987, Chapter XI) that MS(Subplot Error) is the unbiased estimate of $\sigma^2(1 - \rho)$. Also note that $\operatorname{tr}\left[(\Lambda_4'\Lambda_4)^{-1}H_*\right] =$ SS(Times). Thus, restricting the form of Σ leads naturally to

$$T^2 = \text{SS(Times)}/\text{MS(Subplot Error)}.$$

The split plot F statistic for testing main effects in Times is just $T^2/3$. However, by imposing additional structure on Σ one gains degrees of freedom for the denominator of the F statistic.

A similar argument relates the T^2 for parallelism to the F statistic for interaction.

There are a number of ways that one can approach the computational problems involved in one-way MANOVA and profile analysis. The three main computational approaches are through use of flexible interactive statistics packages, structured programs, and matrix manipulation programs. The primary computation involved is finding covariance matrices: one for each treatment, a pooled covariance matrix, and one ignoring treatments. These can be found using a flexible interactive package such as MINITAB that includes some matrix commands. More structured programs require less thought but more reading. For example, BMDP 4V automatically provides the univariate split plot analysis, the multivariate profile analysis, and compromise tests based on the Greenhouse-Geisser (1959) and Huynh-Feldt (1976) adjustments to the degrees of freedom of the univariate tests. The additional reading is needed to identify such things as the definition of Λ used in profile analysis. (The definition given above was just one of an infinite number of equally good choices. While the choice will not effect the standard test statistics, the matrices H_* and E_* do depend on Λ.) Other things that need to be identified in structured programs are the exact definitions of the test statistics. For example, in BMDP 4V, the TRACE statistic is $\operatorname{tr}[HE^{-1}]$ rather than T^2 or V and MXROOT is θ_{max} rather than ϕ_{max}. Of course, with a good matrix manipulation package such as GAUSS you can do any of these computations directly.

I.6   Growth Curves

A natural extension of profile analysis is to develop models for the profiles of the various groups. For example, if the q observations in each row of Y are taken at times $t_1, \ldots, t_q$, we might incorporate a growth curve model

$$\mu_i = \begin{bmatrix} \mu_{i1} \\ \vdots \\ \mu_{iq} \end{bmatrix} = \begin{bmatrix} 1 & t_1 & t_1^2 \\ \vdots & \vdots & \vdots \\ 1 & t_q & t_q^2 \end{bmatrix} \begin{bmatrix} \gamma_{i0} \\ \gamma_{i1} \\ \gamma_{i2} \end{bmatrix}$$
$$= Z\gamma_i \,,$$

which posits that the components of μ_i, when plotted against time, form the parabola $\mu_{ih} = \gamma_{i0} + \gamma_{i1}t_h + \gamma_{i2}t_h^2$. Note that the coefficients of the parabola are allowed to vary with the treatment group but that the design matrix of the growth curve is the same for each treatment.

In general, for a fixed $q \times r$ matrix Z with $r(Z) = r < q$, we assume a linear growth curve model for each μ_i, say

$$\mu_i = Z\gamma_i \,.$$

The multivariate linear model in profile analysis is the multivariate one-way ANOVA

$$Y = XB + e$$

where the cell means parameterization

$$B = \begin{bmatrix} \mu_1' \\ \vdots \\ \mu_a' \end{bmatrix}$$

is used. Incorporating the models for the growth curves gives

$$B = \begin{bmatrix} \gamma_1' Z' \\ \vdots \\ \gamma_a' Z' \end{bmatrix} = \Gamma Z'$$

where

$$\Gamma = \begin{bmatrix} \gamma_1' \\ \vdots \\ \gamma_a' \end{bmatrix} \,.$$

The complete multivariate growth curve model is

$$Y = X\Gamma Z' + e \,. \tag{1}$$

Just as multivariate linear models are a special class of univariate linear models, the growth curve model can also be written as a univariate linear model. Unfortunately, it is not a member of the class that we have been calling multivariate linear models. Multivariate linear models are not allowed to specify structure between the means of the different variables. The optimal estimation results and the distribution theory based on the Wishart distribution depend on applying the same projection operator to each dependent variable vector Y_h. If, as in the growth curve model, one specifies a linear structure for the means across the dependent variables then the best estimates for any variable will incorporate information available from the other variables. For example, $\rho' X \hat{\beta}_h$ will not equal $\rho' M Y_h$ but will depend on the information in the other columns of Y. This is enough to invalidate the usual estimation results and distribution theory. Some alternative form of analysis must be developed.

Model (1) should not be confused with the transformation of $Y = XB + e$ considered in Section 2: $YZ = XBZ + eZ$. The transformed model is still a standard multivariate linear model. It simply reduces the dimensionality from q to r. In both models Z is $q \times r$ with $r(Z) = r < q$, however in model (1) we are multiplying by Z' instead of Z. Thus, the dimensions of the parameter matrices B and Γ are different. In the transformed model, B is $p \times q$ whereas in model (1), Γ is $p \times r$. These distinctions make all the difference in being able to derive optimal estimates and closed form small sample distributions.

We present two methods of analysis for growth curve models. The first is an ad hoc method that transforms the growth curve model into a standard multivariate linear model. The second method is an extension of the first; it uses a multivariate analysis of covariance and has a stronger theoretical foundation.

To transform model (1) into a multivariate linear model, multiply on the right by $Z(Z'Z)^{-1}$ to get

$$YZ(Z'Z)^{-1} = X\Gamma + eZ(Z'Z)^{-1}. \tag{2}$$

Because this is a standard multivariate one-way ANOVA model, estimation and testing can be performed in the usual ways. Note that with

$$Y = \begin{bmatrix} y'_{11} \\ \vdots \\ y'_{an_a} \end{bmatrix},$$

the dependent variable matrix in (2) is

$$YZ(Z'Z)^{-1} = [(Z'Z)^{-1}Z'y_{11}, \ldots, (Z'Z)^{-1}Z'y_{an_a}]'.$$

The rows of the new dependent variable matrix are just the estimated regression coefficients from

$$y_{ij} = Z\gamma_i + e. \tag{3}$$

Writing $\hat{\gamma}_{ij} = (Z'Z)^{-1}Z'y_{ij}$, we have

$$YZ(Z'Z)^{-1} = \begin{bmatrix} \hat{\gamma}'_{11} \\ \vdots \\ \hat{\gamma}'_{an_a} \end{bmatrix}.$$

The essence of the growth curve model is that

$$\mathrm{E}(y_{ij}) = \mu_i = Z\gamma_i$$

so fitting (3) and using the results in the dependent variable matrix of a one-way MANOVA is an intuitive way to arrive at conclusions about Γ.

The main problem with this analysis is that ordinary least squares were used to fit model (3). In fact, $\mathrm{Cov}(y_{ij}) = \Sigma$ so the appropriate estimate of γ_i in (3) is

$$\hat{\gamma}_{ij} = (Z'\Sigma^{-1}Z)^{-1}Z'\Sigma^{-1}y_{ij}.$$

Unfortunately, Σ is not known so the appropriate distance measure for fitting (3) is not available. Any positive definite matrix G can be used to generate estimates

$$\hat{\gamma}_{ij}^{(G)} = (Z'GZ)^{-1}Z'Gy_{ij}.$$

These lead to replacing model (2) with

$$YGZ(Z'GZ)^{-1} = X\Gamma + eGZ(Z'GZ)^{-1}. \tag{4}$$

When Σ is unknown, there is no reason to choose one G rather than another. An intuitively appealing idea is to pick $G = S^{-1}$ in an effort to approximate the optimal values for $\hat{\gamma}_{ij}$. Unfortunately, because S is a function of Y, this leads to a dependent variable matrix in (4) that no longer consists of rows of independent multivariate normal vectors; estimates and tests do not have tractable distributions.

With G fixed, model (4) is a simple multivariate one-way ANOVA. Under the cell means parameterization, the estimate of Γ is

$$\hat{\Gamma} = \begin{bmatrix} \hat{\gamma}'_1 \\ \vdots \\ \hat{\gamma}'_a \end{bmatrix},$$

where,

$$\hat{\gamma}_i = \frac{1}{n_i} \sum_{j=1}^{n_i} \hat{\gamma}_{ij}^{(G)}.$$

Tests of hypotheses for model (4) are performed in the usual way. The most commonly performed test in a one-way MANOVA is equality of means for the various treatments. In this case, the test is of $H_0\colon \gamma_1 = \gamma_2 = \cdots = \gamma_a$. It tests whether the regression coefficients of the growth curves are the same for each treatment. One advantage of analyzing model (4) for some fixed choice of G is that the standard test statistics have known distributions under H_0. Unfortunately, likelihood ratio tests for model (1) have thus far proved to have intractable distributions.

EXAMPLE 1.6.1.   Once again consider the heart rate data of Example 1.5.1. We have seen that standard analysis of variance procedures can be used to examine relationships between drugs. Unfortunately, standard procedures are less applicable for comparing times because observations over time are correlated. If some contrast is of particular interest, it can be examined but such tools as orthogonal contrasts and tabled polynomial contrasts do not retain their attractive properties. In lieu of using orthogonal polynomial contrasts, we fit a polynomial growth curve model. Specifically, we assume a quadratic growth curve model. Recall that heart rates are measured at 2, 7, 12, and 17 minutes after the injection. The model for the time means is

$$
\begin{bmatrix} \mu_{i1} \\ \mu_{i2} \\ \mu_{i3} \\ \mu_{i4} \end{bmatrix} = \begin{bmatrix} 1 & 2 & 4 \\ 1 & 7 & 49 \\ 1 & 12 & 144 \\ 1 & 17 & 289 \end{bmatrix} \begin{bmatrix} \gamma_{i0} \\ \gamma_{i1} \\ \gamma_{i2} \end{bmatrix}
$$
$$
= Z\gamma_i.
$$

$i = 1, 2, 3.$

The simplest choice of G is $G = I$. Let

$$
\begin{aligned}
W_1 &= Z(Z'Z)^{-1} \\
&= \begin{bmatrix} 1.41 & -0.25 & .01 \\ -0.15 & 0.17 & -.01 \\ -0.53 & 0.21 & -.01 \\ 0.27 & -0.13 & .01 \end{bmatrix}.
\end{aligned}
$$

The transformed dependent variable matrix in model (2) is

$$
Y_1 = YW_1
$$

so we fit the multivariate one–way ANOVA

$$
Y_1 = X\Gamma + e .
$$

Note that here Y_1 is *not* the first column of the data matrix Y. It is a transformation of Y. The estimates obtained are

$$
\hat{\Gamma} = \begin{bmatrix} 73.96 & -.093 & -.007 \\ 82.60 & .139 & -.007 \\ 68.19 & 2.561 & -.137 \end{bmatrix}
$$

and
$$S = \begin{bmatrix} 19.447 & -.701 & -.0025 \\ -.701 & .51570 & -.02485 \\ -.0025 & -.02485 & .001379 \end{bmatrix}.$$

Computing the usual test statistics for no treatment effects, we find substantial differences between the three injections. For example, $T^2 = 184.81$ which is huge when compared to the asymptotic null distribution $\chi^2(6)$.

Write
$$Y_1 = [Y_{10}, Y_{11}, Y_{12}].$$

The vector $Y_{10} = (\hat{\gamma}_{10}, \ldots, \hat{\gamma}_{n0})'$ is used for inferences about intercepts; $Y_{11} = (\hat{\gamma}_{11}, \ldots, \hat{\gamma}_{n1})'$ is used for slopes; Y_{12} is used to analyze the coefficients of the quadratic terms in the polynomials. Each of these vectors can be examined in a univariate one-way ANOVA.

The coefficient of the quadratic term for drug B is γ_{32}. The least squares estimate $\hat{\gamma}_{32} = -.137$ has a standard error of .01174 and a t statistic of -11.67. The quadratic term for drug B is clearly important.

We can test whether parabolas are needed to model the placebo and drug A. This is a test of $H_0: \gamma_{12} = \gamma_{22} = 0$ and depends only on Y_{12}. The test is just a univariate ANOVA test. The reduced model is

$$y_{12,ij} = \delta_{i3}\gamma_{i2} + e_{ij}$$

where δ_{i3} is 1 if $i = 3$ and zero otherwise. From the matrix S reported above, we find that $\text{SSE(Full)} = (27)(.001379) = .03723$. For the reduced model, $\text{SSE(R)} = .03821$ so the F statistic is

$$F = \frac{[.03821 - .03723]/2}{.001379} = .355$$

and there is no evidence of the need for a parabola in either the placebo or drug A.

We can also test for equality of the slopes in the placebo and drug A. Because our interest is in the slopes, the dependent variable is Y_{11}. The reduced model is again a one–way ANOVA but now the placebo and drug A are considered as the same treatment. The error sums of squares for the full and reduced models are denoted as $\text{SSE}(F)$ and $\text{SSE}(R)$ and take the values

$$\text{SSE}(F) = 13.924,$$
$$\text{SSE}(R) = 14.193.$$

The F statistic is
$$F = \frac{14.193 - 13.924}{.5157} = .522$$

so there is no evidence of a difference between the placebo and drug A. We can also test whether both slopes are zero. The new reduced model is $y_{11,ij} = \delta_{i3}\gamma_{i1} + e_{ij}$ with

$$\text{SSE}(R) = 14.203$$

and

$$F = \frac{[14.203 - 13.924]/2}{.5157} = .271 \, .$$

There is no evidence of a nonzero slope for the placebo and drug A.

The procedure for testing whether the placebo and drug A act the same can be applied to Y_{10} to determine if there is evidence of a difference in intercepts between the placebo and drug A. If no difference is found, we would have found no evidence of any difference between the placebo and drug A on any of the dependent variables. The statistics are

$$\mathrm{SSE}(F) = 525.07$$

$$\mathrm{SSE}(R) = 898.83$$

$$F = \frac{898.83 - 525.07}{19.45} = 19.22 \, .$$

Comparing this to $F(.995, 1, 27) = 9.34$ or even to the Scheffé critical point $2\,F(.995, 2, 27) = 12.98$ establishes that there is a significant difference at the .005 level between the intercepts for the placebo and drug A.

The heart rates under the placebo seem to be relatively constant at about $\hat{\gamma}_{10} \doteq 74$ beats per minute. Over the course of the 17 minute experiment, drug A yields approximately constant heart rates at $\hat{\gamma}_{20} \doteq 82.5$ beats per minute. Over the course of the experiment, heart rates for drug B can be approximated by the parabola $68.19 + 2.561\,t - .137\,t^2$. Clearly, this is only an approximation. It is unlikely that heart rates would really become negative after thirty-three and a half minutes.

This entire analysis is essentially an exercise in quantifying and evaluating the visual impressions given by Figure I.5. For example, the downward trend seen in the placebo profile is not statistically significant from the current data and analysis. Both $\hat{\gamma}_{11}$ and $\hat{\gamma}_{12}$ are negative so the downward trend is being modeled but neither coefficient is significant, so we do not have firm evidence of a downward trend. Although we only tested that these coefficients were zero in combination with the corresponding values for drug A, the same results occur for the individual tests.

EXERCISE 1.11.   Perform an analysis similar to the one given above using the split plot model. Compare the results to those of the growth curve model.

The use of model (4) to analyze growth curves was apparently first proposed by Potthoff and Roy (1964). An extension of this method based on analysis of covariance and normality has been proposed by Rao (1965, 1966, 1967) and Khatri (1966). Let $W_1 = GZ(Z'GZ)^{-1}$ and let W_2 be a full column rank matrix with $C(W_2) = C(Z)^{\perp}$. As established below in Exercise 1.12, with these choices the matrix $W = [W_1, W_2]$ is nonsingular.

Because we are multiplying by a nonsingular matrix, model (1) is equivalent
to

$$YW = X\Gamma Z'W + eW. \tag{5}$$

Write $Y_1 = YW_1$ and $Y_2 = YW_2$. Note that Y_1 is precisely the dependent
variable matrix in model (4). Using the definition of W, (5) can be rewritten
as

$$[Y_1, Y_2] = [X\Gamma, 0] + [eW_1, eW_2],$$

which is similar to a multivariate linear model. As will be seen later, under
normality the conditional distribution of Y_1 given Y_2 is determined by the
multivariate linear model

$$Y_1 = X\Gamma + Y_2\Psi + e. \tag{6}$$

In particular, this is a multivariate analysis of covariance model. Estimates
and quadratic forms for tests are derived as in Christensen (1987, Chapter IX) so

$$\begin{aligned}
\hat{\Psi} &= [Y_2'(I - M)Y_2]^{-1}Y_2'(I - M)Y_1, \\
X\hat{\Gamma} &= M(Y_1 - Y_2\hat{\Psi}),
\end{aligned}$$

and, with $\Sigma_{1\cdot2}$ denoting the conditional covariance matrix of a row of Y_1,
the estimate of the covariance matrix is based on

$$[n - r(X, Y_2)]S_{1\cdot2} = Y_1'[(I - M) - (I - M)Y_2\{Y_2'(I - M)Y_2\}^{-1}Y_2'(I - M)]Y_1.$$

In general, Γ need not be estimable because X need not have full column
rank. In this section, we consider only the full rank X corresponding to
the cell means parameterization but this is not a substantive restriction.
To test $H_0\colon \gamma_1 = \gamma_2 = \cdots = \gamma_a$, simply test the full model (6) against the
reduced model

$$Y_1 = J\mu' + Y_2\Psi + e.$$

One substantial advantage of the Rao-Khatri modification is that, for
inferences about Γ, the method does not depend on the specific choice
of the matrix W. This can be shown for all aspects of the problem; we
illustrate only that $X\hat{\Gamma}$ does not depend on G. As in profile analysis, let
$E = Y'(I - M)Y$.

$$\begin{aligned}
X\hat{\Gamma} &= M(Y_1 - Y_2\hat{\Psi}) \\
&= M(YW_1 - YW_2\hat{\Psi}) \\
&= MY(W_1 - W_2[Y_2'(I - M)Y_2]^{-1}Y_2'(I - M)Y_1) \\
&= MY(W_1 - W_2[W_2'EW_2]^{-1}W_2'EW_1) \\
&= MY(I - A_2)W_1,
\end{aligned}$$

where $A_2 = W_2(W_2'EW_2)^{-1}W_2'E$. Because $W_2'Z = 0$, Lemma 12.6.2 in
Christensen (1987) implies that $(I - A_2)$ is the oblique projection operator

onto $C(E^{-1}Z)$ along $C(W_2)$. By his Lemma 12.6.3, $E^{-1}Z(Z'E^{-1}Z)^{-1}Z'$ is the projection operator onto $C(E^{-1}Z)$ along $C(W_2)$, so

$$\begin{aligned}
X\hat{\Gamma} &= MYE^{-1}Z(Z'E^{-1}Z)^{-1}Z'W_1 \\
&= MYE^{-1}Z(Z'E^{-1}Z)^{-1}Z'GZ(Z'GZ)^{-1} \\
&= MYE^{-1}Z(Z'E^{-1}Z)^{-1} \\
&= MYS^{-1}Z(Z'S^{-1}Z)^{-1},
\end{aligned}$$

which does not depend on W. Moreover, this is precisely the intuitively appealing estimator based on taking $G = S^{-1}$ in model (4).

It remains to show that (6) is a valid multivariate linear model. A typical row of Y is y'_{ij} so a row of YW is $y'_{ij}W = [y'_{ij}W_1, y'_{ij}W_2]$. Define $[y'_{1ij}, y'_{2ij}] = [y'_{ij}W_1, y'_{ij}W_2]$. We will find the conditional distribution of y_{1ij} given y_{2ij}. Because $[y'_{1ij}, y'_{2ij}]$ is multivariate normal, the conditional distribution is also normal. The conditional mean is the best linear predictor of y_{1ij} based on y_{2ij}. The conditional covariance matrix is the prediction covariance matrix. (These concepts were discussed in Christensen (1987, Chapter VI). They are also discussed in Section III.1) The parameters of the conditional distribution are simple functions of the parameters of the joint distribution. We begin by finding the mean and covariance matrix of the joint distribution. The rows of X can be written as x'_{ij} so $E(y'_{ij}) = x'_{ij}\Gamma Z'$ and as usual $\text{Cov}(y_{ij}) = \Sigma$. It follows that

$$\begin{aligned}
E[y'_{1ij}, y'_{2ij}] &= E(y'_{ij}W) = x'_{ij}\Gamma Z'W \\
&= [x'_{ij}\Gamma ZW_1, x'_{ij}\Gamma Z'W_2] \\
&= [x'_{ij}\Gamma, 0]
\end{aligned}$$

where the last equality follows from the choice of W. Also

$$\begin{aligned}
\text{Cov}\left(\begin{bmatrix} y_{1ij} \\ y_{2ij} \end{bmatrix}\right) &= \text{Cov}(W'y_{ij}) = W'\Sigma W \\
&= \begin{bmatrix} W'_1\Sigma W_1 & W'_1\Sigma W_2 \\ W'_2\Sigma W_1 & W'_2\Sigma W_2 \end{bmatrix}.
\end{aligned}$$

From the theory of best linear predictors

$$y_{1ij}|y_{2ij} \sim N(\Gamma'x_{ij} + \Psi'y_{2ij}, \Sigma_{1\cdot 2}) \tag{7}$$

where $\Psi' = W'_1\Sigma W_2(W'_2\Sigma W_2)^{-1}$ and

$$\Sigma_{1\cdot 2} = W'_1\Sigma W_1 - W'_1\Sigma W_2(W'_2\Sigma W_2)^{-1}W'_2\Sigma W_1.$$

Because the y_{ij}'s are independent, the $W'y_{ij}$'s are also independent. In particular, y_{1ij} is independent of $y_{2i'j'}$ whenever $(i,j) \neq (i',j')$. Thus, the distribution in (7) is also the distribution of y_{1ij} given Y_2. Moreover, because y_{ij} and $y_{i'j'}$ are independent for $(i,j) \neq (i',j')$, the random vector

y_{1ij} given y_{2ij} is independent of any other observations, say $y_{1i'j'}$ given $y_{2i'j'}$. Because both y_{2ij} and $y_{2i'j'}$ can be replaced by Y_2 in the conditioning, it follows that given Y_2, the y_{1ij}'s are i.i.d. with the distribution in (7). This is precisely the definition of the multivariate linear model (6). Because (6) is a conditional normal theory multivariate linear model, the estimates $\hat{\Psi}$ and $X\hat{\Gamma}$ are conditional maximum likelihood estimates. In fact, since $X\hat{\Gamma}$ does not depend on Σ yet maximizes the conditional likelihood for any Σ and since the marginal distribution of Y_2 does not depend on Γ, the conditional maximum likelihood estimate of $X\Gamma$ is also the unconditional MLE.

EXAMPLE 1.6.2.   We now reanalyze the heart rate data using the analysis of covariance growth curve model. Recall from Example 1.6.1, that

$$Z = \begin{bmatrix} 1 & 2 & 4 \\ 1 & 7 & 49 \\ 1 & 12 & 144 \\ 1 & 17 & 289 \end{bmatrix}$$

and we took

$$W_1 = Z(Z'Z)^{-1}.$$

The measurements are taken at equally spaced time intervals so a standard table of polynomial contrasts can be used to obtain the matrix W_2. The cubic contrast with four equally spaced time periods is $(-1, 3, -3, 1)$ and W_2 can be taken as

$$W_2 = \begin{bmatrix} -1 \\ 3 \\ -3 \\ 1 \end{bmatrix}.$$

Computing $Y_1 = YW_1$ and $Y_2 = YW_2$ and fitting the multivariate one–way analysis of covariance model

$$Y_1 = X\Gamma + Y_2\Psi + e$$

gives

$$\hat{\Gamma} = \begin{bmatrix} 74.179 & -.1209 & -.00630 \\ 82.785 & .1162 & -.00640 \\ 68.730 & 2.4926 & -.13528 \end{bmatrix},$$

$$\hat{\Psi} = [.1998, -.02533, .000635],$$

and the error statistic

$$E \equiv E_{1\cdot2} = \begin{bmatrix} 477.88 & -12.95 & -0.22 \\ -12.95 & 13.17 & -0.65 \\ -0.22 & -0.65 & 0.04 \end{bmatrix}$$

with 26 degrees of freedom. Again, X is the full rank design matrix associated with the cell means parameterization so Γ is estimable.

The reduced model for no differences between drugs in the regression coefficients is

$$Y_1 = J\mu' + Y_2\Psi + e$$

with an error statistic of

$$E_0 = \begin{bmatrix} 1470.31 & -162.16 & 8.01 \\ -162.16 & 54.29 & -2.76 \\ 8.01 & -2.76 & 0.14 \end{bmatrix} .$$

For testing the reduced model, the hypothesis statistic is

$$H = E_0 - E = \begin{bmatrix} 992.43 & -149.21 & 8.23 \\ -149.21 & 41.12 & -2.11 \\ 8.23 & -2.11 & 0.11 \end{bmatrix} .$$

The standard test statistics are $U = .064$, $\phi_{max} = 5.43$, $T^2 = 185.3$, $V = 1.434$. There are clear differences due to drugs in the coefficients of the parabolas.

We can now repeat the detailed analysis of the parabolas given in Example 1.6.1. The full and reduced models used in the analysis are all the same except that all now include the covariate matrix Y_2. The specific models used for various tests were discussed in the earlier example.

In looking at the coefficients of the quadratic terms

$$\hat{\gamma}_{32} = -.13528$$

with

$$\text{SE}(\hat{\gamma}_{32}) = .01225$$

and

$$t_{obs} = -11.04 .$$

These results can be obtained by fitting the univariate linear model $Y_{12} = X\gamma_2 + Y_2\psi_2 + e_2$, where Y_{12} is the third component of $Y_1 = [Y_{10}, Y_{11}, Y_{12}]$. The analogous results in Example 1.6.1 were obtained by analyzing the model without the covariate Y_2. This analogy holds between all the models of Example 1.6.1 and the models needed here so only summary statistics are given in the remainder of this example.

For testing H_0: $\gamma_{12} = \gamma_{22} = 0$, the error sums of squares for the full and reduced models are

$$\text{SSE}(F) = .036753 ,$$

and

$$\text{SSE}(R) = .037550 .$$

The F statistic is

$$F = .282$$

which is not significant.

For looking at equality of the slopes in the placebo and drug A,

$$\text{SSE}(F) = 13.166 \ ,$$

$$\text{SSE}(R) = 13.447 \ ,$$

and

$$F = .555 \ .$$

For testing whether the slopes are both zero, to three decimal places we again happen to have

$$\text{SSE}(R) = 13.447$$

but the numerator has two degrees of freedom so the F statistic is half as large,

$$F = .278 \ .$$

Neither test is significant.

To test for differences in intercepts between the placebo and drug A

$$\text{SSE}(F) = 477.88 \ ,$$

$$\text{SSE}(R) = 848.13 \ ,$$

and

$$F = 20.14 \ .$$

Clearly the placebo and drug A have different intercepts.

As in Example 1.6.1, we have found that heart rates are fairly constant for the placebo and drug A. The rates are approximately 74 and 83 beats per minute. Drug B again follows a parabola. The coefficients are remarkably close in the two analyses.

Rao (1965) has suggested that the analysis of covariance procedure be used only with columns of Y_2 that are important contributors to model (6). Formal tests can be made for the various columns but these depend on the choice of G.

EXAMPLE 1.6.3.   To test whether Y_2 contributes to the model using the heart rate data with $G = I$, the error matrix from Example 1.6.1 is the appropriate E_0 and the error matrix from Example 1.6.2 is E. Subtracting these gives

$$H = \begin{bmatrix} 47.19 & -5.976 & .15005 \\ -5.976 & .7584 & -.01897 \\ .15005 & -.01897 & .00048 \end{bmatrix}$$

and multiplying by E^{-1} leads to

$$T^2 = 4.54 \ .$$

The comparison value for T^2 is 3 so there is no overwhelming evidence of the importance of Y_2. The hypothesis matrix H has only one degree of freedom so McKeon's approximate distribution for T^2 is exact. As in Sections 3 and 4, the formal test is based on comparing

$$\frac{T^2}{dfE} \frac{dfE + 1 - q}{q} = \frac{4.54}{26} \frac{24}{3} = 1.40$$

to an $F(q, dfE + 1 - q)$ distribution. The result is far from significant. This is consistent with the fact that the results of our two analyses of this data are not very different.

EXERCISE 1.12.
a) Show that the covariance matrix $\Sigma_{1 \cdot 2}$ does not depend on the choice of the matrix G in W_1.
b) Show that W is nonsingular if and only if $C(W_1) \cap C(W_2) = \{0\}$.
c) Show that $C(W_1) \cap C(W_2) = \{0\}$.
Hint: Since G is positive definite, so is G^{-1}. Recall that $v = 0$ if and only if $v'Gv = 0$.

I.7   Testing for Additional Information

In some multivariate linear model problems, it is of interest to examine whether all the information about XB can be obtained from a subset of the dependent variables. If so, the other variables provide no additional information beyond that available in the subset. Let y be the vector of dependent variables, and partition it as $y' = (y_1', y_2')$. Our interest is in determining whether y_1 contains all the useful information or whether there is useful information in y_2.

Assume a linear model

$$y_i' = x_i'B + \varepsilon_i'$$

$i = 1, \ldots, n$, with $\varepsilon_i \sim N(0, \Sigma)$, and for $i \neq j$, ε_i and ε_j independent. Partition each y_i as $y_i' = (y_{i1}', y_{i2}')$ and partition B and Σ in conformance with the y_i's to give

$$(y_{i1}', y_{i2}') = x_i'[B_1, B_2] + (\varepsilon_{i1}', \varepsilon_{i2}')$$

where

$$\operatorname{Cov}\left(\begin{bmatrix} \varepsilon_{i1} \\ \varepsilon_{i2} \end{bmatrix}\right) = \begin{bmatrix} \Sigma_{11} & \Sigma_{12} \\ \Sigma_{21} & \Sigma_{22} \end{bmatrix}.$$

Let

$$Y_i = \begin{bmatrix} y_{1i}' \\ \vdots \\ y_{ni}' \end{bmatrix}$$

$i = 1, 2$. In matrix form, the linear model is

$$[Y_1, Y_2] = X[B_1, B_2] + [e_1, e_2].$$

The procedure for evaluating additional information is based on the conditional distribution of y_{i2} given y_{i1} and is similar to the analysis of covariance method used in the previous section. The conditional distribution is normal with mean equal to the best linear predictor of y_{i2} based on y_{i1} and covariance matrix equal to the covariance of the prediction error. Thus

$$y_{i2} \mid y_{i1} \sim N\left(B_2' x_i + \Sigma_{21}\Sigma_{11}^{-1}(y_{i1} - B_1' x_i), \Sigma_{22} - \Sigma_{21}\Sigma_{11}^{-1}\Sigma_{12}\right).$$

$i = 1, \ldots, n$. Conditioning on Y_1, the random vectors $y_{i2}, i = 1, \ldots n$ are still independent so the conditional distribution also defines a multivariate linear model. Let

$$e_{2 \cdot 1} = \begin{bmatrix} \xi_1' \\ \vdots \\ \xi_n' \end{bmatrix}$$

where the ξ_i's are i.i.d. $N(0, \Sigma_{22} - \Sigma_{21}\Sigma_{11}^{-1}\Sigma_{12})$. Then

$$\begin{aligned} Y_2 &= XB_2 + (Y_1 - XB_1)\Sigma_{11}^{-1}\Sigma_{12} + e_{2 \cdot 1} \\ &= X(B_2 - B_1\Sigma_{11}^{-1}\Sigma_{12}) + Y_1(\Sigma_{11}^{-1}\Sigma_{12}) + e_{2 \cdot 1} \qquad (1) \\ &\equiv X\Delta + Y_1\Gamma + e_{2 \cdot 1} \ . \end{aligned}$$

Our interest is in whether Y_2 contains any additional information about XB beyond that available in Y_1. If $X\Delta = 0$, the conditional distribution of Y_2 depends only on Σ; XB is not involved. Thus Y_2 provides no additional information on XB. In particular, if $\Delta = (B_2 - B_1\Sigma_{11}^{-1}\Sigma_{12}) = 0$ then knowledge of B_1 determines XB. The linear structure involved in the components $Y_2 = XB_2 + e_2$ is determined entirely through the dependence between Y_1 and Y_2.

EXAMPLE 1.7.1.   Suppose $Y = XB + e$ is a multivariate one-way analysis of variance. The matrix X merely identifies the treatment group for each observation. If the conditional distribution of Y_2 does not involve X then the conditional distribution does not depend on the treatment groups. It follows that for the purpose of modeling treatment means, there is no additional information in Y_2.

The actual test procedure is straightforward. The hypothesis H_0: $X\Delta = 0$ can be tested by comparing the fit of model (1) to that of

$$Y_2 = Y_1\Gamma + e_{2 \cdot 1} \ . \qquad (2)$$

Alternatively, if Δ is estimable, analysis of covariance methods can be used to obtain $\hat{\Delta}$ and a test can be based on the estimate. In practice, the idea that Δ could really be zero is rather farfetched. The important thing is that the test provides a way of evaluating whether Y_2 is worth bothering about. If the reduced model (2) fits nearly as well as (1) then the $X\Delta$ structure is not really needed to explain Y_2. If Y_2 has almost no relationship to $C(X)$ except through Y_1, an analysis of Y_1 should provide essentially all the information on the relationship between (Y_1, Y_2) and $C(X)$

EXAMPLE 1.7.2.   Typically, the method of testing for additional information is of most value when the number of dependent variables q is quite large. It provides a method of *reducing the dimensionality* of the problem. The current example is restricted to demonstrating some elementary computations using the heart rate data of Example 1.5.1. In particular, we test whether the first two measurements add any information beyond that contained in the last two measurements.

The error matrix for the test can be obtained from a multivariate analysis of covariance (MACOVA). The components of the MACOVA are available in the matrix E reported in Example 1.5.1. The error matrix for the analysis of covariance model is

$$E = \begin{bmatrix} 479.5 & 483.7 \\ 483.7 & 610.5 \end{bmatrix}$$

$$- \begin{bmatrix} 363.0 & 237.5 \\ 475.6 & 319.7 \end{bmatrix} \begin{bmatrix} 540.8 & 366.4 \\ 366.4 & 381.0 \end{bmatrix}^{-1} \begin{bmatrix} 363.0 & 475.6 \\ 237.5 & 319.7 \end{bmatrix}$$

$$= \begin{bmatrix} 235.21 & 164.30 \\ 164.30 & 192.19 \end{bmatrix}.$$

The reduced model involves only the covariates and not the analysis of variance structure. To obtain the reduced model error matrix E_0, one can regress y_1 on y_3, y_4 and y_2 on y_3, y_4 without including intercepts. The diagonal elements of E_0 are the sums of the squared residuals. The off-diagonal element is the sum of the cross-products of the two sets of residuals. These computations yield

$$E_0 = \begin{bmatrix} 422.747 & 205.842 \\ 205.842 & 206.801 \end{bmatrix}.$$

The hypothesis matrix is

$$H = E_0 - E = \begin{bmatrix} 187.539 & 41.538 \\ 41.538 & 14.610 \end{bmatrix}.$$

and

$$HE^{-1} = \begin{bmatrix} 1.60460 & -1.15564 \\ .30659 & -0.18608 \end{bmatrix}.$$

Under the null hypothesis, HE^{-1} should approximate a scalar multiple of the identity matrix. The observed matrix HE^{-1} is nothing like that. In particular, $T^2 = 25 \operatorname{tr}\left[HE^{-1}\right] = 35.46$, while the comparison value for T^2, as introduced in Section 2, is only 6. Formal tests of $H_0\colon \Delta = 0$ are highly significant. The first two time measurements are *not* extraneous.

The analysis of additional information can be generalized to examine whether Y_2 contains additional information for specific purposes. For example, if X is the design matrix for a one-way ANOVA, we can test whether Y_2 contains additional information for the purpose of distinguishing treatment groups. (This is different from modeling treatment means.) Rather than testing model (1) against the reduced model (2), test model (1) against the analysis of covariance model without treatment groups

$$Y_2 = J\mu' + Y_1\Gamma + e_{2.1}.\tag{3}$$

By assuming a parameterization of model (1) in which

$$X = [J, X_*]$$

and

$$B = \begin{bmatrix} B_{10} & B_{20} \\ B_{11} & B_{21} \end{bmatrix},$$

it is easily seen that this is a test of whether Y_2 contains any additional information about the part of the model that distinguishes treatment groups, $X_*[B_{11}, B_{21}]$.

It is of interest to note that if Y_2 consists of only one column, the test of model (3) versus model (1) is a standard univariate F test. In particular, sequential (stepwise) evaluation of the individual variables is quite simple. Moreover, the F test can be constructed from the diagonal elements of E^{-1} and $(H + E)^{-1}$ that correspond to Y_2; here E and H are defined for testing the multivariate linear models $Y = XB + e$ and $Y = J\mu' + e$. This provides a simple way to compute all of the F statistics needed for a stepwise evaluation. Variable selection methods in discriminant analysis are based on testing the additional information available for distinguishing treatment groups, cf. Section II.2.

EXERCISE 1.13.   Find the F test in terms of E^{-1} and $(H + E)^{-1}$.

EXERCISE 1.14.   Test whether the first two time variates in the heart rate data are needed for distinguishing the treatment groups.

I.8   Additional Exercises

EXERCISE 1.8.1.   Jolicoeur and Mosimann (1960) give data on the length, width, and height of painted turtle shells. The carapace dimensions of twenty-four females and twenty-four males are given in Table I.3. Use Hotelling's T^2 statistic to test whether there is a sex difference in shell dimensions. Is there a significant sex difference between any of the individual dimensions? Use plots to check the validity of your assumptions.

TABLE I.3. Carapace Dimensions

Female			Male		
Length	Width	Height	Length	Width	Height
98	81	38	93	74	37
103	84	38	94	78	35
103	86	42	96	80	35
105	86	42	101	84	39
109	88	44	102	85	38
123	92	50	103	81	37
123	95	46	104	83	39
133	99	51	106	83	39
133	102	51	107	82	38
133	102	51	112	89	40
134	100	48	113	88	40
136	102	49	114	86	40
138	98	51	116	90	43
138	99	51	117	90	41
141	105	53	117	91	41
147	108	57	119	93	41
149	107	55	120	89	40
153	107	56	120	93	44
155	115	63	121	95	42
155	117	60	125	93	45
158	115	62	127	96	45
159	118	63	128	95	45
162	124	61	131	95	46
177	132	67	135	106	47

EXERCISE 1.8.2.   Analyze the repeated measures data given by Danford, Hughes, and McNee (1960) in *Biometrics* on pages 562, 563.

EXERCISE 1.8.3.   Box (1950) gives data on the weights of three groups of rats. One group was given thyroxin in their drinking water, one thiouracil, and the third group was a control. Weights are measured in grams at weekly intervals. The data are given in Table I.4

a) Perform a multivariate one-way analysis of variance.

b) Evaluate the validity of the assumptions.

c) My eight-year-old son told me that the true means for the control group are $(60, 80, 100, 120, 140)'$. Use Hotelling's T^2 to test the validity of his claim.

d) Do a profile analysis.

e) Analyze the data using the analysis of covariance method for growth curve analysis.

TABLE I.4. Rat Weights

| | | Thyroxin | | |
Time 0	Time 1	Time 2	Time 3	Time 4
59	85	121	156	191
54	71	90	110	138
56	75	108	151	189
59	85	116	148	177
57	72	97	120	144
52	73	97	116	140
52	70	105	138	171

| | | Thiouracil | | |
Time 0	Time 1	Time 2	Time 3	Time 4
61	86	109	120	129
59	80	101	111	122
53	79	100	106	133
59	88	100	111	122
51	75	101	123	140
51	75	92	100	119
56	78	95	103	108
58	69	93	114	138
46	61	78	90	107
53	72	89	104	122

| | | Control | | |
Time 0	Time 1	Time 2	Time 3	Time 4
57	86	114	139	172
60	93	123	146	177
52	77	111	144	185
49	67	100	129	164
56	81	104	121	151
46	70	102	131	153
51	71	94	110	141
63	91	112	130	154
49	67	90	112	140
57	82	110	139	169

EXERCISE 1.8.4.   Smith, Gnanadesikan, and Hughes (1962) provide data on characteristics of the urine of young men. The men are categorized into four groups based on their degree of obesity. The four variables given in Table I.5 consist of a covariate $x = 10^3((\text{specific gravity}) - 1)$ and three dependent variables, $y_1 = $ pigment creatinine, $y_2 = $ chloride, and $y_3 = $

chlorine.

a) Do a one-way MANOVA on this data ignoring the covariate.

b) Do a one-way multivariate ACOVA (MACOVA) on this data.

TABLE I.5. Excretory Characteristics

	Group I				Group II		
x	y_1	y_2	y_3	x	y_1	y_2	y_3
24	17.6	5.15	7.5	31	18.1	9.00	14.5
32	13.4	5.75	7.1	23	19.7	5.30	12.5
17	20.3	4.35	2.3	32	16.9	9.85	8.0
30	22.3	7.55	4.0	20	23.7	3.60	4.9
30	20.5	8.50	2.0	18	19.2	4.05	0.2
27	18.5	10.25	2.0	23	18.0	4.40	3.6
25	12.1	5.95	16.8	31	14.8	7.15	12.0
30	12.0	6.30	14.5	28	15.6	7.25	5.2
28	10.1	5.45	0.9	21	16.2	5.30	10.2
24	14.7	3.75	2.0	20	14.1	3.10	8.5
26	14.8	5.10	0.4	15	17.5	2.40	9.6
27	14.4	4.05	3.8	26	14.1	4.25	6.9
				24	19.1	5.80	4.7
				16	22.5	1.55	3.5

	Group III				Group IV		
x	y_1	y_2	y_3	x	y_1	y_2	y_3
18	17.0	4.55	1.9	32	12.5	2.90	22.5
10	12.5	2.65	0.7	25	8.7	3.00	19.5
33	21.5	6.50	8.3	28	9.4	3.40	1.3
25	22.2	4.85	9.3	27	15.0	5.40	20.0
35	13.0	8.75	13.0	23	12.9	4.45	1.0
33	13.0	5.20	18.3	25	12.1	4.30	5.0
31	10.9	4.75	10.5	26	13.2	5.00	3.0
34	12.0	5.85	14.5	34	11.5	3.40	5.1
16	22.8	2.85	3.3				
31	16.5	6.55	6.3				
28	18.4	6.60	4.9				

EXERCISE 1.8.5.   Box (1950) presents data on the weight loss of a fabric due to abrasion. Two fillers were used in three proportions. Some of the fabric was given a surface treatment. Weight loss was recorded after 1000, 2000, and 3000 revolutions of a machine designed to test abrasion resistance. The data are given in Table I.6. Perform a multivariate analysis of variance, a profile analysis, and fit a simple linear growth curve model. Check your assumptions.

TABLE I.6. Abrasion Resistance

| Surf. | | | 25% | | | 50% | | | 75% | | |
Treat.	Fill	1000	2000	3000	1000	2000	3000	1000	2000	3000
	A	194	192	141	233	217	171	265	252	207
	A	208	188	165	241	222	201	269	283	191
Yes										
	B	239	127	90	224	123	79	243	117	100
	B	187	105	85	243	123	110	226	125	75
	A	155	169	151	198	187	176	235	225	166
	A	173	152	141	177	196	167	229	270	183
No										
	B	137	82	77	129	94	78	155	76	91
	B	160	82	83	98	89	48	132	105	67

(The "Proportions" heading spans the three percentage groups.)

EXERCISE 1.8.6.   Show that if $W \sim W(n, \Sigma, 0)$ then

$$AWA' \sim W(n, A\Sigma A', 0).$$

EXERCISE 1.8.7.   Show that if $W_1, \cdots W_r$ are independent with $W_i \sim W(n_i, \Sigma, 0)$ then

$$\sum_{i=1}^{r} W_i \sim W\left(\sum_{i=1}^{r} n_i, \Sigma, 0\right).$$

EXERCISE 1.8.8.   Show that if $W \sim W(n, \Sigma, 0)$, then

$$\frac{\lambda' W \lambda}{\lambda' \Sigma \lambda} \sim \chi^2(n, 0).$$

EXERCISE 1.8.9.   For $i = 1, 2, 3$, let $y_i \sim N\left(\mu + (i-2)\xi, \Sigma\right)$ where Σ is known and y_1, y_2, and y_3 are independent. Find the maximum likelihood estimates of μ and ξ.

EXERCISE 1.8.10.   Based on the multivariate linear model

$$Y = XB + e, \; \mathrm{E}(e) = 0, \; \mathrm{Cov}(\varepsilon_i, \varepsilon_j) = \delta_{ij}\Sigma,$$

find a 99% prediction interval for $y_0'\xi$ where y_0 is an independent observation that is distributed $N(B'x_0, \Sigma)$.

EXERCISE 1.8.11.   Let $y_1, y_2, \cdots, y_n$ be i.i.d. $N(X\beta, \Sigma)$ where $\Sigma_{n \times n}$ is unknown. Show that the maximum likelihood estimate of $X\beta$ is

$$X\hat{\beta} = X(X'E^{-1}X)^{-}X'E^{-1}\bar{y}..$$

EXERCISE 1.8.12.   Consider the multivariate linear model $Y = XB + e$ and the parametric function $\Lambda'BZ$ where Z is a $q \times r$ matrix of rank r. Find simultaneous confidence intervals for all parameters of the form $\zeta'\Lambda'BZ\xi$.

EXERCISE 1.8.13.   Use Lemma 1.2.3 to show that if A is nonnegative definite and B is positive definite, then AB^{-1} is nonnegative definite. Hint: Show that the nonnegative definite matrix $B^{-1/2}AB^{-1/2}$ has the same eigenvalues as AB^{-1}.

EXERCISE 1.8.14.   a) Show that

$$\frac{d[A(s)B(s)]}{ds} = \frac{dA(s)}{ds}B(s) + A(s)\frac{dB(s)}{ds}.$$

b) Show that

$$\frac{d\Sigma^{-1}(s)}{ds} = -\Sigma^{-1}(s)\frac{d\Sigma(s)}{ds}\Sigma^{-1}(s).$$

c) Show that

$$\frac{d\log|\Sigma(s)|}{ds} = tr\left[\Sigma^{-1}(s)\frac{d\Sigma(s)}{ds}\right].$$

Hints: For a), consider $A(s)B(s)$ elementwise. For b), use a) and the fact that $0 = dI/ds = d\left[\Sigma(s)\Sigma^{-1}(s)\right]/ds$. For c), write $\Sigma = P\,Diag(\phi_i)\,P'$ and show that both sides equal

$$\sum_{i=1}^{q} \frac{d\phi_i(s)}{ds}\frac{1}{\phi_i(s)}.$$

For the right-hand side use (a) and the fact that $0 = dI/ds = dPP'/ds$.

Chapter II

Discrimination and Allocation

Consider the eight populations of people determined by all combinations of sex (male, female) and age (adult, adolescent, child, infant). These are commonly used distinctions, but the populations are not clearly defined. It is not obvious when infants become children, when children become adolescents, nor when adolescents become adults. On the other hand, most people can clearly be identified as members of one of these eight groups. It might be of interest to see whether one can *discriminate* between these populations on the basis of, say, various aspects of their blood chemistry. The discrimination problem is sometimes referred to as the problem of *separation*. Another potentially interesting problem is trying to predict the population of a new individual given only the information on their blood chemistry. The problem of predicting the population of a new case is referred to as the problem of *allocation*. Other names for this problem are *identification* and *classification*.

Most books on multivariate analysis contain extensive discussions of discrimination and allocation. The author can particularly recommend the treatments in Anderson (1984), Johnson and Wichern (1988) and Seber (1984). In addition, Hand (1981) and Lachenbruch (1975) have written monographs on the subject. As is so often the case in statistics, the first modern treatment of these problems was by Sir Ronald A. Fisher, cf. Fisher (1936, 1938). The discussion in this chapter is closely related to methods associated with the multivariate normal distribution. An alternative approach is based on logistic regression, cf. Christensen (1990), Press (1982), Press and Wilson (1978), or Seber (1984). There are also a variety of nonparametric methods available.

Discrimination seems to be a purely descriptive endeavor. The observations are vectors in $\mathbf{R}^q$. All observations come from known populations. Discriminate analysis uses the observations to partition $\mathbf{R}^q$ into regions, each uniquely associated with a particular population. Given a partition, it is easy to allocate future observations. An observation y is allocated to population r if y falls into the region of $\mathbf{R}^q$ associated with the r'th population. The difficulty lies in developing a rational approach to partitioning $\mathbf{R}^q$.

Just as a solution to the discrimination problem implicitly determines an allocation rule, a solution to the allocation problem implicitly solves the discrimination problem. The set of all y values to be allocated to population r determines the region associated with population r.

Our discussion will be centered on the allocation problem. We present allocation rules based on Mahalanobis's distance, maximum likelihood, and Bayes theorem. An advantage of the Mahalanobis distance method is that the method is based solely on the means and covariances of the population distributions. The other methods require knowledge of the entire distribution in the form of a density. Not surprisingly, the Mahalanobis, the maximum likelihood, and the Bayes rules are similar for normally distributed populations.

In general, consider the situation in which there are t populations and q variables $y_1, \ldots, y_q$ with which to discriminate among them. In particular, if $y = (y_1, \ldots, y_q)'$ is an observation from the i'th population, we assume that either the mean and covariance matrix of the population are known, say

$$\mathrm{E}(y) = \mu_i$$

and

$$\mathrm{Cov}(y) = \Sigma_i$$

or that the density of the population distribution, say

$$f(y|i)$$

is known. An important special case is where the covariance matrix is the same for all populations, say

$$\Sigma = \Sigma_1 = \cdots = \Sigma_t \,.$$

In practice, neither the density nor the mean nor the covariance matrix will be known. These must be estimated using data from the various populations. We assume that a random sample of n_i observations are available from the i'th population. The j'th observation from the i'th population is denoted $y_{ij} = (y_{ij,1}, \ldots, y_{ij,q})'$. Note that

$$\mathrm{E}(y_{ij}) = \mu_i$$

and

$$\mathrm{Cov}(y_{ij}) = \Sigma_i \,.$$

It is important to recognize that in the special case of equal covariance matrices, the data follow a multivariate one-way ANOVA model.

Section II.1 deals with the general allocation problem. The special case of equal covariance matrices is dealt with in Section II.2. Section II.3 examines the relationship between MANOVA and discrimination and introduces discrimination coordinates that are useful in visualizing the discrimination procedure.

II.1   The General Allocation Problem

In this section we discuss allocation rules based on Mahalanobis's distance, maximum likelihood, and Bayes theorem. These rules are based on populations with either known means and covariances or known densities. The section ends with a discussion of estimated versions of these rules.

MAHALANOBIS'S DISTANCE

As discussed in Christensen (1987, Section XIII.1), the Mahalanobis distance
$$D^2 = (y - \mu)'\Sigma^{-1}(y - \mu)$$
is a frequently used measure of how far a random vector is from the center of its distribution. In the allocation problem, we have a random vector y and t possible distributions from which it could arise. A reasonable allocation procedure is to assign y to the population that minimizes the observed Mahalanobis distance. In other words, allocate y to population r if

$$(y - \mu_r)'\Sigma_r^{-1}(y - \mu_r) = \min_i (y - \mu_i)'\Sigma_i^{-1}(y - \mu_i). \tag{1}$$

MAXIMUM LIKELIHOOD

If the densities $f(y|i)$ are known for each population, the population index "i" is the only unknown parameter. Given an observation y, the likelihood function is
$$L(i) = f(y|i)$$
which is defined on the set $i = 1, \ldots, t$. The maximum likelihood allocation rule assigns y to population r if

$$L(r) = \max_i L(i)$$

or equivalently if
$$f(y|r) = \max_i f(y|i).$$

If the observations have a multivariate normal distribution, the maximum likelihood rule is very similar to the Mahalanobis distance rule. From Christensen (1987, Section I.2), the likelihoods (densities) are

$$L(i) = f(y|i) = (2\pi)^{-q/2}|\Sigma_i|^{-1/2}\exp[-(y - \mu_i)'\Sigma_i^{-1}(y - \mu_i)/2],$$

$i = 1, \ldots, t$. The logarithm is a monotone increasing function so maximizing the log-likelihood is equivalent to maximizing the likelihood. The log-likelihood is

$$\ell(i) = \log(L(i)) = -\frac{q}{2}\log(2\pi) - \frac{1}{2}\log(|\Sigma_i|) - \frac{1}{2}(y - \mu_i)'\Sigma_i^{-1}(y - \mu_i).$$

If we drop the constant term $-\frac{q}{2}\log(2\pi)$ and minimize the negative of the log-likelihood rather than maximizing the log-likelihood we see that the maximum likelihood rule for normally distributed populations is: assign y to population r if

$$\log(|\Sigma_r|)+(y-\mu_r)'\Sigma_r^{-1}(y-\mu_r) = \min_i\{\log(|\Sigma_i|)+(y-\mu_i)'\Sigma_i^{-1}(y-\mu_i)\}. \quad (2)$$

Note that the only difference between the maximum likelihood rule and the Mahalanobis rule is the inclusion of the term $\log(|\Sigma_i|)$. Both the Mahalanobis rule and the maximum likelihood rule involve quadratic functions of y. Methods related to these rules are often referred to as *quadratic discrimination* methods.

BAYESIAN METHODS

We will discuss two procedures for Bayesian allocation. One is an intuitive rule. The other is a formal procedure based on costs of misclassification. It will also be shown that the intuitive rule can be arrived at by the formal procedure.

Bayesian allocation methods presuppose that for each population i there exists a prior probability, say $p(i)$, that the new observation y comes from that population. Typically, these prior probabilities are arrived at either subjectively or through the use of the maximum entropy principle, cf. Berger (1985, Section 3.4). The maximum entropy principle dictates that the $p(i)$'s should be chosen to minimize the amount of information contained in them. This is achieved by selecting

$$p(i) = 1/t, \quad (3)$$

$i = 1,\ldots,t.$

Given the prior probabilities and the data y, the posterior probability that y came from population i can be computed using Bayes theorem, cf. Berger (1985, Section 4.2). The posterior probability is

$$p(i|y) = f(y|i)p(i) \Big/ \sum_{j=1}^{t} f(y|j)p(j). \quad (4)$$

A simple intuitive allocation rule is to assign y to the population with the highest posterior probability. In other words, assign y to population r if

$$p(r|y) = \max_i p(i|y).$$

Note that the denominator in (4) does not depend on i so the allocation rule is equivalent to choosing r such that

$$f(y|r)p(r) = \max_i\{f(y|i)p(i)\}.$$

In the important special case in which the $p(i)'s$ are all equal, this corresponds to maximizing $f(y|i)$, i.e., choosing r so that

$$f(y|r) = \max_i f(y|i) \,.$$

Thus, for equal initial probabilities, the intuitive Bayes allocation rule is the same as the maximum likelihood allocation rule. In particular, if the populations are normal, the Bayes rule with equal prior probabilities is based on (2).

To develop a formal Bayesian analysis requires knowledge of the costs of correct classification (allocation) and of misclassification. Let $c(j|i)$ be the cost of classifying y into population j when y is in fact from population i. Note that $c(i|i)$ is the cost of correct classification. For any $j \neq i$, $c(j|i)$ is a cost for misclassification. Typically, we would take $c(i|i) \leq c(j|i)$ for all j.

The expected cost of classifying y into population i (given the data) is simply

$$C(i|y) = \sum_{j=1}^{t} c(i|j)p(j|y) \,.$$

The formal Bayes allocation rule is to assign y to population r if

$$C(r|y) = \min_i C(i|y) \,.$$

Now consider the special case where

$$c(i|i) = 0 \tag{5}$$

and

$$c(j|i) = c \qquad \text{for} \qquad j \neq i \,. \tag{6}$$

This cost structure leads to the intuitive Bayes rule discussed above. Clearly, the formal Bayes rule is equivalent to choosing r such that

$$c - C(r|y) = \max_i \{c - C(i|y)\} \,.$$

Note two things. First, by the definition of probability

$$\sum_{j=1}^{t} p(j|y) = 1 \,.$$

Second, from the assumed cost structure,

$$C(i|y) = c \sum_{j \neq i} p(j|y) \,.$$

Using these two facts,

$$c - C(i|y) \;=\; c\sum_{j=1}^{t} p(j|y) - c\sum_{j\neq i} p(j|y)$$

$$=\; c\,p(i|y)\,.$$

Thus, the formal Bayes rule allocates y to population r if

$$c\,p(r|y) = \max_i c\,p(i|y)$$

or equivalently if

$$p(r|y) = \max_i p(i|y)\,.$$

This is precisely the intuitive rule given earlier.

Combining our Bayesian results we see that assuming equal prior probabilities as in (3) and the classification cost structure of (5) and (6), the formal Bayes rule is identical to the maximum likelihood rule. In particular, for normally distributed populations, the Bayes rule is given by (2).

Estimated Parameters

One serious problem with the allocation rules (1) and (2) is that typically the values μ_i and Σ_i are unknown. In practice, allocation is often based on estimated means and covariances. The estimated Mahalanobis distance rule is that an observation y is allocated to population r if

$$(y - \bar{y}_{r\cdot})' S_r^{-1}(y - \bar{y}_{r\cdot}) = \min_i (y - \bar{y}_{i\cdot})' S_i^{-1}(y - \bar{y}_{i\cdot})$$

where

$$S_i = \sum_{j=1}^{n_i}(y_{ij} - \bar{y}_{i\cdot})(y_{ij} - \bar{y}_{i\cdot})' \Big/ (n_i - 1)\,.$$

An estimated maximum likelihood-Bayes allocation rule is to assign y to population r if

$$\log(|S_r|) + (y - \bar{y}_{r\cdot})' S_r^{-1}(y - \bar{y}_{r\cdot}) = \min_i\{\log(|S_i|) + (y - \bar{y}_{i\cdot})' S_i^{-1}(y - \bar{y}_{i\cdot})\}\,.$$

With estimated parameters, the Bayes allocation rule is really only a quasi-Bayesian rule. The allocation is Bayesian but the estimation of $f(y|i)$ is not. Geisser's (1971) suggestion of using the Bayesian predictive distribution as an estimate of $f(y|i)$ has been shown to be optimal under frequentist criteria by Aitchison (1975), Murray (1977), and Levy and Perng (1986). In particular, for normal data, treating the maximum likelihood estimates as the mean and covariance matrix of the normal gives an inferior estimate for the distribution of new observations. The appropriate frequentist distribution is a multivariate t with the same location vector and scale matrix. See Geisser (1977) for a discussion of these issues in relation to discrimination.

In any case, plugging in estimates of μ_i and Σ_i requires that good estimates be available. Friedman (1989) has proposed an alternative estimation technique for use with small samples.

Finally, in examining the data, the results of the allocation are generally not enough to look at. Typically, one is interested not only in the population to which a case is allocated, but also in the clarity of the allocation. It is desirable to know whether a case y is clearly in one population or whether it could have come from two or more populations. The posterior probabilities from the Bayesian method address these questions in the simplest fashion. Similar information can be gathered from examining the entire likelihood function or the entire set of Mahalanobis distances. The idea of clarity of allocation is illustrated in Example 2.2.1. An alternative approach to this problem is presented in Section 3.

II.2   Equal Covariance Matrices

If the populations are all multivariate normal with the same covariance matrix, say $\Sigma = \Sigma_1 = \cdots = \Sigma_t$, then the Mahalanobis distance rule (1) and the maximum likelihood–Bayes allocation rule (2) are identical. The maximum likelihood-Bayes allocation rule assigns y to the population r that satisfies

$$\log(|\Sigma|) + (y - \mu_r)'\Sigma^{-1}(y - \mu_r) = \min_i\{\log(|\Sigma|) + (y - \mu_i)'\Sigma^{-1}(y - \mu_i)\}\,.$$

However, the term $\log(|\Sigma|)$ is the same for all populations so the rule is equivalent to choosing the population r that satisfies

$$(y - \mu_r)'\Sigma^{-1}(y - \mu_r) = \min_i(y - \mu_i)'\Sigma^{-1}(y - \mu_i)\,.$$

This is precisely the Mahalanobis distance rule.

In practice, estimates must be substituted for Σ and the μ_i's. With equal covariance matrices the data fit the multivariate one-way ANOVA model of Section I.5 so the standard estimates $\bar{y}_{i\cdot}$, $i = 1,\ldots,t$ and $S = E/(n-t)$ are reasonable. The allocation rule is: assign y to population r if

$$(y - \bar{y}_{r\cdot})'S^{-1}(y - \bar{y}_{r\cdot}) = \min_i(y - \bar{y}_{i\cdot})'S^{-1}(y - \bar{y}_{i\cdot})\,.$$

Recall that in a one-way ANOVA, the estimated covariance matrix is a weighted average of the individual estimates, i.e.,

$$S = \sum_{i=1}^{t}(n_i - 1)S_i/(n - t)\,.$$

Although $(y - \bar{y}_{i\cdot})'S^{-1}(y - \bar{y}_{i\cdot})$ is a quadratic function of y, the allocation only depends on a linear function of y. Note that

$$(y - \bar{y}_{i\cdot})'S^{-1}(y - \bar{y}_{i\cdot}) = y'S^{-1}y - 2\bar{y}_{i\cdot}'S^{-1}y + \bar{y}_{i\cdot}'S^{-1}\bar{y}_{i\cdot}\,.$$

The term $y'S^{-1}y$ is the same for all populations. Subtracting this constant and dividing by -2, the allocation rule can be rewritten as: assign y to population r if

$$y'S^{-1}\bar{y}_{r\cdot} - \frac{1}{2}\bar{y}'_{r\cdot}S^{-1}\bar{y}_{r\cdot} = \max_i \left\{ y'S^{-1}\bar{y}_{i\cdot} - \frac{1}{2}\bar{y}_{i\cdot}S^{-1}\bar{y}_{i\cdot} \right\}.$$

This is based on a linear function of y, so methods related to this allocation rule are often referred to as *linear discrimination* methods.

The methods discussed explicitly in this chapter are all related to the normal distribution. If the true distributions $f(y|i)$ are elliptically symmetric both the quadratic and linear methods work well. Moreover, the linear discrimination method is generally quite robust; it even seems to work quite well for discrete data. See Lachenbruch, Sneeringeer, and Revo (1973), Lachenbruch (1975), and Hand (1983) for details.

CROSS–VALIDATION

It is of considerable interest to be able to evaluate the performance of allocation rules. Depending on the populations involved, there is generally some level of misclassification that is unavoidable. If the distributions that determine the allocation rules are known, one can simply classify random samples from the various populations to see how often the data are misclassified. This provides simple yet valid estimates of the error rates. Unfortunately, things are rarely this straightforward. In practice, the data available are used to estimate the distributions of the populations. If the same data are also used to estimate the error rates, a bias is introduced. Typically, this double dipping in the data overestimates the performance of the allocation rules. The method of estimating error rates by reclassifying the data is often called the *resubstitution method.*

To avoid the bias of the resubstitution method, *cross-validation* is often used, cf. Geisser (1977) and Lachenbruch (1975). In its most common form, cross-validation involves leaving out one data point, estimating the allocation rule from the remaining data, and then classifying the deleted case using the estimated rule. Every data point is left out in turn. Error rates are estimated by the proportions of misclassified cases. This version of cross-validation is also known as the *jackknife.* The computation of the cross-validation error rates can be simplified by the use of updating formula similar to those discussed in Christensen (1987, Chapter XIII).

Alternatively, cross-validation can be performed by leaving out any randomly selected group of observations, estimating the rules, and allocating the observations that were left out. This second approach requires a great deal of data to be effective. Large amounts of data are needed to estimate both the rules and the error rates.

While cross-validation has less bias than the resubstitution method, it typically has a considerably larger variance. If the number of observations is much larger than the number of parameters to be estimated, resubstitution is often adequate for estimating the error rates. When the number of parameters is large relative to the number of observations the bias becomes unacceptably high. Under normal theory, the parameters involved are simply the means and covariances for the populations. Thus the key issue is the number of variables used in the discrimination relative to the number of observations. (While we have not discussed nonparametric discrimination, in the context of this discussion nonparametric methods should be considered as highly parametric methods.)

The bootstrap has also been suggested as a tool for estimating error rates. It often has both small bias and small variance but it is computationally intensive and handles large biases poorly. The interested reader is referred to Efron (1983) and the report of the Panel Discussion on Discriminate Analysis etc. in *Statistical Science* (1989).

EXAMPLE 2.2.1.   Aitchison and Dunsmore (1975) present data on Cushing's syndrome, a medical condition characterized by overproduction of cortisol by the adrenal cortex. Twenty-one individuals where identified as belonging to one of three types: *adenoma, bilateral hyperplasia,* and *carcinoma.* The amounts of tetrahydrocortisone and pregnanetriol excreted in the urine were measured. The data are given in Table II.1. We wish to discriminate between the three types of Cushing's syndrome based on the urinary excretion data. A quick glance at Table II.1 establishes that none of the data groups seems to be from a bivariate normal distribution. Note also that the pregnanetriol value for case 4 was possibly misrecorded. It is the only value that is nonzero in the hundredths place. In the absence of information to the contrary, we will treat Table II.1 as correct, cf. Exercise 2.4.1. Following Aitchison and Dunsmore, the analysis is performed on the logarithms of the data. The log data are plotted in Figure II.1.

Performing a discriminant analysis involves nothing more complicated than estimating the means and covariance matrices for the three populations. These are given below.

Variable	Group Means			
				Grand
	a	b	c	Mean
log(Tet)	1.0433	2.0073	2.7097	1.8991
log(Preg)	−.60342	−.20604	1.5998	.11038

Covariance Matrix for Adenoma

	log(Tet)	log(Preg)
log(Tet)	0.1107	0.1239
log(Preg)	0.1239	4.0891

TABLE II.1. Data on Cushing's Syndrome

Case	Type	Tetra	Preg
1	*a*	3.1	11.70
2	*a*	3.0	1.30
3	*a*	1.9	0.10
4	*a*	3.8	0.04
5	*a*	4.1	1.10
6	*a*	1.9	0.40
7	*b*	8.3	1.00
8	*b*	3.8	0.20
9	*b*	3.9	0.60
10	*b*	7.8	1.20
11	*b*	9.1	0.60
12	*b*	15.4	3.60
13	*b*	7.7	1.60
14	*b*	6.5	0.40
15	*b*	5.7	0.40
16	*b*	13.6	1.60
17	*c*	10.2	6.40
18	*c*	9.2	7.90
19	*c*	9.6	3.10
20	*c*	53.8	2.50
21	*c*	15.8	7.60

Covariance Matrix for Bilateral Hyperplasia

	log(Tet)	log(Preg)
log(Tet)	0.2119	0.3241
log(Preg)	0.3241	0.7203

Covariance Matrix for Carcinoma

	log(Tet)	log(Preg)
log(Tet)	0.5552	−0.2422
log(Preg)	−0.2422	0.2885

We begin by assuming equal covariance matrices and equal prior probabilities. Equal covariance matrices implies a linear discriminant analysis. Equal prior probabilities corresponds to a maximum likelihood analysis. All results are based on the pooled covariance matrix given below.

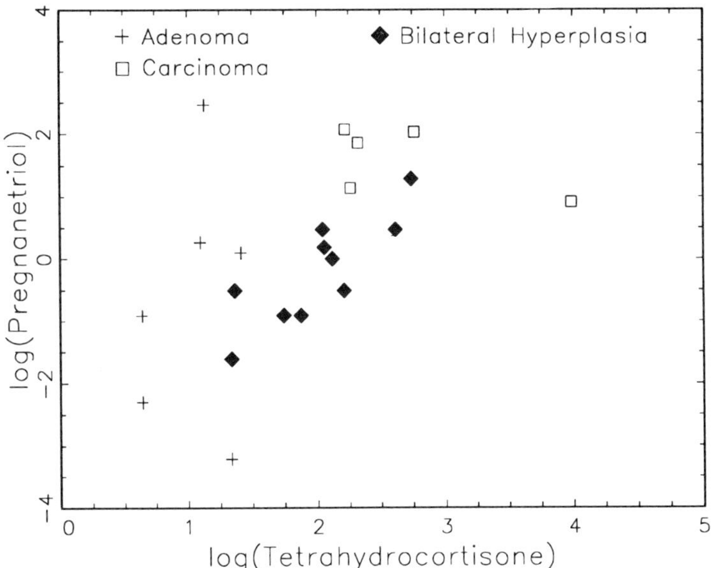

FIGURE II.1. Cushing's Syndrome

Pooled Covariance Matrix

	log(Tet)	log(Preg)
log(Tet)	0.2601	0.1427
log(Preg)	0.1427	1.5601

The results of the linear discriminant analysis are summarized in Table
II.2. Based on resubstitution, four cases are misallocated, all from group
b. Based on single-case cross-validation, three additional cases are misallocated. Cases 8 and 9 are consistently classified as belonging to group a
and cases 12 and 16 are classified as c. In addition, when they are left out
of the fitting process, cases 1 and 4 are allocated to groups c and b respectively while case 19 is misallocated as b. It is interesting to note that linear
discrimination has a hard time deciding whether case 19 belongs to type b
or c. A more detailed discussion of these cases is given later.

The results of a discriminant analysis without the assumption of equal covariance matrices is given in Table II.3. The analysis is based on equal costs
and prior probabilities; hence it is again a maximum likelihood analysis.
Again, the quadratic discrimination procedure requires means and covariance matrices for each group but the covariance matrices are not pooled.
Using the resubstitution method, only two cases are misallocated: 9 and
12. Under cross-validation, cases 1, 8, 19, and 20 are also misclassified.

Careful inspection of Table II.1 and Figure II.1 sheds light on both
the linear and quadratic procedures. From Figure II.1, there seems to

TABLE II.2. Linear Discrimination

Allocated to Group	Resubstitution True Group			Cross-Validation True Group		
	a	b	c	a	b	c
a	6	2	0	4	2	0
b	0	6	0	1	6	1
c	0	2	5	1	2	4

Case		Group	Resubstitution Probability			Cross-Validation Probability		
			a	b	c	a	b	c
1	**	a	.80	.14	.05	.17	.30	.54
2		a	.83	.16	.01	.80	.19	.01
3		a	.96	.04	.00	.94	.06	.00
4	**	a	.61	.39	.00	.12	.88	.00
5		a	.60	.37	.03	.56	.41	.03
6		a	.96	.04	.00	.96	.04	.00
7		b	.08	.71	.21	.09	.69	.22
8	**	b	.64	.35	.00	.75	.25	.00
9	**	b	.64	.35	.01	.69	.30	.01
10		b	.10	.69	.22	.11	.67	.23
11		b	.06	.77	.17	.07	.75	.19
12	**	b	.00	.20	.80	.00	.12	.88
13		b	.10	.64	.26	.11	.62	.27
14		b	.19	.75	.06	.21	.73	.06
15		b	.29	.68	.04	.31	.65	.04
16	**	b	.01	.42	.58	.01	.36	.63
17		c	.02	.26	.73	.02	.31	.67
18		c	.02	.26	.72	.04	.36	.61
19	**	c	.03	.43	.53	.03	.49	.47
20		c	.00	.02	.98	.00	.14	.86
21		c	.00	.10	.89	.00	.12	.88

TABLE II.3. Quadratic Discrimination

Allocated to Group	Resubstitution True Group			Cross-Validation True Group		
	a	b	c	a	b	c
a	6	1	0	5	2	0
b	0	8	0	0	7	2
c	0	1	5	1	1	3

Case		Group	Resubstitution Probability			Cross-Validation Probability		
			a	b	c	a	b	c
1	**	a	.61	.00	.39	.24	.00	.76
2		a	1.00	.00	.00	1.00	.00	.00
3		a	.91	.09	.00	.81	.19	.00
4		a	1.00	.00	.00	.96	.04	.00
5		a	.92	.08	.00	.87	.13	.00
6		a	1.00	.00	.00	1.00	.00	.00
7		b	.00	1.00	.00	.00	1.00	.00
8	**	b	.42	.58	.00	.62	.38	.00
9	**	b	.61	.39	.00	.93	.07	.00
10		b	.00	1.00	.00	.00	1.00	.00
11		b	.00	1.00	.00	.00	1.00	.00
12	**	b	.00	.28	.72	.00	.12	.88
13		b	.01	.99	.00	.01	.98	.01
14		b	.02	.98	.00	.03	.97	.00
15		b	.04	.96	.00	.05	.95	.00
16		b	.00	.96	.04	.00	.94	.06
17		c	.00	.01	.99	.00	.01	.99
18		c	.00	.00	1.00	.00	.00	1.00
19	**	c	.00	.40	.60	.00	1.00	.00
20	**	c	.00	.00	1.00	.00	1.00	.00
21		c	.00	.05	.95	.00	.08	.92

be almost no evidence that the covariance matrices are equal. Adenoma displays large variability in log(Pregnanetriol), very small variability in log(Tetrahydrocortisone), and almost no correlation between the variables. Carcinoma is almost the opposite. It has large variability in log(Tet) and small variability in log(Preg). Carcinoma seems to have a negative correlation. Bilateral hyperplasia displays substantial variability in both variables with a positive correlation. These conclusions are also visible from the estimated covariance matrices. Given that the covariance structure seems to differ from group to group, linear discrimination does surprisingly well when evaluated by resubstitution. Recall that linear discriminant analysis has been found to be rather robust. Of course, quadratic discrimination does a much better job for this data.

The fact that the assessments based on cross-validation are much worse than those based on resubstitution is due largely to the existence of influential observations. The mean of group c and especially the covariance structure of group c are dominated by the large value of log(Tet) for case 20. Case 20 is not misclassified by the linear analysis because its effect on the covariance structure is minimized by the pooling of covariance estimates over groups. In cross-validated quadratic discrimination, its effect on the covariance of group c is eliminated so case 20 seems more consistent with group b. The large log(Preg) value of case 1 is also highly influential. With case 1 dropped out and case 20 included, case 1 is more consistent with carcinoma than with adenoma. The reason that cases 8 and 9 are misclassified is simply that they tend to be consistent with group a, cf. Figure II.1. In examining Table II.1, a certain symmetry can be seen involving cases 12 and 19. Because of case 19, when case 12 is unassigned it looks more like group c than its original group. Similarly, because of case 12, when 19 is unassigned it looks more like group b than group c under quadratic discrimination. Case 19 is essentially a toss up under linear discrimination. Cases 4 and 16 are misclassified under linear discrimination because they involve very unusual data. Case 4 has an extremely small pregnanetriol value and case 16 has a very large tetrahydrocortisone value for being part of the bilateral hyperplasia group.

In a data set this small, it seems unreasonable to drop influential observations. If we cannot believe the data, there is little hope of being able to arrive at a reasonable analysis. If further data bear out the covariance tendencies visible in Figure II.1, the best analysis is provided by quadratic discrimination. It must be acknowledged that the error rates obtained by resubstitution are unreliable. They are generally biased towards underestimating the true error rates and they may be particularly bad for this data. Quadratic discrimination simply provides a good description of the data. There is probably insufficient data to produce good predictions. Christensen (1990) uses this data to illustrate logistic discrimination.

One interesting problem in linear allocation is the choice of variables. Including variables that have no ability to discriminate between populations can only muddy the issues involved. By analogy with multiple regression, one might expect to find advantages to allocation procedures based solely on variables with high discriminatory power. In multiple regression, methods for eliminating independent variables are either directly based on or closely related to testing whether exclusion of the variables hurts the regression model. In other words, a test is performed of whether, given the included variables, the excluded variables contain any additional information for prediction. In discrimination and allocation, methods for eliminating discriminatory variables such as *stepwise discrimination* are based on testing whether, given the included variables, the excluded variables contain any additional information for discrimination between the populations. We have noted that linear discrimination is closely related to the multivariate one-way ANOVA model. Tests of additional information can be performed as in Section I.7. In particular, they are typically performed by testing a one-way ACOVA model such as (1.7.1) against the no treatment effects ACOVA model (1.7.3).

EXAMPLE 2.2.2.   We now illustrate the process of stepwise discrimination using data given by Lubischew (1962). He considered the problem of discriminating between three populations of flea-beatles within the genus Chaetocnema. Six variables were given: y_1 – the width, in microns, of the first joint of the first tarsus, y_2 – the same measurement for the second joint, y_3 – the maximum width, in microns, of the aedeagus in the fore part, y_4 – the front angle, in units of 7.5 degrees, of the aedeagus, y_5 – the maximum width of the head, in .01 millimeter units, between the external edges of the eyes, and y_6 – the width of the aedeagus from the side, in microns. In addition, Lubischew mentions that $r_{12} \equiv y_1/y_2$ is very good for discriminating between one of the species and the other two. The vector of dependent variables is taken as $y' = (y_1, y_2, y_3, y_4, y_5, y_6, r_{12})$. Stepwise discrimination is carried out by testing for additional information in the one-way MANOVA.

Evaluating the assumptions of a one-way MANOVA with three groups and 7 dependent variables is a daunting task. There are three 7×7 covariance matrices that should be roughly similar. To wit, there are $\binom{7}{2} = 21$ bivariate scatter plots to check for elliptical patterns. If the capability exists for the user, there are $\binom{7}{3} = 35$ three-dimensional plots to check. There are $3(7) = 21$ normal plots to evaluate the marginal distributions and at least some linear combinations of the variables should be evaluated for normality. Of course, if y_1 and y_2 are multivariate normal, the constructed variable r_{12} cannot be. However, it may be close enough for our purposes.

If the assumptions break down, it is difficult to know how to proceed. After any transformation, everything needs to be reevaluated with no guarantee that things will have improved. It seems like the best bet for a transformation is some model-based system similar to the Box and Cox (1964) method, cf. Andrews, Gnanadesikan, and Warner (1971).

For the most part, in this example, we will cross our fingers and hope for the best. In other words, we will rely on the robustness of the procedure. While it is certainly true that the P values used in stepwise discriminant analysis should typically not be taken at face value (this is true for almost any statistical modeling technique), the P values can be viewed as simply a one to one transformation of the test statistics. Thus, decisions based on P values are based on the relative sizes of comparable test statistics. The test statistics are reasonable even without the assumption of multivariate normality so, from this point of view, multivariate normality is not a crucial issue.

The assumption of equal covariance matrices is a stickier issue. From Christensen (1987, Section III.2), univariate test statistics are based on the squared length of the vector of differences between the optimal estimates under the full model and the optimal estimates under the reduced model. If the reduced model is (nearly) correct, the difference vector should be near zero. For the multivariate linear model, the hypothesis statistic H consists of the squared length for each variable and inner products between distinct variables. Small matrices H are consistent with the reduced model and large matrices H are consistent with the alternative. Even with unequal covariance matrices for the groups, the least squares estimates are reasonable so valid conclusions can be based on the size of H. In fact, with unequal covariance matrices, least squares estimates are still optimal for the full one-way MANOVA model. To see this, write the model as a univariate linear model and, as in Section I.1, use Theorem 10.4.5 from Christensen (1987). The problem with unequal covariance matrices is twofold. First, it is difficult to evaluate explicitly what we mean by large and small matrices H. Second, the inner products (and thus the lengths) are being evaluated with the Euclidean inner product, which is less than optimal.

Although the properties of formal tests can be greatly affected by the invalidity of the MANOVA assumptions, crude but valid evaluations can still be made based on the test statistics. This is often the most that we have any right to expect from multivariate procedures. For univariate models, Scheffé (1959, Chapter 10) gives an excellent discussion of the effects of invalid assumptions on formal tests.

The three species of flea-beetles considered will be referred to as simply A, B, and C and indexed as 1, 2, and 3 respectively. There are 21 observations on species A with

$$\bar{y}_1' = (183.1, 129.6, 51.2, 146.2, 14.1, 104.9, 1.41)$$

and

$$
S_1 = \begin{bmatrix}
147.5 & 66.64 & 18.53 & 15.08 & -5.21 & 14.21 & .406 \\
66.64 & 51.25 & 11.55 & 2.48 & -1.81 & 3.09 & -.044 \\
18.53 & 11.55 & 4.99 & 5.85 & -.524 & 5.49 & .017 \\
15.08 & 2.48 & 5.85 & 31.66 & -.969 & 15.63 & .090 \\
-5.21 & -1.81 & -.524 & -.969 & .791 & -1.99 & -.021 \\
14.21 & 3.09 & 5.49 & 15.63 & -1.99 & 38.23 & .078 \\
.406 & -.044 & .017 & .090 & -.021 & .078 & .0036
\end{bmatrix}.
$$

Species B has 31 observations with

$$
\bar{y}_2' = (201.0, 119.3, 48.9, 124.6, 14.3, 81.0, 1.69)
$$

and

$$
S_2 = \begin{bmatrix}
222.1 & 63.40 & 22.60 & 30.37 & 4.37 & 29.47 & .926 \\
63.40 & 44.16 & 7.91 & 11.82 & .337 & 11.47 & -.100 \\
22.60 & 7.91 & 5.52 & 5.69 & .005 & 4.23 & .075 \\
30.37 & 11.82 & 5.69 & 21.37 & -.327 & 11.70 & .088 \\
4.37 & .337 & .005 & -.327 & 1.21 & 1.27 & .029 \\
29.47 & 11.47 & 4.23 & 11.70 & 1.27 & 79.73 & .085 \\
.926 & -.100 & .075 & .088 & .029 & .085 & .009
\end{bmatrix}.
$$

For species C, there are 22 observations with

$$
\bar{y}_3' = (138.2, 125.1, 51.6, 138.3, 10.1, 106.6, 1.11)
$$

and

$$
S_3 = \begin{bmatrix}
87.33 & 44.55 & 20.53 & 19.17 & -.736 & 15.29 & .301 \\
44.55 & 73.04 & 15.71 & 14.02 & -.390 & 21.23 & -.267 \\
20.53 & 15.71 & 8.06 & 8.21 & -.294 & 4.97 & .027 \\
19.17 & 14.02 & 8.21 & 2.16 & -.502 & 7.93 & .027 \\
-.736 & -.390 & -.294 & -.502 & .944 & .277 & -.002 \\
15.29 & 21.23 & 4.97 & 7.93 & .277 & 34.25 & -.061 \\
.301 & -.267 & .027 & .027 & -.002 & -.061 & .0046
\end{bmatrix}.
$$

The pooled estimate of the covariance is a weighted average of S_1, S_2, and S_3 with approximately 50% more weight on S_2 than on the other estimates.

Although typically, backward elimination is to be preferred to forward selection in stepwise procedures, it is illustrative to demonstrate forward selection on this data set. We will begin by making a very rigorous requirement for inclusion: variables will be included if the P value for adding them is .01 or less.

The first step in forward selection consists of performing the univariate one-way ANOVA F tests for each variable.

Step 1: Statistics for Entry, $df = 2, 71$

Variable	F_{obs}	$\Pr[F > F_{obs}]$
y_1	160.339	0.0001
y_2	12.499	0.0001
y_3	9.659	0.0002
y_4	134.353	0.0001
y_5	129.633	0.0001
y_6	101.314	0.0001
r_{12}	351.292	0.0001

The P values are all sufficiently small to warrant inclusion of the variables. By far the largest F statistic and thus the smallest P value, is for r_{12} so this is the first variable included for use in discrimination. Note that r_{12} is the variable constructed by Lubischew.

The second and all subsequent steps of the procedure involve performing a one-way analysis of covariance for each variable not yet included. For the second step, the sole covariate is r_{12} and a test is made for treatment effects in the analysis of covariance model. For the dependent variables y_1 through y_6, the results are given below.

Step 2: Statistics for Entry, $df = 2, 70$

Variable	F_{obs}	$\Pr[F > F_{obs}]$
y_1	9.904	0.0002
y_2	8.642	0.0004
y_3	6.386	0.0028
y_4	87.926	0.0001
y_5	30.549	0.0001
y_6	28.679	0.0001

The largest F statistic is for y_4 and the corresponding P value is less than .01, so y_4 is included for discrimination.

At the third step, both r_{12} and y_4 are used as covariates in a one-way analysis of covariance. Again, the F tests for treatment differences are performed.

Step 3: Statistics for Entry, $df = 2, 69$

Variable	F_{obs}	$\Pr[F > F_{obs}]$
y_1	2.773	0.0694
y_2	3.281	0.0436
y_3	6.962	0.0018
y_5	24.779	0.0001
y_6	3.340	0.0412

Variable y_5 is included for discrimination. Note the large difference between the F statistic for y_5 and that for the other variables. There is an order of magnitude difference between the abilities of the r_{12}, y_4, and y_5 to discriminate and the abilities of the other variables. Considering the questionable validity of formal tests, this is an important point. It should also be mentioned that this conclusion is based on one sequence of models. There is a possibility that other sequences would lead to different conclusions about the relative importance of the variables. In fact, it would be desirable to check all models or, better yet, have an algorithm to identify the best models.

Step 4 simply adds weight to our conclusions of the previous paragraph. In performing the analysis of covariance with three covariates, none of the variables considered have the very large F statistics seen earlier.

Step 4: Statistics for Entry, $df = 2, 68$

Variable	F_{obs}	$\Pr[F > F_{obs}]$
y_1	1.985	0.1453
y_2	2.567	0.0842
y_3	3.455	0.0372
y_6	3.359	0.0406

Any rule that terminates forward selection when all P values exceed .0371 will stop the selection process at Step 4. In particular, our stringent stopping rule based on P values of .01 terminates here.

In practice, it is much more common to use a stopping rule based on P values of .05, .10, or .15. By any of these rules we would add variable y_3 and continue checking variables. This leads to Step 5 and the corresponding F statistics.

Step 5: Statistics for Entry, $df = 2, 67$

Variable	F_{obs}	$\Pr[F > F_{obs}]$
y_1	7.040	0.0017
y_2	8.836	0.0004
y_6	3.392	0.0395

Surprisingly, adding y_3 has changed things dramatically. While the F statistic for y_6 is essentially unchanged, the F values for y_1 and y_2 have more than tripled. Of course, we are still not seeing the huge F statistics that were encountered earlier but, apparently, one can discriminate much better with y_3 and either y_1 or y_2 than would be expected from the performance of any of these variables individually. This is precisely the sort of thing that is very easily missed by forward selection procedures and one of the main reasons why they are considered to be poor methods for model selection. Forward selection does have advantages. In particular, it is cheap and it is able to accommodate huge numbers of variables.

The stepwise procedure finishes off with two final steps. Variable y_2 was added in the previous step. The results from Step 6 are given below.

Step 6: Statistics for Entry, $df = 2, 66$

Variable	F_{obs}	$\Pr[F > F_{obs}]$
y_1	0.827	0.4418
y_6	3.758	0.0285

Variable y_6 is added if our stopping rule is not extremely stringent. This leaves just y_1 to be evaluated.

Step 7: Statistics for Entry, $df = 2, 65$

Variable	F_{obs}	$\Pr[F > F_{obs}]$
y_1	0.907	0.4088

By any standard y_1 would not be included. Of course, r_{12} is the ratio of y_1 and y_2 so it is not surprising that there is no need for all three variables. A forward selection procedure that does not include r_{12} would simply include all of the variables.

We have learned that r_{12}, by itself, is a powerful discriminator. The variables r_{12}, y_4, and y_5, when taken together, have major discriminatory powers. Variable y_3, taken together with either y_1 or y_2 and the previous three variables, may provide substantial help in discrimination. Finally, y_6 may also contribute to distinguishing among the populations. Most of these conclusions are visible from the table given below that summarizes the results of the forward selection.

Summary of Forward Selection

Step	Variable Entered	F_{obs}	$\Pr[F > F_{obs}]$
1	r_{12}	351.292	0.0001
2	y_4	87.926	0.0001
.3	y_5	24.779	0.0001
4	y_3	3.455	0.0372
5	y_2	8.836	0.0004
6	y_6	3.758	0.0285

It is also of interest to see the results of a multivariate analysis of variance for all of the variables included at each step. For example, after Step 3, variables r_{12}, y_4, and y_5 were included for discrimination. The likelihood ratio test statistic for no group effects in the one-way MANOVA is $U = .0152$. This is a very small, hence very significant, number. The following table lists the results of such tests for each step in the process.

MANOVA Tests

Step	Variable Entered	LRTS U_{obs}	$\Pr[U < U_{obs}]$
1	r_{12}	0.09178070	0.0001
2	y_4	0.02613227	0.0001
3	y_5	0.01520881	0.0001
4	y_3	0.01380601	0.0001
5	y_2	0.01092445	0.0001
6	y_6	0.00980745	0.0001

Based on their P values, all of the variables added had substantial discriminatory power. Thus, it is not surprising that the U statistics decrease as each variable is added.

In practice, decisions about the practical discriminatory power of variables should not rest solely on the P values. After all, the P values are often unreliable. Other methods, such as the graphical methods presented in the next section, should be used in determining the practical usefulness of results based on multivariate normal distribution theory.

II.3   Linear Discrimination Coordinates

As mentioned earlier, one is typically interested in the clarity of classification. This can be investigated by examining the posterior probabilities, the entire likelihood function, or the entire set of Mahalanobis distances. It is done by computing the allocation measures for each element of the data set. The allocation measure can be estimated either by the entire data set or the data set having deleted the case currently being allocated. To many people, the second approach is more appealing.

An alternative approach to examining the clarity of discrimination is through the use of linear discrimination coordinates. This approach derives from the work of Fisher (1938) and Rao (1948, 1952). It consists of redefining the coordinate system in $\mathbf{R}^q$ in such a way that the different treatment groups in the one–way ANOVA have, in some sense, maximum separation in each coordinate. The clarity of discrimination can then be examined visually by inspecting one, two, or three-dimensional plots of the data. In these plots, cases are identified by their populations. If the new coordinate system is effective, observations from the same population should be clustered together and distinct populations should be well separated.

It is standard practice to redefine the coordinate system by taking linear combinations of the original variables. It is also standard practice to define the new coordinate system sequentially. In particular, the first coordinate is chosen to maximize the separation between the groups. The second coordinate maximizes the separation between the groups given that the second linear combination is uncorrelated with the first. The third maximizes the separation given that the linear combination is uncorrelated with the first two. Subsequent coordinates are defined similarly. In the discussion below, we assume a constant covariance matrix for the t groups. It remains to define what precisely is meant by "maximum separation of the groups."

Recall that with equal covariance matrices, the data available in a discriminant analysis fits a multivariate one-way ANOVA

$$Y = XB + e.$$

Thus,

$$E = \sum_{i=1}^{t} \sum_{j=1}^{n_i} (y_{ij} - \bar{y}_{i\cdot})(y_{ij} - \bar{y}_{i\cdot})'$$

and

$$H = \sum_{i=1}^{t} n_i (\bar{y}_{i\cdot} - \bar{y}_{\cdot\cdot})(\bar{y}_{i\cdot} - \bar{y}_{\cdot\cdot})'.$$

Also define

$$H_* = \sum_{i=1}^{t} (\bar{y}_{i\cdot} - \bar{y}_{\cdot\cdot})(\bar{y}_{i\cdot} - \bar{y}_{\cdot\cdot})'.$$

The linear discrimination coordinates are based on E and either H or H_*. We will examine the use of H in detail. Some comments will also be made on the motivation for using H_*.

For any vector $y = (y_1, \ldots, y_q)'$, the first linear discrimination coordinate is defined by

$$y' a_1$$

where the vector a_1 is chosen so that the univariate one-way ANOVA model

$$(Y a_1) = X(B a_1) + (e a_1)$$

has the largest possible F statistic for testing equality of group effects. Intuitively, the linear combination of the variables that maximizes the F statistic must have the greatest separation between groups. The degrees of freedom are not affected by the choice of a_1 so we need to find a_1 that maximizes

$$\frac{(Y a_1)' \left(M - \frac{1}{n} J_n^n\right)(Y a_1)}{(Y a_1)'(I - M)(Y a_1)}$$

or equivalently

$$\frac{a_1' H a_1}{a_1' E a_1}.$$

A one-dimensional plot of the n elements of $Y a_1$ shows the maximum separation between groups that can be achieved in a one-dimensional plot.

The second linear discrimination coordinate is

$$y' a_2$$

such that

$$\frac{a_2' H a_2}{a_2' E a_2}$$

is maximized subject to the constraint that for any i and j, the estimated covariance between $y_{ij}' a_1$ and $y_{ij}' a_2$ is zero. The covariance condition can be rewritten as

$$a_1' S a_2 = 0$$

or equivalently as

$$a_1' E a_2 = 0\,.$$

Another way of thinking of this condition is that a_1 and a_2 are orthogonal in the inner product space defined using the matrix E. (Inner products are discussed in the paragraphs following Definition B.50 in Christensen (1987).)

A one-dimensional plot of $Y a_2$ illustrates visually the separation in the groups. Even more productively, the n ordered pairs that are the rows of $Y(a_1, a_2)$ can be plotted to illustrate the discrimination achieved by the first two linear discrimination coordinates.

For $h = 3, \ldots, q$ the h'th linear discriminant coordinate is

$$y' a_h$$

where

$$a_h' H a_h / a_h' E a_h$$

is maximized subject to the covariance condition

$$a_h' E a_i = 0 \qquad i = 1, 2, \ldots, h - 1\,.$$

Note that, using the inner product for $\mathbf{R}^q$ based on E, this defines an orthogonal system of coordinates, i.e., $a_1, \ldots, a_q$ define an orthogonal basis for $\mathbf{R}^q$ using the inner product defined by E.

Unfortunately, the discrimination coordinates are not uniquely defined. Given a vector a_h, any scalar multiple of a_h also satisfies the requirements listed above. One way to avoid the nonuniqueness is to impose another condition. The most commonly used extra condition is that $a_h' E a_h = 1$, so that $a_1, \ldots, a_q$ is an orthonormal basis for $\mathbf{R}^q$ under the inner product defined by E.

Before going into the details of actually finding the linear discrimination coordinates, we illustrate their use. It will be established below that the linear discrimination coordinate vectors a_i, $i = 1, \ldots, q$ are eigenvectors of $E^{-1} H$. Moreover, the appropriate metric for examining variables transformed into the linear discrimination coordinates is the standard Euclidean metric. This allows simple visual inspection of the transformed data. Writing $A = [a_1, \ldots, a_q]$, the mapping Y into YA gives the data matrix in the linear discrimination coordinates.

EXAMPLE 2.3.1.   Consider again the heart rate data of Example 1.5.1. The data structure needed for development of linear discrimination coordinates is the same as for a one-way MANOVA. We have already examined these data for multivariate normality and equal covariance matrices. The data seem to satisfy the assumptions.

The linear discrimination coordinates are defined by the matrix of eigenvectors of $E^{-1}H$. This is

$$A = \frac{1}{10} \begin{bmatrix} .739 & .382 & .581 & .158 \\ -.586 & -.323 & -.741 & .543 \\ -.353 & -.234 & .792 & -.375 \\ .627 & -.184 & -.531 & -.218 \end{bmatrix}.$$

Recall that E and H were given in Example 1.5.1. The columns of A define four new data vectors Ya_1, Ya_2, Ya_3, and Ya_4. If we perform an analysis of variance on each variable we get F statistics for discriminating between groups. All have 2 degrees of freedom in the numerator and 27 in the denominator.

Variable	F
Ya_1	74.52
Ya_2	19.47
Ya_3	0.0
Ya_4	0.0

As advertised, the F statistics are nonincreasing. The first two F statistics clearly establish that there are group differences in the first two coordinates. The last two F statistics are zero because with three groups there are two degrees of freedom for treatments and H is a 4×4 matrix of rank 2. Only two of the linear discrimination coordinates can have positive F statistics. This issue is discussed in more detail later in the section.

The big advantage of linear discrimination coordinates is that they allow us to plot the data in ways that let us visualize the separation in the groups. Figure II.2 shows a series of one dimensional plots that display the first discrimination coordinate values for each population. Figure II.3 is an alternative method of plotting the first discrimination coordinate. It consists of a scatter plot of Ya_1 versus the index of the populations. Note that the degree of separation is substantial and about the same for all three groups. The edges of the middle group are close to the edges of the other groups. The placebo has one observation that is consistent with drug A.

Figure II.4 is similar to Figure II.2 except that it plots the data in the second discrimination coordinate. Note that in the second coordinate it is very difficult to distinguish between drugs A and B. The Placebo is separated from the other groups but there is more overlap around the edges than was present in the first coordinate.

Figure II.5 is a scatter plot of the data in the first two discrimination coordinates. Together, the separation is much clearer than in either of the

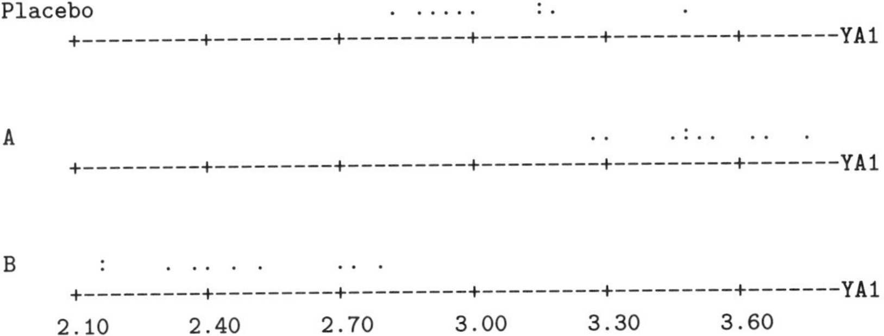

FIGURE II.2. One-Dimensional Plots of the Heart Rate Data in the First Linear Discrimination Coordinate

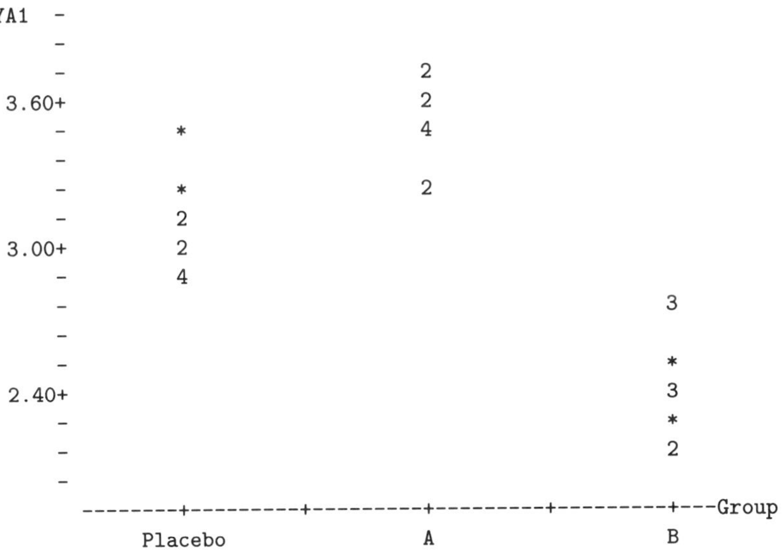

FIGURE II.3. Scatter Plot of the Heart Rate Data in the First Discrimination Coordinate

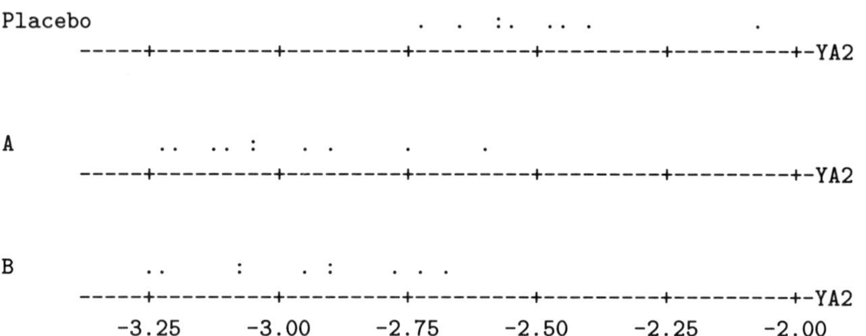

FIGURE II.4. One-Dimensional Plots of the Heart Rate Data in the Second Discrimination Coordinate

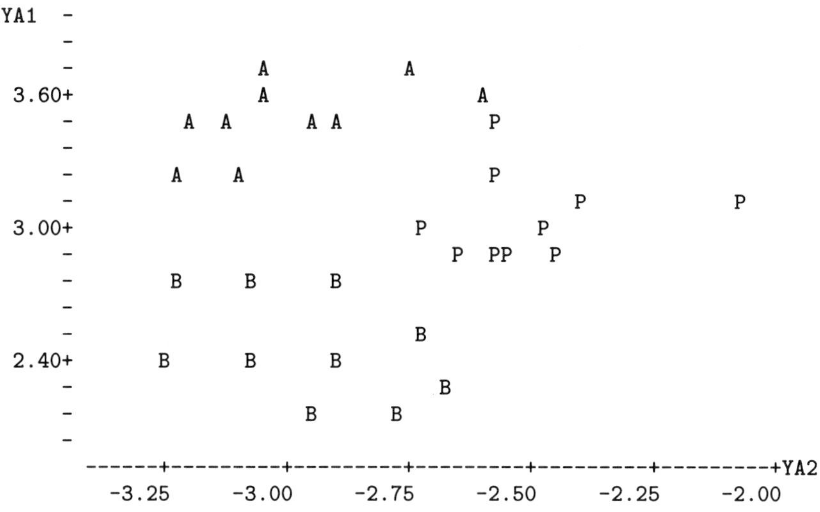

FIGURE II.5. Scatter Plot of the Heart Rate Data in the First Two Linear Discrimination Coordinates

individual coordinates. There is still one observation from drug A that is difficult to distinguish from the Placebo group but, other than that, the groups are very well separated. That the one observation from drug A is similar to the Placebo is a conclusion based on the Euclidean distance of the point from the centers of the groups for drug A and the Placebo. It is not clear that Euclidean distances are appropriate but that will be shown later in this section.

The vectors $a_1, \ldots, a_q$ can be taken as eigenvectors of the matrix $E^{-1}H$.

Before showing this we prove a result similar to Theorem B.15 in Christensen (1987). Theorem B.15 states that given any symmetric matrix, say W, there exists an orthonormal basis for $\mathbf{R}^n$ consisting of eigenvectors of W. Lemma 2.3.2 states that, relative to an inner product on $\mathbf{R}^q$ defined by an arbitrary positive definite matrix E, there exists an orthonormal basis for $\mathbf{R}^q$ consisting of eigenvectors of $E^{-1}H$ where H is an arbitrary symmetric matrix. Note that although we have continued to use the symbols H and E and our immediate interest is in application of these results to the specific matrices H and E defined earlier, the result does not depend on the choice of these matrices except as indicated in the lemma. The series of results given below will also be used in Chapter III to derive principal components. In Chapter III, E and H will not be the error and hypothesis matrices from a multivariate linear model.

Lemma 2.3.2.   Let E be any $q \times q$ positive definite matrix and let H be any symmetric $q \times q$ matrix. Then there exists a $q \times q$ diagonal matrix Λ and a matrix A such that

$$E^{-1}HA = A\Lambda$$

and

$$A'EA = I\,.$$

Observe that the columns of A must be eigenvectors and the elements of Λ must be eigenvalues of $E^{-1}H$.

PROOF.   Define $E^{1/2}$ as in Lemma 1.2.3. By Christensen's Theorem B.15 there exists B such that

$$E^{-1/2}HE^{-1/2}B = B\Lambda \tag{1}$$

with

$$I = BB' = B'B\,.$$

Let $A = E^{-1/2}B$ then multiplying (1) on the left by $E^{-1/2}$ gives

$$E^{-1}HA = A\Lambda$$

and

$$A'EA = B'E^{-1/2}EE^{-1/2}B = B'B = I\,.$$

$\square$

We will later need the following result.

Corollary 2.3.3.   If A is $q \times q$, $E^{-1}HA = A\Lambda$ and $A'EA = I$ then

$$E^{-1} = AA'.$$

PROOF.   If $A'EA = I$ and both E and A are $q \times q$, then A must be nonsingular. It follows that

$$I = (A'EA)^{-1} = A^{-1}E^{-1}(A')^{-1}.$$

Multiplying on the left by A and on the right by A' gives

$$AA' = AA^{-1}E^{-1}(A')^{-1}A' = E^{-1}.$$

$\square$

The argument that the linear discrimination coordinates can be taken as eigenvectors of $E^{-1}H$ has similarities to the proof of Lemma 2.3.2 and also to the argument in Section I.2 that relates Roy's $\phi_{\max}$ test statistic to the maximum eigenvalue of HE^{-1}. Once again, the argument does not depend on the specific choices of H and E.

Proposition 2.3.4.   Let E be any $q \times q$ positive definite matrix and let H be any $q \times q$ symmetric matrix. The vectors $a_1, \ldots, a_q$ satisfy

$$\frac{a_1'Ha_1}{a_1'Ea_1} = \sup_a \frac{a'Ha}{a'Ea},$$

and for $i > 1$,

$$a_i'Ea_j = 0 \qquad j = 1, \ldots, i-1,$$

and

$$\frac{a_i'Ha_i}{a_i'Ea_i} = \sup_a \left\{ \frac{a'Ha}{a'Ea} \middle| a'Ea_j = 0 \qquad j = 1, \ldots, i-1 \right\}$$

if and only if for $i = 1, \ldots, q$, a_i is an eigenvector of $E^{-1}H$ corresponding to an eigenvalue ϕ_i where $\phi_1 \geq \cdots \geq \phi_q$ and where, for $i > 1$, $a_iEa_j = 0$, $j = 1, \ldots, i-1$.

PROOF.   The idea of the proof is to transform the difficult problem of looking at $a'Ha/a'Ea$ to a simpler problem of looking at $c'\Lambda c$ where Λ is a diagonal matrix. The conditions $a'Ea_j = 0$ are transformed into conditions

$c'c_j = 0$. The proof is by induction. The inductive step requires the assumption of an induction hypothesis that is more detailed than one might expect. We begin by examining the transformation.

Define $E^{1/2}$ as in Lemma 1.2.3. Write

$$\Lambda = D(\phi_1, \ldots, \phi_q)$$

and pick B and A as in the proof of Lemma 2.3.2. Note that

$$E^{-1/2}HE^{-1/2} = B\Lambda B'$$

and recall that

$$E^{-1}HA = A\Lambda.$$

For any vector a, define

$$b = E^{1/2}a$$

and

$$c = B'b = B'E^{1/2}a.$$

The transformation from a to c is invertible, i.e.,

$$\begin{aligned} a &= E^{-1/2}Bc \\ &= Ac. \end{aligned}$$

Observe that

$$\begin{aligned} \frac{a'Ha}{a'Ea} &= \frac{b'E^{-1/2}HE^{-1/2}b}{b'b} = \frac{b'B\Lambda B'b}{b'b} \\ &= \frac{b'B\Lambda B'b}{b'BB'b} = \frac{c'\Lambda c}{c'c}. \end{aligned}$$

Moreover, if a and a_* are two vectors, by the choice of B

$$a'Ea_* = b'b_* = b'BB'b_* = c'c_*.$$

Thus, properties related to $a'Ha/a'Ea$ and $a'Ea_*$ can be examined by investigating $c'\Lambda c$ and $c'c_*$.

Finally, we need to establish that a is an eigenvector of $E^{-1}H$ corresponding to some value ϕ if and only if c is an eigenvector of Λ corresponding to ϕ. First, if $E^{-1}Ha = a\phi$ then $\Lambda c = A^{-1}(A\Lambda)c = A^{-1}(E^{-1}HA)c = A^{-1}E^{-1}Ha = A^{-1}a\phi = A^{-1}Ac\phi = c\phi$. Conversely, if $\Lambda c = c\phi$, then $E^{-1}Ha = E^{-1}HAc = A\Lambda c = Ac\phi = a\phi$.

We begin by showing that $a_1'Ha_1/a_1'Ea_1$ maximizes $a'Ha/a'Ea$ if and only if a_1 is an eigenvector of $E^{-1}H$ corresponding to ϕ_1. A vector a_1 maximizes $a'Ha/a'Ea$ if and only if the corresponding vector c_1 maximizes

$$\frac{c'\Lambda c}{c'c} = \sum_{j=1}^{q} c_j^2 \phi_j \bigg/ \sum_{j=1}^{q} c_j^2.$$

Because this is a weighted average of the ϕ_j's and the ϕ_j's are ordered from largest to smallest, the maximum value is ϕ_1. Given the previously demonstrated relationship between eigenvectors of $E^{-1}H$ and Λ, the proof is complete if we show that c_1 maximizes $c'\Lambda c/c'c$ if and only if c_1 is an eigenvector of Λ corresponding to ϕ_1. If c_1 is an eigenvector of Λ corresponding to ϕ_1, i.e., if

$$\Lambda c_1 = c_1 \phi_1 ,$$

then

$$c_1' \Lambda c_1 / c_1' c_1 = \phi_1$$

so c_1 maximizes $c'\Lambda c/c'c$. Conversely, suppose $\phi_1 = \cdots = \phi_s > \phi_{s+1}$, then to maximize $c'\Lambda c/c'c$ a vector $c = (k_1, \ldots, k_q)$ must have $k_{s+1} = \cdots = k_q = 0$. Such vectors clearly satisfy $\Lambda c = c\phi_1$.

We complete the proof by induction. Suppose that for $i = 1, \ldots, h-1$

$$\frac{a_i' H a_i}{a_i' E a_i} = \sup_a \left\{ \frac{a' H a}{a' E a} \middle| a' E a_j = 0 \quad j = 1, \ldots, i-1 \right\}$$

with

$$a_i' E a_j = 0 \quad j = 1, \ldots, i-1$$

if and only if a_i is an eigenvector of $E^{-1}H$ corresponding to ϕ_i with $a_i' H a_j = 0$, $j = 1, \ldots, i-1$. We need to prove that

$$\frac{a_h' H a_h}{a_h' E a_h} = \sup_a \left\{ \frac{a' H a}{a' E a} \middle| a' E a_j = 0 \quad j = 1, \ldots, h-1 \right\}$$

with

$$a_h' E a_j = 0 \quad j = 1, \ldots, h-1$$

if and only if a_h is an eigenvector of $E^{-1}H$ corresponding to ϕ_h with $a_h' E a_j = 0$, $j = 1, \ldots, h-1$. From the equivalence established earlier, a_h maximizes $a'Ha/a'Ea$ subject to the conditions if and only if the corresponding vector c_h maximizes $c'\Lambda c/c'c$ subject to the conditions $c'c_j = 0$, $j = 1, \ldots, h-1$ and a_h is an eigenvector with $a_h' E a_j = 0$ if and only if c_h is an eigenvector with $c_h' c_j = 0$.

Suppose $\phi_r = \cdots = \phi_h = \cdots = \phi_s$ with either $r = 1$ or $\phi_{r-1} > \phi_r$ and either $s = q$ or $\phi_s > \phi_{s+1}$. Write $c = (k_1, \ldots, k_q)'$ and $c_i = (k_{i1}, \ldots, k_{iq})'$. Think of the element in the j'th row of a vector as corresponding to ϕ_j. As additional parts of the induction hypothesis, assume that the terms in c_i corresponding to eigenvalues less than ϕ_i must be zero and that if c is orthogonal to the c_i's the terms in c corresponding to eigenvalues greater than ϕ_h must be zero. Specifically, for $i = 1, \ldots, h-1$ if $\phi_j < \phi_i$, then $k_{ij} = 0$ and that if $c'c_i = 0$, $i = 1, \ldots, h-1$ and $\phi_j > \phi_h$ then $k_j = 0$. Note that the first of these conditions holds for c_1 and that the second condition also holds because it does not apply to c_1.

The second of the assumptions implies that for c orthogonal to $c_1, \ldots, c_{h-1}$,

$$c' \Lambda c / c' c = \sum_{j=r}^{q} k_j^2 \phi_j \bigg/ \sum_{j=r}^{q} k_j^2 \, .$$

This is a weighted average of the values ϕ_j, $j = r, \ldots, q$. The maximum value is $\phi_r = \phi_h$. As before, c_h will attain the maximum if and only if c_h is an eigenvector. If c_h is an eigenvector, i.e., $\Lambda c_h = c_h \phi_h$, then the maximum is attained. Conversely, a maximum is only attained if $k_{s+1} = \cdots = k_q = 0$. Thus, an orthogonal maximizing vector c_h must have $k_{h1} = \cdots = k_{hr-1} = k_{hs+1} = \cdots = k_{hq} = 0$. Clearly, any such vector satisfies $\Lambda c = c \phi_h$ and thus is an eigenvector of ϕ_h. In particular, c_h can be any eigenvector that satisfies $c_h' c_j = 0$, $j = 1, \ldots, h - 1$.

To complete the proof we need to prove that our additional induction hypotheses hold for $i = h$. We have already established that $k_{h,s+1} = \cdots = k_{hq} = 0$ which is precisely the condition that if $\phi_j < \phi_h$ then $k_{hj} = 0$. We also need to show that if $c' c_i = 0$, $i = 1, \ldots, h$, and $\phi_j > \phi_{h+1}$ then $k_j = 0$. Equivalently we need to show that if $h < s$, $k_1 = \cdots = k_{r-1} = 0$ and if $h = s$, $k_1 = \cdots = k_s = 0$. If $h < s$, there is nothing to prove; the result follows from the induction hypothesis. For $h = s$, the two induction hypotheses and the argument in the previous paragraph give: (1) for $i = r, \ldots, s$, $k_{is+1} = \cdots = k_{iq} = 0$ and (2) $k_1 = \cdots = k_{r-1} = 0$. Writing $d_i = (k_{ir}, \ldots, k_{is})$ and $d = (k_r, \ldots, k_s)$, we see that for $i = r, \ldots, s$, $c' c_i = 0$ if and only if $d' d_i = 0$. In particular, $d_r, d_{r+1}, \ldots, d_s$ is an orthogonal basis for $\mathbf{R}^{s-r+1}$. Any other vector d that is orthogonal to $d_r, \ldots, d_s$ must be the zero vector. Thus, if $c' c_i = 0$, $i = r, \ldots, s$, then $k_r = \cdots = k_s = 0$. This is precisely what we needed to prove. $\square$

We have established a practical procedure for finding linear discriminant coordinates and for transforming the original data into the discrimination coordinates. Simply find A such that

$$E^{-1} H A = A \Lambda$$

where $\Lambda = D(\phi_1, \ldots, \phi_q)$ and

$$A' E A = I \, .$$

This is not difficult to do using a good matrix manipulation computer program. The transformed data is

$$Z = Y A \, .$$

As illustrated in Example 2.3.1, the n rows of Z can be plotted in a variety of ways to examine the efficacy of the different coordinates for discrimination. The columns of Z corresponding to the largest eigenvalues show the clearest discrimination because they maximize the one-way ANOVA F test. The estimated covariance matrix of a transformed vector $A'y$ is

$$\widehat{\mathrm{Cov}}(A'y) = A'SA = \frac{1}{n-t}A'EA = \frac{1}{n-t}I$$

so the Euclidean metric is appropriate for evaluating relationships between data points. This is important in that it allows intuitive evaluation of plots.

With t groups there are at most $t-1$ coordinates that are valuable for discrimination. If $t > q$, this is not of much interest; but if $t \leq q$, this means that some coordinates have no discriminatory power. Recall that

$$H = Y'\left(M - \frac{1}{n}J_n^n\right)Y$$

and that $r\left(M - \frac{1}{n}J_n^n\right) = t - 1$. It follows that for the $q \times q$ matrix H

$$r(H) \leq \min(q, t-1).$$

Any choice of linear discrimination coordinates corresponds to a set of eigenvectors for $E^{-1}H$. Write $A = (a_1, \ldots, a_q)$ and partition A as $A = (A_*, A_0)$ where the columns of A_* correspond to the nonzero eigenvalues of $E^{-1}H$ and the columns of A_0 correspond to the eigenvalue zero. If a_h is a column of A_0 then as will be seen below

$$a_h'Ha_h/a_h'Ea_h = 0. \tag{2}$$

In other words, if the data is transformed to the h'th discrimination coordinate, i.e., Ya_h, then a one-way ANOVA applied to the transformed data gives an F statistic of zero for testing differences between groups. Thus, the coordinate is useless in discrimination. This result (2) is an immediate consequence of the following lemma.

Lemma 2.3.5. $HA_0 = 0$

PROOF.   The columns of A_0 are eigenvectors of $E^{-1}H$ corresponding to the eigenvalue 0. Hence $E^{-1}HA_0 = 0$. Multiplying on the left by E gives the result. $\square$

Sometimes the linear discrimination coordinates are used to allocate new observations. Generally, this is done when the coordinates are chosen as in Lemma 2.3.2. Suppose the first s coordinates are to be used. Let

$$A_s = [a_1, \ldots, a_s].$$

The standard allocation rule is to assign y to population r if

$$(y - \bar{y}_{r\cdot})'[A_s A_s'](y - \bar{y}_{r\cdot}) = \min_i (y - \bar{y}_{i\cdot})'[A_s A_s'](y - \bar{y}_{i\cdot}). \qquad (3)$$

Note that

$$(y - \bar{y}_{i\cdot})'[A_s A_s'](y - \bar{y}_{i\cdot}) = \sum_{j=1}^{s}[(y - \bar{y}_{i\cdot})'a_j]^2$$

so the allocation is based on the squared values of the first s linear discrimination coordinates of the vector $(y - \bar{y}_{i\cdot})$. Once again we see that the Euclidean metric is appropriate for the transformed variables. If the coordinates are not chosen as in Lemma 2.3.2, i.e., if the vectors $a_1, \ldots, a_q$ are not an orthonormal set in the appropriate inner product, then taking a simple sum of squares is not appropriate. Thus, the restriction on the choice of coordinates is imposed.

This allocation procedure is closely related to the Mahalanobis distance method. If $s = q$, then $A_s = A$ and by Corollary 2.3.3, $AA' = E^{-1}$. The allocation rule is based on the distances

$$(y - \bar{y}_{i\cdot})'E^{-1}(y - \bar{y}_{i\cdot}).$$

The estimated covariance matrix is $S^{-1} = (n-t)E^{-1}$ so these distances are simply a constant multiple of the estimated Mahalanobis distances used in Section 2. The distances differ by merely a constant multiple; therefore the allocation rules are identical.

Recall that there are at most $\min(q, t-1)$ useful linear discrimination coordinates. Eigenvectors of $E^{-1}H$ that correspond to the eigenvalue zero are not useful for discrimination. It makes little sense to choose s greater than the number of nonzero eigenvalues of $E^{-1}H$. For reasons of collinearity (cf. Christensen, 1987, Chapter XIV), it might make sense to choose s less than the number of nonzero eigenvalues of $E^{-1}H$ when some of those nonzero eigenvalues are very close to zero. In fact, choosing the number of linear discrimination coordinates is reminiscent of choosing the number of principal components to use in principal component regression. (Recall the similarities between variable selection in discrimination and variable selection in regression.)

We now show that if s is chosen to be exactly the number of nonzero eigenvalues, i.e., $s = r(E^{-1}H)$, then the linear discrimination coordinate allocation rule based on (3) is precisely the same as the Mahalanobis distance rule. If $r(E^{-1}H) = q$, then $A_s = A$ and we have already shown the result. We need consider only the case where $r(E^{-1}H) \le t - 1 < q$. Using notation from earlier in the section, $A = [A_*, A_0]$. With $s = r(E^{-1}H)$, we have $A_s = A_*$. Before proving the equivalence of allocation rules, we need the following lemma.

Lemma 2.3.6.   If $t - 1 < q$, then for any $i = 1, \ldots, t$,

$$A_0'(\bar{y}_{i\cdot} - \bar{y}_{\cdot\cdot}) = 0.$$

PROOF.   Recall that

$$H = \sum_{j=1}^{t} n_j (\bar{y}_{j\cdot} - \bar{y}_{\cdot\cdot})(\bar{y}_{j\cdot} - \bar{y}_{\cdot\cdot})'$$

$$= [\sqrt{n_1}(\bar{y}_{1\cdot} - \bar{y}_{\cdot\cdot}), \ldots, \sqrt{n_t}(\bar{y}_{t\cdot} - \bar{y}_{\cdot\cdot})] \begin{bmatrix} \sqrt{n_1}(\bar{y}_{1\cdot} - \bar{y}_{\cdot\cdot})' \\ \vdots \\ \sqrt{n_t}(\bar{y}_{t\cdot} - \bar{y}_{\cdot\cdot})' \end{bmatrix}.$$

By Proposition B.51 in Christensen (1987)

$$C(H) = C([\sqrt{n_1}(\bar{y}_{1\cdot} - \bar{y}_{\cdot\cdot}), \ldots, \sqrt{n_t}(\bar{y}_{t\cdot} - \bar{y}_{\cdot\cdot})]).$$

It follows that for $i = 1, \ldots, t$

$$\bar{y}_{i\cdot} - \bar{y}_{\cdot\cdot} \in C(H)$$

and for some vector d

$$\bar{y}_{i\cdot} - \bar{y}_{\cdot\cdot} = Hd.$$

By Lemma 2.3.5

$$0 = A_0' Hd = A_0'(\bar{y}_{i\cdot} - \bar{y}_{\cdot\cdot}).$$

$\square$

The equivalence of the allocation rules is established in the following proposition.

Proposition 2.3.7.   If $t - 1 < q$, then for any y

$$(y - \bar{y}_{i\cdot})' E^{-1} (y - \bar{y}_{i\cdot}) = (y - \bar{y}_{i\cdot})' [A_* A_*'](y - \bar{y}_{i\cdot}) + (y - \bar{y}_{\cdot\cdot})' [A_0 A_0'](y - \bar{y}_{\cdot\cdot}) \quad (4)$$

and the Mahalanobis distance allocation rule is identical to the linear discrimination coordinate allocation rule with $s = r(E^{-1}H)$.

PROOF.   We begin by arguing that if equation (4) holds, the two allocation rules are identical. As mentioned earlier, the Mahalanobis rule minimizes $(y - \bar{y}_{i\cdot})' E^{-1} (y - \bar{y}_{i\cdot})$ with respect to i. The term $(y - \bar{y}_{\cdot\cdot})' [A_0 A_0'](y - \bar{y}_{\cdot\cdot})$, on the right of (4), does not depend on i so the Mahalanobis rule minimizes $(y - \bar{y}_{i\cdot})' [A_* A_*'](y - \bar{y}_{i\cdot})$ with respect to i. However, this is simply the linear discrimination coordinate rule for $s = r(E^{-1}H)$.

We now prove equation (4). From Corollary 2.3.3 and the partition of A

$$E^{-1} = AA' = A_* A'_* + A_0 A'_0$$

so

$$(y - \bar{y}_{i\cdot})' E^{-1} (y - \bar{y}_{i\cdot}) = (y - \bar{y}_{i\cdot})' A_* A'_* (y - \bar{y}_{i\cdot}) + (y - \bar{y}_{i\cdot})' A_0 A'_0 (y - \bar{y}_{i\cdot}).$$

It suffices to show that

$$(y - \bar{y}_{i\cdot})' A_0 A'_0 (y - \bar{y}_{i\cdot}) = (y - \bar{y}_{\cdot\cdot})' A_0 A'_0 (y - \bar{y}_{\cdot\cdot})$$

or

$$A'_0 (y - \bar{y}_{i\cdot}) = A'_0 (y - \bar{y}_{\cdot\cdot}).$$

Clearly,

$$\begin{aligned} A'_0 (y - \bar{y}_{i\cdot}) &= A'_0 (y - \bar{y}_{\cdot\cdot} - \bar{y}_{i\cdot} + \bar{y}_{\cdot\cdot}) \\ &= A'_0 (y - \bar{y}_{\cdot\cdot}) - A'_0 (\bar{y}_{i\cdot} - \bar{y}_{\cdot\cdot}). \end{aligned}$$

By Lemma 2.3.6, $A'_0 (\bar{y}_{i\cdot} - \bar{y}_{\cdot\cdot}) = 0$ and the proof is complete. $\square$

Our motivation for the choice of linear discrimination coordinates has been based entirely on maximizing analysis of variance F statistics. An alternative motivation, based on population rather than sample values, leads to slightly different results. Consider a linear combination of the dependent variable vector y, say $a'y$. It follows that $\mathrm{Var}(a'y) = a'\Sigma a$ and, depending on its population, $\mathrm{E}(a'y) = a'\mu_i$. Define

$$\Omega = \sum_{i=1}^{t} (\mu_i - \bar{\mu}_{\cdot})(\mu_i - \bar{\mu}_{\cdot})'.$$

The value

$$\frac{a'\Omega a}{a'\Sigma a}$$

can be viewed as a measure of the variability between the population means $a'\mu_i$ relative to the variance of $a'y$. Choosing a to maximize this measure may be a reasonable way to choose linear discrimination coordinates. Both Ω and Σ are unknown parameters and must be estimated. The covariance matrix can be estimated with S and Ω can be estimated with

$$H_* = \sum_{i=1}^{t} (\bar{y}_{i\cdot} - \bar{y}_{\cdot\cdot})(\bar{y}_{i\cdot} - \bar{y}_{\cdot\cdot})'.$$

The sample version of $a'\Omega a / a'\Sigma a$ is

$$\frac{a' H_* a}{a' S a}.$$

The error statistic E is a constant multiple of S so it is equivalent to work with

$$\frac{a'H_*a}{a'Ea}.$$

The subsequent development of linear discrimination coordinates follows as in our discussion based on $a'Ha/a'Ea$.

II.4   Additional Exercises

EXERCISE 2.4.1.   Reanalyze the Cushing's syndrome data of Example 2.2.1 using a value of .40 for the pregnanetriol reading associated with case 4.

EXERCISE 2.4.2.   Consider the data of Example 1.5.1. Suppose a person has heart rate measurements of $y = (84, 82, 80, 69)'$.
a) Using normal theory linear discrimination, what is the estimated maximum likelihood allocation for this person?
b) Using normal theory quadratic discrimination, what is the estimated maximum likelihood allocation for this person?
c) If the two drugs have equal prior probabilities but the placebo is twice as probable as the drugs, what is the estimated maximum posterior probability allocation?
d) Suppose the costs of correct classification are zero, the costs of misclassification depend only on the true population, and the cost of misclassifying an individual who actually has taken drug B is twice that of the other populations. Using the prior probabilities of (c), what is the estimated Bayesian allocation?
e) What is the optimal allocation using only the first two linear discrimination coordinates?

EXERCISE 2.4.3.   In the motion picture *Diary of a Mad Turtle* the main character, played by Richard Benjamin Kingsley, claims to be able to tell a female turtle by a quick glance at her carapace. Based on the data of Exercise 1.8.1, do you believe that it is possible to accurately identify a turtle's sex based on its shell? Explain. Include graphical evaluation of the linear discrimination coordinates.

EXERCISE 2.4.4.   Using the data of Exercise 1.8.3, do a stepwise discriminant analysis to distinguish between the thyroxin, thiouracil, and control rat populations based on their weights at various times. To which group is a rat with the following series of weights most likely to belong: $(56, 75, 104, 114, 138)$?

EXERCISE 2.4.5.   Lachenbruch (1975) presents information on four groups of junior technical college students from greater London. The information consists of summary statistics for the performance of the groups on arithmetic, English, and form relations tests that were given in the last year of secondary school. The four groups are Engineering, Building, Art, and Commerce students. The sample means are:

	Engineering	Building	Art	Commerce
Arithmetic (y_1)	27.88	20.65	15.01	24.38
English (y_2)	98.36	85.43	80.31	94.94
Form Relations (y_3)	33.60	31.51	32.01	26.69
Sample Size	404	400	258	286

The pooled estimate of the covariance matrix is

$$S_p = \begin{bmatrix} 55.58 & 33.77 & 11.66 \\ 33.77 & 360.04 & 14.53 \\ 11.66 & 14.53 & 69.21 \end{bmatrix}.$$

What advice could you give to a student planning to go to a junior technical college who just achieved scores of $(22, 90, 31)'$?

EXERCISE 2.4.6.   Suppose the concern in Exercise 2.4.5 is minimizing the cost to society of allocating students to the various programs of study. The great bureaucrat in the sky, who works on the top floor of the tallest building in Whitehall, has determined that the costs of classification are as follows:

Cost		Optimal Study Program			
		Engineering	Building	Art	Commerce
Allocated	Engineering	1	2	8	2
Study	Building	4	2	7	3
Program	Art	8	7	4	4
	Commerce	4	3	5	2

Evaluate the program of study that the bureaucrat thinks is appropriate for the student from Exercise 2.4.5.

EXERCISE 2.4.7.   Show that the Mahalanobis distance is invariant under affine transformations $z = Ay + b$ of the random vector y when A is nonsingular.

EXERCISE 2.4.8.   Let y be an observation from one of two populations that have means of μ_1 and μ_2 and common covariance matrix Σ. Define $\lambda' = (\mu_1 - \mu_2)'\Sigma^{-1}$.
a) Show that under linear discrimination, y is allocated to population 1 if and only if

$$\lambda'y - \lambda'\frac{1}{2}(\mu_1 + \mu_2) > 0.$$

b) Show that if y is from population 1,

$$E(\lambda'y) - \lambda'\frac{1}{2}(\mu_1 + \mu_2) > 0$$

and if y is from population 2,

$$E(\lambda'y) - \lambda'\frac{1}{2}(\mu_1 + \mu_2) < 0.$$

EXERCISE 2.4.9.   Consider a two-group allocation problem in which the prior probabilities are $p(1) = p(2) = .5$ and the sampling distributions are exponential, i.e.,

$$f(y|i) = \theta_i e^{-\theta_i y}, \quad y \geq 0.$$

Find the optimal allocation rule. Assume a cost structure where $c(i|j)$ is zero for $i = j$ and one otherwise. The *total probability of misclassification* for an allocation rule is precisely the Bayes risk of the allocation rule under this cost structure. Let $\delta(y)$ be an allocation rule. The frequentist risk for the true population j is $R(j, \delta) = \int c(\delta(y)|j)f(y|j)dy$ and the Bayes risk is $r(p, \delta) = \sum_{j=1}^{t} R(j, \delta)p(j)$. See Berger (1985, Section 1.3) for more on risk functions. Find the total probability of misclassification for the optimal rule.

EXERCISE 2.4.10.   Suppose that the distributions for two populations are bivariate normal with the same covariance matrix. For $p(1) = p(2) = .5$ find the value of the correlation coefficient that minimizes the total probability of misclassification. The total probability of misclassification is defined in Exercise 2.4.9.

Chapter III

Principal Components and Factor Analysis

Suppose that observations are available on q variables. In practice, q is often quite large. If, for example, $q = 25$, it can be very difficult to grasp the relationships between the many variables. It might be convenient if the 25 variables could be reduced to a more manageable number. Clearly, it is easier to work with 4 or 5 variables than with 25. Of course, one cannot reasonably expect to get a substantial reduction in dimensionality without some loss of information. We want to minimize that loss. Assuming that a reduction in dimensionality is desirable, how can it be performed efficiently? One reasonable method is to choose a small number of linear combinations of the dependent variables based on their ability to reproduce the entire set of variables. In effect, we want to create a few new variables that are best able to predict the original variables. *Principal component analysis* finds linear combinations of the original variables that are best linear predictors of the full set of variables. This predictive approach to *dimensionality reduction* seems intuitively reasonable. We emphasize this interpretation of principal component analysis rather than the traditional motivation of finding linear combinations that account for most of the variability in the data. The predictive approach is mentioned in Rao (1973, p. 591). Seber (1984) takes an approach that is essentially predictive. Seber's discussion is derived from Okamoto and Kanazawa (1968). Schervish (1986) gives an explicit derivation in terms of prediction. Other approaches, that are not restricted to linear combinations of the dependent variables, are discussed by Gnanadesikan (1977, Section 2.4) and Li and Chen (1985). Jolliffe (1986) gives a thorough discussion with many examples.

Principal components are similar in spirit to the linear discrimination coordinates discussed in Section II.3. Principal components actually form a new coordinate system for $\mathbf{R}^q$. These coordinates are defined sequentially so that they are mutually orthogonal in an appropriate inner product and have successively less ability to predict the original dependent variables. In practice, only the first few coordinates are used to represent the entire vector of dependent variables.

Section 1 presents a review of best linear prediction and gives results used not only in deriving principal components but also in later chapters. Section 2 presents several alternative derivations for theoretical principal components including both predictive and nonpredictive motivations. Section 3 examines the use of sample principal components. The final section of this chapter examines *Factor Analysis*. Although many people consider principal component analysis a special case of factor analysis, in fact their theoretical bases are quite different. Nevertheless, we discuss both subjects in the same chapter.

III.1   Properties of Best Linear Predictors

Best linear prediction is a key concept in the development of many of the topics in statistics. It is introduced in Christensen (1987, Chapter VI) and used again in his Chapter XII. In this volume, it is used in the discussions of principal components, time series analysis, the Kalman filter, and Kriging.

We need to establish general properties of best linear predictors that are analogous to results for conditional expectations. Let $y = (y_1, \ldots, y_q)'$ and $x = (x_1, \ldots, x_{p-1})'$. Denote

$$\begin{aligned} \mathrm{E}(y) &= \mu_y & \mathrm{E}(x) &= \mu_x \\ \mathrm{Cov}(y) &= V_{yy} & \mathrm{Cov}(x) &= V_{xx} \end{aligned}$$

and

$$\mathrm{Cov}(y, x) = V_{yx} = V_{xy}'.$$

Recall from Christensen (1987, Section VI.3 and Exercise 6.3) that the best linear predictor (BLP) of y is defined to be the linear function $f(x)$ that minimizes

$$\mathrm{E}\left[(y - f(x))'(y - f(x))\right].$$

The best linear predictor, also called the linear expectation, is

$$\hat{E}(y|x) \equiv \mu_y + \beta'(x - \mu_x)$$

where β is a solution

$$V_{xx}\beta = V_{xy}.$$

In general, the linear expectation $\hat{E}(y|x)$ is neither the conditional expectation of y given x nor an estimate of the conditional expectation; it is a different concept. The conditional expectation $\mathrm{E}(y|x)$ is the best predictor of y based on x. $\hat{E}(y|x)$ is the best *linear* predictor. Conditional expectations require knowledge of the entire multivariate distribution. Linear expectations depend only on the mean vector and covariance matrix. For some families of multivariate distributions, of which the multivariate normal is the best known, the linear expectation and the conditional expectation happen to be the same. This is similar to the fact that for multivariate normals best linear unbiased estimates are also best within the broader class of (possibly nonlinear) unbiased estimates. Linear expectations have a number of properties that are similar to those of conditional expectations. Many of these properties will be explored in the current section. The notation $\hat{E}(y|x)$ for the linear expectation has been used for at least thirty-five years, cf. Doob (1953).

Recall from Christensen (1987, Section VI.3) that $\hat{E}(y|x)$ is a function of x, that $\mathrm{E}(y) = \mathrm{E}[\hat{E}(y|x)]$, and that the *prediction error covariance matrix* is

$$
\begin{aligned}
\mathrm{Cov}&(y - \hat{E}(y|x)) \\
&= E[(y - \mu_y) - V_{yx}V_{xx}^{-}(x - \mu_x)][(y - \mu_y) - V_{yx}V_{xx}^{-}(x - \mu_x)]' \\
&= V_{yy} - V_{yx}V_{xx}^{-}V_{xy}.
\end{aligned}
$$

To simplify the discussion, it is assumed below that appropriate inverses exist. In particular,

$$
\beta = V_{xx}^{-1}V_{xy}
$$

so that the linear expectation is unique. First, we show that linear expectation is a linear operator.

Proposition 3.1.1.   Let A be a $r \times q$ matrix and let a be an $r \times 1$ vector. The best linear predictor of $Ay + a$ based on x is

$$
\hat{E}(Ay + a|x) = A\hat{E}(y|x) + a\,.
$$

EXERCISE 3.1.   Prove Proposition 3.1.1.

If we predict a random variable y from a set of random variables that includes y, then the prediction is just y. It is convenient to state this result in terms of the x vector.

Proposition 3.1.2.   For $x = (x_1, \ldots, x_{p-1})'$,

$$
\hat{E}(x_i|x) = x_i\,.
$$

EXERCISE 3.2.   Prove Proposition 3.1.2.
Hint: By definition, $\hat{E}(x_i|x) = \mu_i + \beta'(x - \mu_x)$ where $V_{xx}\beta = V_{xx_i}$. In this case, V_{xx_i} is the i'th column of V_{xx}.

Propositions 3.1.1 and 3.1.2 lead to the following corollary.

Corollary 3.1.3.   If $\beta \in \mathbf{R}^{p-1}$, then

$$
\hat{E}(x'\beta|x) = x'\beta\,.
$$

The next result is that a nonsingular linear transformation of the predictors does not change the linear expectation.

Proposition 3.1.4.   Let A be a $(p-1) \times (p-1)$ nonsingular matrix and let a be a vector in $\mathbf{R}^{p-1}$, then

$$\hat{E}(y|Ax + a) = \hat{E}(y|x).$$

PROOF.   Note that $\mathrm{Cov}(y, Ax + a) = \mathrm{Cov}(y, Ax)$, $\mathrm{Cov}(Ax + a) = \mathrm{Cov}(Ax)$, and $(Ax + a) - \mathrm{E}(Ax + a) = A(x - \mu_x)$. Then

$$
\begin{aligned}
\hat{E}(y|Ax + a) &= \mu_y + \mathrm{Cov}(y, Ax)[\mathrm{Cov}(Ax)]^{-1}A(x - \mu_x) \\
&= \mu_y + V_{yx}A'[AV_{xx}A']^{-1}A(x - \mu_x) \\
&= \mu_y + V_{yx}A'A'^{-1}V_{xx}^{-1}A^{-1}A(x - \mu_x) \\
&= \mu_y + V_{yx}V_{xx}^{-1}(x - \mu_x) \\
&= \hat{E}(y|x).
\end{aligned}
$$

$\square$

The next proposition involves predictors that are uncorrelated with the random vector to be predicted.

Proposition 3.1.5.   If $\mathrm{Cov}(y, x) = 0$, then

$$\hat{E}(y|x) = \mu_y.$$

PROOF. $\hat{E}(y|x) = \mu_y + \beta'(x - \mu_x)$ where $V_{xx}\beta = V_{xy}$. If $V_{xy} = 0$, the vector $\beta = 0$ is the solution, thus giving the result. $\square$

Again, all of these results are analogous to results for condition expectations, cf. Christensen (1987, Appendix D). In Proposition 3.1.5, the condition $\mathrm{Cov}(y, x) = 0$ is analogous to the idea, for conditional expectations, that y and x are independent. In Proposition 3.1.4, the idea that A is nonsingular in the transformation $Ax + a$ corresponds to taking an invertible transformation of the conditioning variable. Because of these analogies, any proofs that depend only on the five results given above have corresponding proofs for conditional expectations. This observation generalizes results from best linear predictors to best predictors. As mentioned in Christensen (1987, Section VI.3) and earlier in this section, the best predictor of y

based on x is $E(y|x)$. The reason for not using best predictors is that they require knowledge of the joint distribution of the random variables. Best linear predictors require knowledge only of the first and second moments. For Gaussian processes best linear predictors are also best predictors.

EXERCISE 3.3.   Assume that

$$\begin{bmatrix} y \\ x \end{bmatrix} \sim N\left(\begin{bmatrix} \mu_y \\ \mu_x \end{bmatrix}, \begin{bmatrix} V_{yy} & V_{yx} \\ V_{xy} & V_{xx} \end{bmatrix} \right)$$

and that the covariance matrix is nonsingular so that a density exists for the joint distribution. Show that

$$y|x \sim N\big(\hat{E}(y|x), \mathrm{Cov}\big[y - \hat{E}(y|x)\big]\big) \,.$$

In applications to the time domain models of Chapter V, it is occasionally convenient to allow x to be an infinite vector $x = (x_1, x_2, \ldots)'$. For infinite vectors, the condition $V_{xx}\beta = V_{xy}$ can be written rigorously as $\sum_{j=1}^{\infty} \sigma_{ij}\beta_j = \sigma_{iy}$ for $i = 1, 2, \ldots$ where $V_{xx} = [\sigma_{ij}]$ and $V_{xy} = [\sigma_{iy}]$. It is not difficult to see that all of the propositions given above continue to hold.

Finally, to derive joint prediction properties of principal components and later the Kalman filter, some additional results are required. The first involves linear expectations based on a predictor consisting of two vectors with zero correlation. (I am not aware of any corresponding result for conditional expectations given two independent vectors.)

To handle three vectors simultaneously, we need some additional notation. Consider a partition of y, say $y' = (y_1', y_2')$. Denote

$$\mathrm{Cov}(y) = \mathrm{Cov}\left(\begin{array}{c} y_1 \\ y_2 \end{array} \right) = \begin{bmatrix} V_{11} & V_{12} \\ V_{21} & V_{22} \end{bmatrix}$$

and

$$\mathrm{Cov}(y_i, x) = V_{ix} \qquad i = 1, 2 \,.$$

Also, let

$$E(y_i) = \mu_i \qquad i = 1, 2 \,.$$

The main result is given below.

Proposition 3.1.6.   If $\mathrm{Cov}(y_1, x) = 0$ then

$$\hat{E}(y_2|y_1, x) = \hat{E}(y_2|x) + \hat{E}(y_2|y_1) - \mu_2 \,.$$

PROOF.   By definition

$$\hat{E}(y_2|y_1, x) = \mu_2 + (V_{21}, V_{2x}) \begin{bmatrix} V_{11} & V_{1x} \\ V_{x1} & V_{xx} \end{bmatrix}^{-1} \left(\begin{array}{c} y_1 - \mu_1 \\ x - \mu_x \end{array} \right) \,. \qquad (1)$$

By assumption $V_{1x} = 0$ and it follows that

$$\begin{bmatrix} V_{11} & V_{1x} \\ V_{x1} & V_{xx} \end{bmatrix}^{-1} = \begin{bmatrix} V_{11} & 0 \\ 0 & V_{xx} \end{bmatrix}^{-1} = \begin{bmatrix} V_{11}^{-1} & 0 \\ 0 & V_{xx}^{-1} \end{bmatrix} . \tag{2}$$

Substituting (2) into (1) gives

$$\begin{aligned} \hat{E}(y_2|y_1, x) &= \mu_2 + V_{21}V_{11}^{-1}(y_1 - \mu_1) + V_{2x}V_{xx}^{-1}(x - \mu_x) \\ &= \hat{E}(y_2|y_1) + \hat{E}(y_2|x) - \mu_2 . \end{aligned}$$

$\square$

To simplify notation let

$$e(y_1|x) \equiv y_1 - \hat{E}(y_1|x) .$$

This is the prediction error from predicting y_1 using x. Our primary application of Proposition 3.1.6 will be through the following lemma.

Lemma 3.1.7.

$$\mathrm{Cov}\big(e(y_1|x), x\big) = 0 .$$

PROOF.

$$\begin{aligned} \mathrm{Cov}\,(e(y_1|x), x) &= \mathrm{Cov}(y_1 - \hat{E}(y_1|x), x) \\ &= \mathrm{Cov}((y_1 - \mu_1) - V_{1x}V_{xx}^{-1}(x - \mu_x), x) \\ &= \mathrm{Cov}(y_1 - \mu_1, x) - V_{1x}V_{xx}^{-1}\mathrm{Cov}(x - \mu_x, x) \\ &= V_{1x} - V_{1x}V_{xx}^{-1}V_{xx} \\ &= 0 . \end{aligned}$$

$\square$

Lemma 3.1.7 leads to the key result.

Proposition 3.1.8.

$$\hat{E}(y_2|y_1, x) = \hat{E}(y_2|x) + \mathrm{Cov}\,(y_2, e(y_1|x))\,[\mathrm{Cov}\,(e(y_1|x))]^{-1}\,e(y_1|x) .$$

PROOF.   First note that $\hat{E}(y_1|x)$ is a linear function of x. Thus, we can write

$$\begin{bmatrix} y_1 - \hat{E}(y_1|x) \\ x \end{bmatrix} = \begin{bmatrix} I & B \\ 0 & I \end{bmatrix} \begin{bmatrix} y_1 \\ x \end{bmatrix}$$

for some matrix B. The upper triangular matrix

$$\begin{bmatrix} I & B \\ 0 & I \end{bmatrix}$$

is nonsingular so by Proposition 3.1.4

$$\hat{E}(y_2|y_1, x) = \hat{E}\big(y_2|y_1 - \hat{E}(y_1|x), x\big) = \hat{E}\big(y_2|e(y_1|x), x\big).$$

The result follows from Proposition 3.1.6, Lemma 3.1.7 and observing that by definition

$$\hat{E}\left(y_2|e(y_1|x)\right) = \mu_2 + \mathrm{Cov}\left(y_2, e(y_1|x)\right)\left[\mathrm{Cove}(y_1|x)\right]^{-1} e(y_1|x)$$

so subtracting μ_2 from both sides gives

$$\hat{E}\left(y_2|e(y_1|x)\right) - \mu_2 = \mathrm{Cov}\left(y_2, e(y_1|x)\right)\left[\mathrm{Cove}(y_1|x)\right]^{-1} e(y_1|x).$$

$$\square$$

Note that if x is vacuous, Proposition 3.1.8 is just the standard definition of $\hat{E}(y_2|y_1)$.

The prediction error covariance matrix can be written in a form analogous to results in Christensen (1987, Chapter IX).

Proposition 3.1.9.

$$\mathrm{Cov}\big(y_2 - \hat{E}(y_2|y_1, x)\big) = \mathrm{Cov}\big(y_2 - \hat{E}(y_2|x)\big)$$
$$- \mathrm{Cov}\big(y_2, e(y_1|x)\big)\left[\mathrm{Cov}\big(e(y_1|x)\big)\right]^{-1}\mathrm{Cov}\big(e(y_1|x), y_2\big).$$

PROOF.   To simplify notation let $e \equiv e(y_1|x)$. By Lemma 3.1.7, $\mathrm{Cov}(x, e) = 0$. Moreover, because $\hat{E}(y_2|x)$ is a linear function of x, $\mathrm{Cov}(\hat{E}(y_2|x), e) = 0$ and $\mathrm{Cov}(y_2 - \hat{E}(y_2|x), e) = \mathrm{Cov}(y_2, e) - \mathrm{Cov}(\hat{E}(y_2|x), e) = \mathrm{Cov}(y_2, e)$. Using Proposition 3.1.8 and this fact about covariances gives

$$\mathrm{Cov}(y_2 - \hat{E}(y_2|y_1, x))$$
$$= \mathrm{Cov}\big([y_2 - \hat{E}(y_2|x)] - \mathrm{Cov}(y_2, e)[\mathrm{Cov}(e)]^{-1}e\big)$$
$$= \mathrm{Cov}\big(y_2 - \hat{E}(y_2|x)\big) + \mathrm{Cov}(y_2, e)[\mathrm{Cov}(e)]^{-1}\mathrm{Cov}(e, y_2)$$
$$\quad - \mathrm{Cov}\big(y_2 - \hat{E}(y_2|x), e\big)[\mathrm{Cov}(e)]^{-1}\mathrm{Cov}(e, y_2)$$
$$\quad - \mathrm{Cov}(y_2, e)[\mathrm{Cov}(e)]^{-1}\mathrm{Cov}\big(e, y_2 - \hat{E}(y_2|x)\big)$$
$$= \mathrm{Cov}\big(y_2 - \hat{E}(y_2|x)\big) + \mathrm{Cov}(y_2, e)[\mathrm{Cov}(e)]^{-1}\mathrm{Cov}(e, y_2)$$
$$\quad - 2\mathrm{Cov}(y_2, e)[\mathrm{Cov}(e)]^{-1}\mathrm{Cov}(e, y_2)$$
$$= \mathrm{Cov}\big(y_2 - \hat{E}(y_2|x)\big) - \mathrm{Cov}(y_2, e)[\mathrm{Cov}(e)]^{-1}\mathrm{Cov}(e, y_2)$$

which proves the result. $\square$

Proposition 3.1.10.

$$\mathrm{Cov}[y, y - \hat{E}(y|x)] = \mathrm{Cov}[y - \hat{E}(y|x)].$$

PROOF.

$$
\begin{aligned}
\mathrm{Cov}[y - \hat{E}(y|x)] &= V_{yy} - V_{yx}V_{xx}^{-1}V_{xy} \\
\mathrm{Cov}[y, y - \hat{E}(y|x)] &= \mathrm{Cov}[y] - \mathrm{Cov}[y, \hat{E}(y|x)] \\
&= V_{yy} - \mathrm{Cov}[y, \mu_y + V_{yx}V_{xx}^{-1}(x - \mu_x)] \\
&= V_{yy} - V_{yx}V_{xx}^{-1}V_{xy}.
\end{aligned}
$$

$\square$

III.2   The Theory of Principal Components

In this section we give several derivations of principal components. First, principal components are derived as a sequence of orthogonal linear combinations of the dependent variable vector y. Each linear combination has maximum capability to predict the full set of dependent variables subject to the condition that each combination is orthogonal to the previous linear combinations. In this sequence, orthogonality is defined using the inner product determined by Σ, the covariance matrix of y. Second, it is shown that the first r principal components have maximum capability to predict y among all sets of r linear combinations of y. Thus, if r linear combinations of y are to be analyzed instead of the full vector y, the first r principal components of y are linear combinations from which y can be most nearly reconstructed. The section closes with a discussion of alternate derivations of principal components and of principal components based on the correlation matrix.

SEQUENTIAL PREDICTION

Let $y = (y_1, \ldots, y_q)'$ be a vector in $\mathbf{R}^q$. We wish to define new coordinates $a_1'y, a_2'y, \ldots, a_q'y$. These coordinates are to have certain statistical properties. If we think of y as a random vector with $\mathrm{E}(y) = \mu$ and $\mathrm{Cov}(y) = \Sigma$ where Σ is positive definite, then the random variables $a_1'y, \ldots, a_q'y$ give the random vector y represented in the new coordinate system. The coordinate vectors $a_1, \ldots, a_q$ are to be chosen so that they are orthogonal in the inner product defined by Σ, i.e.,

$$a_i'\Sigma a_j = 0 \qquad i \neq j.$$

This condition implies that the corresponding random variables are uncorrelated, i.e.,

$$\mathrm{Cov}(a_i'y, a_j'y) = a_i'\Sigma a_j = 0 \qquad i \neq j.$$

The new coordinates are to provide sequentially optimal predictions of y given the orthogonality conditions. Thus, a_1 is chosen to minimize

$$\mathrm{E}\{[y - \hat{E}(y|a'y)]'[y - \hat{E}(y|a'y)]\} \tag{1}$$

and for $i > 1$, a_i is chosen to minimize (1) subject to the condition that $a_i'\Sigma a_j = 0$, $j = 1, \ldots, i-1$.

If we restate the problem, we can use Proposition 2.3.4 to solve it. Note that

$$\begin{aligned}
\mathrm{E}\{[y - \hat{E}(y|a'y)]'[y - \hat{E}(y|a'y)]\} &= \mathrm{tr}(\mathrm{E}\{[y - \hat{E}(y|a'y)][y - \hat{E}(y|a'y)]'\}) \\
&= \mathrm{tr}\{\mathrm{Cov}[y - \hat{E}(y|a'y)]\}.
\end{aligned}$$

Thus, minimizing (1) is identical to minimizing the trace of the prediction error covariance matrix. Write

$$V_a = \mathrm{Cov}[y - \hat{E}(y|a'y)].$$

Given $a_1, \ldots, a_{r-1}$, we wish to find a_r such that

$$\mathrm{tr}(V_{a_r}) = \inf_a \{\mathrm{tr}(V_a) | a'\Sigma a_i = 0, \ i = 1, \ldots, r-1\}.$$

It is surprising that eigenvectors $a_1, \ldots, a_q$ of Σ corresponding to the ordered eigenvalues $\phi_1 \geq \cdots \geq \phi_q > 0$ solve this optimal prediction problem. To see this first note that

$$V_a = \mathrm{Cov}[y - \hat{E}(y|a'y)] = \Sigma - \Sigma a(a'\Sigma a)^{-1}a'\Sigma$$

so

$$\begin{aligned}
\mathrm{tr}(V_a) &= \mathrm{tr}(\Sigma) - \mathrm{tr}[\Sigma a(a'\Sigma a)^{-1}a'\Sigma] \\
&= \mathrm{tr}(\Sigma) - \mathrm{tr}[(a'\Sigma a)^{-1}a'\Sigma\Sigma a] \\
&= \mathrm{tr}(\Sigma) - \frac{a'\Sigma^2 a}{a'\Sigma a}.
\end{aligned}$$

Thus minimizing $\mathrm{tr}(V_a)$ is equivalent to maximizing $a'\Sigma^2 a/a'\Sigma a$. It suffices to find $a_1, \ldots, a_q$ such that

$$\frac{a_1'\Sigma^2 a_1}{a_1'\Sigma a_1} = \sup_a \frac{a'\Sigma^2 a}{a'\Sigma a},$$

and for $i = 2, \ldots, q$

$$a_i'\Sigma a_j = 0, \ j = 1, \ldots, i-1,$$

with

$$\frac{a_i'\Sigma^2 a_i}{a_i'\Sigma a_i} = \sup_a \left\{ \frac{a'\Sigma^2 a}{a'\Sigma a} \,\middle|\, a'\Sigma a_j = 0 \quad j = 1, \ldots, i-1 \right\}.$$

By taking $H = \Sigma^2$ and $E = \Sigma$, this is just a special case of the problem solved in Proposition 2.3.4. An optimal vector a_i is an eigenvector of $E^{-1}H = \Sigma^{-1}\Sigma^2 = \Sigma$ that corresponds to the i'th largest eigenvalue ϕ_i of Σ.

To evaluate how well each coordinate does in predicting y, observe that

$$\begin{aligned}
\operatorname{tr}(V_{a_i}) &= \operatorname{tr}(\Sigma) - a_i'\Sigma^2 a_i / a_i'\Sigma a_i \\
&= \sum_{j=1}^{q} \phi_j - \phi_i^2/\phi_i \\
&= \sum_{j=1}^{q} \phi_j - \phi_i .
\end{aligned}$$

Thus, because the ϕ_i's are decreasing, each coordinate does no better at predicting y than the previous components.

JOINT PREDICTION

It is also of interest to evaluate the overall predictive ability of a set of principal components, say, $a_1'y, \ldots, a_r'y$ that provide optimal sequential prediction. Using Propositions 3.1.5 and 3.1.9, a simple inductive proof yields

$$\operatorname{tr}\{\operatorname{Cov}[y - \hat{E}(y|a_1'y, \ldots, a_r'y)]\}\} = \sum_{j=r+1}^{q} \phi_j . \tag{2}$$

Together, the first r principal components do a good job of predicting y if the ratio

$$\frac{\operatorname{tr}\{\operatorname{Cov}[y - \hat{E}(y|a_1'y, \ldots, a_r'y)]\}}{\operatorname{tr}\{\operatorname{Cov}[y]\}} = \frac{\displaystyle\sum_{j=r+1}^{q} \phi_j}{\displaystyle\sum_{j=1}^{q} \phi_j}$$

is very small. This occurs if $\phi_{r+1}, \ldots, \phi_q$ are all very small relative to $\sum_{j=1}^{q} \phi_j$. While the user must decide how much information can be sacrificed to the goal of dimensionality reduction, Johnson and Wichern (1988, p. 343) suggest that for many purposes the principal components form an effective substitute for the original variables when the ratio is .2 or less for large q and $r = 1$, 2 or 3.

EXERCISE 3.4.   Prove equation (2).

As a matter of fact, $a_1'y, \ldots, a_r'y$ do as good or better at predicting y than any other r linear combinations of y. Let B be a $q \times r$ matrix of rank r. Then $B'y$ is a vector consisting of r linear combinations of y. The best linear predictor of y based on $B'y$ has

$$
\begin{aligned}
\mathrm{Cov}[y - \hat{E}(y|B'y)] &= \Sigma - \Sigma B(B'\Sigma B)^{-1}B'\Sigma \\
&= \Sigma^{1/2}(I - \Sigma^{1/2}B(B'\Sigma B)^{-1}B'\Sigma^{1/2})\Sigma^{1/2} \\
&= \Sigma^{1/2}(I - M_{\Sigma^{1/2}B})\Sigma^{1/2}
\end{aligned}
$$

where $M_{\Sigma^{1/2}B}$ is the perpendicular projection operator onto the column space of $\Sigma^{1/2}B$, under the standard Euclidean inner product. It will be shown below (by the reader) that the prediction error satisfies

$$
\begin{aligned}
E\big\{ [y - \hat{E}(y|B'y)]'[y - \hat{E}(y|B'y)]\big\} &= \mathrm{tr}\big\{\mathrm{Cov}[y - \hat{E}(y|B'y)]\big\} \qquad (3) \\
&= \mathrm{tr}\big\{\Sigma^{1/2}(I - M_{\Sigma^{1/2}B})\Sigma^{1/2}\big\} \\
&= \mathrm{tr}\big\{(I - M_{\Sigma^{1/2}B})\Sigma\big\} \\
&\geq \sum_{j=r+1}^{q} \phi_j \, .
\end{aligned}
$$

By equation (2), the r linear combinations given by the principal components achieve the lower bound so the first r principal components are not only sequentially optimal predictors, but also jointly optimal predictors.

The inequality (3) can be established using the following lemma.

Lemma 3.2.1.   Let $\phi_1 \geq \cdots \geq \phi_q > 0$ be the eigenvalues of Σ. Let $v_1, \ldots, v_r$ be orthonormal vectors in $\mathbf{R}^q$. Then

a) $\displaystyle \sum_{j=q-r+1}^{q} \phi_j \leq \sum_{j=1}^{r} v_j'\Sigma v_j \leq \sum_{j=1}^{r}\phi_j \, .$

b) If M is a perpendicular projection operator on $\mathbf{R}^q$ (standard inner product) and $r(M) = r$, then

$$
\sum_{j=q-r+1}^{q} \phi_j \leq \mathrm{tr}\{M\Sigma\} \leq \sum_{j=1}^{r}\phi_j \, .
$$

EXERCISE 3.5.   Prove Lemma 3.2.1.
Hints: For a) first consider the special case Σ diagonal and use some ideas from the proof of Proposition 2.3.4. Then use the eigenvector-eigenvalue decomposition $\Sigma = P\Lambda P'$ where $P'P = I_q$ and Λ is diagonal. For b) use the orthonormal basis decomposition for a perpendicular projection operator, i.e., $M = OO'$ with $O'O = I_r$.

To see that (3) is a result of Lemma 3.2.1, note that $I - M_{\Sigma^{1/2}B}$ is a perpendicular projection operator of rank $q - r$, and that part (b) of the lemma leads immediately to (3).

In applications, the correlation between the i'th principal component $a_i'y$ and the h'th dependent variable y_h is often cited. The correlations for $h = 1, \ldots, q$ are frequently used in trying to develop an interpretation for the i'th principal component. The idea is to recognize some common characteristic(s) of the y_h's that have large correlations with $a_i'y$. The principal component is then interpreted as an underlying factor that measures this characteristic. This procedure is really a factor analytic use of principal components and is open to criticisms similar to those made of factor analysis, cf. Section 4. In particular, such interpretations are not only subjective (subjectivity is unavoidable) but they are apparently unverifiable (a much more serious problem). Regardless of the appropriate use of these correlations, they have a simple mathematical form. Note that because a_i is an eigenvector of Σ

$$\text{Cov}(y, a_i'y) = \Sigma a_i = \phi_i a_i \, .$$

Thus, for the h component of y

$$\text{Cov}(y_h, a_i'y) = \phi_i a_{ih}$$

where

$$a_i' = (a_{i1}, \ldots, a_{iq}) \, .$$

In addition

$$\text{Var}(a_i'y) = a_i'\Sigma a_i = \phi_i a_i' a_i$$

and

$$\text{Var}(y_h) = \sigma_{hh}$$

so

$$\text{Corr}(y_h, a_i'y) = \phi_i a_{ih} / \sqrt{\sigma_{hh} \phi_i a_i' a_i} \, .$$

Often the vectors a_i are chosen so that $a_i' a_i = 1$. This generates a simplification in the formula for the correlation.

OTHER DERIVATIONS OF PRINCIPAL COMPONENTS

Although the methodology of principal components was first proposed by Karl Pearson (1901), principal component analysis in its modern form was originated by Hotelling (1933). It is curious that, although Hotelling was originally interested in a prediction problem (actually a factor analysis problem), he transformed his problem into one of finding vectors $a_1, \ldots, a_r$ of fixed length such that $a_i'y$ has maximum variance subject to the condition that $\text{Cov}(a_i'y, a_j'y) = 0$, $j = 1, \ldots, i - 1$. This is the form in which principal component analysis is traditionally presented.

Apparently, it was intuitively clear to Hotelling that his linear combinations $a_i'y$ should have maximum predictive capability. In fact, the solution to Hotelling's problem is identical to the solution just given for the prediction problem. Unfortunately, that fact is not obvious to many of us who do not share Hotelling's keen insight into multivariate analysis. We now prove that this equivalence is true. To begin, we solve a slightly different problem and then show that the solution to Hotelling's problem and the alternative problem are the same.

Suppose we want to find vectors $a_1, \ldots, a_q$ that satisfy three properties: $a_i'a_i = K$ for all i and some constant K, $\mathrm{Var}(a_1'y)$ is maximized, and given that $a_i'a_j = 0$, $j = 1, \ldots, i-1$, $\mathrm{Var}(a_i'y)$ is maximized. The condition that $a_i'a_i = K$ is necessary because multiplying a_i by a constant changes the variance. If we allow different size vectors, we could never attain a maximum variance. Alternatively, we could allow different size vectors, but maximize the standardized variance. In other words, maximizing $\mathrm{Var}(a'y)$ for a fixed size vector is identical to maximizing $\mathrm{Var}(a'y)/a'a = a'\Sigma a/a'a$. Using this equivalence, our problem is to find a_1 such that

$$a_1'\Sigma a_1/a_1'a_1 = \sup_a a'\Sigma a/a'a$$

and for $i = 2, \ldots, q$ find a_i such that

$$\frac{a_i'\Sigma a_i}{a_i'a_i} = \sup_a \left\{ \frac{a'\Sigma a}{a'a} \,\middle|\, a'a_j = 0 \quad j = 1, \ldots, i-1 \right\}$$

and

$$a_i'a_j = 0 \quad j = 1, \ldots, i-1 \,.$$

Using Proposition 2.3.4 with $E = I$ and $H = \Sigma$, just as in the sequential prediction problem the vectors a_i are eigenvectors of Σ with respect to the eigenvalues $\phi_1 \geq \cdots \geq \phi_q$.

Hotelling's problem was slightly different. For $i = 2, \ldots, q$ and $j = 1, \ldots, i-1$, Hotelling wanted $\mathrm{Cov}(a_i'y, a_j'y) = a_i'\Sigma a_j = 0$ instead of $a_i'a_j = 0$. We want to show that the eigenvectors that solve our modified problem also solve Hotelling's problem. We can show this inductively. Clearly, a_1 is the same for either problem. Now suppose the first $i-1$ eigenvectors solve both problems. We need to show that if a_i solves our modified problem it will also solve Hotelling's problem. The key point is that for any vector a and $j = 1, \ldots, i-1$

$$a'\Sigma a_j = \phi_j a'a_j$$

because a_j is an eigenvector of Σ. It follows that with Σ positive definite,

$$a'\Sigma a_j = 0 \quad \text{if and only if} \quad a'a_j = 0 \,.$$

Thus

$$\left\{ \left. \frac{a'\Sigma a}{a'a} \right| a'a_j = 0 \quad j = 1, \ldots, i-1 \right\}$$
$$= \left\{ \left. \frac{a'\Sigma a}{a'a} \right| a'\Sigma a_j = 0 \quad j = 1, \ldots, i-1 \right\}$$

so any a_i that is a solution to our modified problem also satisfies

$$\frac{a_i'\Sigma a_i}{a_i'a_i} = \sup_a \left\{ \left. \frac{a'\Sigma a}{a'a} \right| a'\Sigma a_j = 0 \quad j = 1, \ldots, i-1 \right\}$$

with

$$a_i'\Sigma a_j = 0 \quad j = 1, \ldots, i-1 .$$

This establishes that eigenvectors of Σ provide solutions to Hotelling's problem just as they do for the modified problem and the prediction problems.

Finally, principal components can be related to ellipsoids. If $y \sim N(\mu, \Sigma)$, the set of points that have constant likelihood (the same value of the density) fall on ellipsoids defined by Σ^{-1}. An ellipsoid defined by Σ^{-1} is

$$\{a | a'\Sigma^{-1}a = c\}$$

where c is some constant. The vectors $a_1, \ldots, a_q$ are the directions of the axes of the ellipsoid. The principal axis is the longest vector a on the ellipsoid. The subsequent axes are the longest vectors on the ellipsoid that are orthogonal (in the standard inner product) to the previous axes. Any vector a can be made to fit on the ellipsoid by standardizing it, i.e., $(\sqrt{c}/\sqrt{a'\Sigma^{-1}a})a$ is always on the ellipsoid because $(\sqrt{c}/\sqrt{a'\Sigma^{-1}a})a'\Sigma^{-1}(\sqrt{c}/\sqrt{a'\Sigma^{-1}a})a = c$. The principal axis is the longest vector of the form $(\sqrt{c}/\sqrt{a'\Sigma^{-1}a})a$. Because c is a constant, a vector a_1 is in the direction of the principal axis if and only if

$$\frac{a_1'a_1}{a_1'\Sigma^{-1}a_1} = \sup_a \frac{a'a}{a'\Sigma^{-1}a} .$$

Similarly, vectors in the directions of the other axes must satisfy

$$\frac{a_i'a_i}{a_i'\Sigma^{-1}a_i} = \sup_a \left\{ \left. \frac{a'a}{a'\Sigma^{-1}a} \right| a'a_j = 0 \quad j - 1, \ldots, i-1 \right\}$$

and

$$a_i'a_j = 0 \quad j = 1, \ldots, i-1 .$$

Again, to show that the eigenvectors of Σ provide a solution to this problem, we solve a related problem in which we require

$$\frac{a_i'a_i}{a_i'\Sigma^{-1}a_i} = \sup_a \left\{ \left. \frac{a'a}{a'\Sigma^{-1}a} \right| a'\Sigma^{-1}a_j = 0 \quad j = 1, \ldots, i-1 \right\}$$

and

$$a_i'\Sigma^{-1}a_j = 0\,.$$

The related problem is solved using Proposition 2.3.4 with $E = \Sigma^{-1}$ and $H = I$. The eigenvectors of $E^{-1}H = \Sigma$ provide solutions. The equivalence of the solutions to the ellipsoid problem and our related problem is based on the facts that $\Sigma a = \phi a$ if and only if $\frac{1}{\phi}a = \Sigma^{-1}a$ and that for an eigenvector a_j of Σ

$$a'\Sigma^{-1}a_j = 0 \quad \text{if and only if} \quad a'a_j = 0\,.$$

PRINCIPAL COMPONENTS BASED ON THE CORRELATION MATRIX

In our discussion of principal components, we have used minimization of the trace of the prediction error covariance matrix as a criterion for optimal prediction. This is precisely the sum of the prediction error variances for each component of y, i.e.,

$$\sum_{h=1}^{q} \text{Var}\left(y_h - \hat{E}(y_h|a'y)\right)\,.$$

If an individual variable, say y_h, has a very small variance, then any linear combination $a'y$ generates a very small prediction error variance for y_h because the prediction error variance is never greater than the original variance. An optimal linear combination $a'y$ minimizes the sum of all the prediction variances so prediction based on $a'y$ may lead to incongruities. Variables that happen to be measured on scales with small absolute variability are almost ignored in favor of predicting variables that are measured with large variability. Moreover, because the variance of any individual variable can be made arbitrarily large or small simply by multiplying the variable by a constant, the principal components are subject to the whims of measurement scale. An obvious way to avoid this problem is to standardize the variance of the individual variables. Suppose $\Sigma = [\sigma_{hh'}]$ and let $D = \text{Diag}(\sigma_{11},\ldots,\sigma_{qq})$, then the variable $z = D^{-1/2}y$ has

$$\text{Cov}(z) = D^{-1/2}\Sigma D^{-1/2}$$

which is the correlation matrix for y. Because $\text{Cov}(z)$ is a matrix that has 1's down the main diagonal, the principal components for predicting $z = (z_1,\ldots,z_q)$ will give comparable weight to predicting each of the individual variables z_h. Moreover, because z is a nonsingular linear transformation of y, no information is lost in analyzing z in place of y. The transformation of y to z is simply an implicit way of defining a new criterion for optimal prediction of y. Schervish (1986) gives an explicit derivation of principal components based on the correlation matrix.

III.3   Sample Principal Components

In practice, the covariance matrix Σ is unknown so the principal components cannot be computed. However, if a sample $y_1, \ldots, y_n$ of observations on y is available, sample principal components can be computed from either the sample covariance matrix

$$S = \sum_{i=1}^{n} (y_i - \bar{y}_\cdot)(y_i - \bar{y}_\cdot)'/(n-1)$$

or the sample correlation matrix

$$R = D^{-1/2} S D^{-1/2}$$

where $S = [s_{ij}]$ and $D = \mathrm{Diag}(s_{11}, \ldots, s_{qq})$. Most often the correlation matrix seems to be the appropriate choice.

It is convenient to use S or R to define an inner product and to choose $a_1, \ldots, a_q$ as an orthonormal set of eigenvectors corresponding to the eigenvalues $\phi_1 \geq \cdots \geq \phi_q$ of S or R respectively. Write $A = [a_1, \ldots, a_q]$ and, for $r \leq q$, $A_r = [a_1, \ldots, a_r]$. A vector w, rewritten in the principal component coordinate system is $A'w$. Write the entire data set as

$$Y = \begin{bmatrix} y_1' \\ \vdots \\ y_n' \end{bmatrix}.$$

Using S, the data in the principal component coordinate system are

$$YA.$$

If principal components are based on the correlation matrix R, the rescaled data are

$$\begin{bmatrix} z_1' \\ \vdots \\ z_n' \end{bmatrix} = Z = YD^{-1/2}$$

and the data in the principal component coordinate system are

$$ZA.$$

The point of principal component analysis is to reduce dimensionality. If the smallest ordered eigenvalues of S, $\phi_{r+1}, \ldots, \phi_q$ are small, a random vector y with covariance matrix S can be predicted well by $A_r'y$. If the entire data set is transformed in this way, a principal component observation matrix is obtained

$$YA_r = [Ya_1, \ldots, Ya_r]$$

where

$$Y a_i = \begin{bmatrix} a_i' y_1 \\ \vdots \\ a_i' y_n \end{bmatrix}.$$

The elements of the vector $Y a_i$ consist of the i'th principal component applied to each of the n observation vectors. The principal component observation matrix combines these vectors for each of the first r principal components.

The analysis of the data can be performed on the principal component observations with a minimal loss of information. This includes various plots and formal statistical techniques for the analysis of a sample from one population.

THE SAMPLE PREDICTION ERROR

A question that arises immediately is just how much information is lost by using r principal components rather than the entire data set. Based on the substitutions $\bar{y}. = \mu$ and $S = \Sigma$, the total error of prediction from using the first r principal components is

$$\sum_{i=1}^{n} \left[y_i - \hat{E}(y|A_r' y_i) \right]' \left[y_i - \hat{E}(y|A_r' y_i) \right]$$

where

$$\hat{E}(y|A_r' y_i) = \bar{y}. + S A_r (A_r' S A_r)^{-1} A_r' (y_i - \bar{y}.).$$

The total error can be rewritten in terms of the eigenvalues of S, $\phi_1 \geq \cdots \geq \phi_q$. Note that

$$\begin{aligned} y_i - \hat{E}(y|A_r' y_i) &= (y_i - \bar{y}.) - S A_r (A_r' S A_r)^{-1} A_r' (y_i - \bar{y}.) \\ &= [I - S A_r (A_r' S A_r)^{-1} A_r'](y_i - \bar{y}.). \end{aligned}$$

Let $P_r = A_r (A_r' S A_r)^{-1} A_r' S$; this is an oblique projection operator. In particular, $P_r' S = S P_r$. Using this fact and basic results on traces, observe that

$$\begin{aligned} &\sum_{i=1}^{n} \left[y_i - \hat{E}(y|A_r' y.) \right]' \left[y_i - \hat{E}(y|A_r' y_i) \right] \\ &= \operatorname{tr} \left\{ \sum_{i=1}^{n} \left[y_i - \hat{E}(y|A_r' y_i) \right] \left[y_i - \hat{E}(y|A_r' y_i) \right]' \right\} \\ &= \operatorname{tr} \left\{ \sum_{i=1}^{n} [I - P_r'](y_i - \bar{y}.)(y_i - \bar{y}.)'[I - P_r']' \right\} \end{aligned}$$

$$
\begin{aligned}
&= \; \mathrm{tr}\left\{ [I - P_r'] \left[\sum_{i=1}^{n} (y_i - \bar{y}.)(y_i - \bar{y}.)' \right] [I - P_r'] \right\} \\
&= \; \mathrm{tr}\left\{ [I - P_r'][(n-1)S][I - P_r] \right\} \\
&= \; \mathrm{tr}\left\{ (n-1)\left[S - SA_r(A_r'SA_r)^{-1}A_r'S \right] \right\} \\
&= \; (n-1)\mathrm{tr}\left\{ S - SA_r(A_r'SA_r)^{-1}A_r'S \right\}.
\end{aligned}
$$

From the subsection on joint prediction

$$
\mathrm{tr}\{ S - SA_r(A_r'SA_r)^{-1}A_r'S \} = \mathrm{tr}\{ \mathrm{Cov}[y - \hat{E}(y|A_r'y)] \}
$$

where $\mathrm{Cov}(y) = S$. By equation (3.2.2)

$$
\sum_{i=1}^{n} \left[y_i - \hat{E}(y|A_r'y_i) \right]' \left[y_i - \hat{E}(y|A_r'y_i) \right] = (n-1) \sum_{j=r+1}^{q} \phi_j.
$$

To evaluate the quality of prediction, the total error of prediction using r components can be compared to the maximum possible prediction
error. The maximum possible prediction error can be viewed as using zero
principal components. The maximum is

$$
\begin{aligned}
\sum_{i=1}^{n} (y_i - \bar{y}.)'(y_i - \bar{y}.) &= \; \mathrm{tr}\left\{ \sum_{i=1}^{n} (y_i - \bar{y}.)(y_i - \bar{y}.)' \right\} \\
&= \; (n-1)\mathrm{tr}\{S\} \\
&= \; (n-1)\sum_{j=1}^{q} \phi_j.
\end{aligned}
$$

The value

$$
100 \sum_{j=r+1}^{q} \phi_j \Big/ \sum_{j=1}^{q} \phi_j
$$

is the percentage of the maximum prediction error left unexplained by
$\hat{E}(y_i|A_r'y_i)$, $i = 1, \ldots, n$. Alternatively,

$$
100 \sum_{j=1}^{r} \phi_j \Big/ \sum_{j=1}^{q} \phi_j
$$

is the percent of the maximum prediction error accounted for by $A_r'y$.

Using Principal Components

Although the analysis of data can be performed on the principal component observations with a minimal loss of information, why accept any loss
of information? Two possibilities come to mind. First, if q is very large,

an analysis of all q variables may be untenable. If one must reduce the dimensionality before any work can proceed, principal components are a reasonable place to begin. However, it should be kept in mind that principal components are based on linear combinations of y and linear predictors of y. If the important structure in the data is nonlinear, principal components can totally miss that structure. A second reason for giving up information is when you do not trust all of the information. Principal component regression was discussed in Christensen (1987, Chapter XIV) as a method for dealing with near collinearity of the design matrix in regression analysis. The main idea was that, with errors in the design matrix, directions corresponding to small eigenvalues are untrustworthy. In the present context, we might say that any statistical relationships depending on directions that do not provide substantial power of prediction are questionable.

Principal components are well designed for data reduction within a given population. If there are samples available from several populations with the same covariance matrix, then the optimal data reduction will be the same for every group and can be estimated using the pooled covariance matrix. Note that this essentially requires doing a one-way MANOVA prior to the principal component analysis. If an initial MANOVA is required, you may wonder why one would bother to reduce the data having already done a significant analysis on the unreduced set.

As a general principle, to reduce the dimensionality of data successfully you need to know ahead of time that it can be reduced. Whether it can be reduced depends on the goal of the analysis. The work involved in figuring out whether data reduction can be accomplished often negates the value of doing it. The situation is similar to that associated with Simpson's paradox in contingency table analysis, cf. Christensen (1990, Section III.1). Valid inferences cannot always be obtained from a collapsed contingency table. To know whether valid inferences can be obtained one needs to analyze the full table first. Having analyzed the full table, there may be little point in collapsing to a smaller dimensional table. Often, it would be convenient to reduce a data set using principal components and then do a MANOVA on the reduced data. Unfortunately, about the only way to find out if that approach is reasonable is to examine the results of a MANOVA on the entire data set. For example, with two populations it is possible that the difference in the mean vectors of the two groups has the same direction as the smallest principal component. Thus a reduction of dimensionality will eliminate precisely that dimension in which the difference in the populations exists. Not only is this problem potentially devastating for MANOVA but also for the related problems of discrimination and allocation. Jolliffe (1986, Section 9.1) discusses this problem in more detail.

Data reduction is also closely related to a more nebulous idea, the identification of underlying factors that determine the observed data. For example, the dependent variable vector y may consist of a battery of tests on a variety of subjects. One may seek to explain scores on the entire set of tests using a few key factors such as general intelligence, quantitative reasoning, verbal reasoning, etc. It is common practice to examine the principal components and try to interpret them as measuring some sort of underlying factor. Such interpretations are based on examination of the relative sizes of the elements of a_i. Although factor identification is commonly performed, it is, at least in some circles, quite controversial.

EXAMPLE 3.3.1.   One of the well-traveled data sets in multivariate analysis is from Jolicoeur and Mosimann (1960) on the shell (carapace) sizes of painted turtles. Aspects of this data have been examined by Morrison (1976) and Johnson and Wichern (1988). The data are given in Exercise 1.8.1. The analysis is based on $10^{3/2}$ times the natural logs of the height, width, and length of the shells. Because all of the measurements are taken on a common scale, it may be reasonable to examine the sample covariance matrix rather than the sample correlation matrix. The point of this example is to illustrate the type of analysis commonly used in identifying factors. No claim is made that these procedures are reasonable.

For 24 males, the covariance matrix is

$$S = \begin{bmatrix} 6.773 & 6.005 & 8.160 \\ 6.005 & 6.417 & 8.019 \\ 8.160 & 8.019 & 11.072 \end{bmatrix}.$$

The eigenvalues and corresponding eigenvectors for S are given below.

	c_1	c_2	c_3
$10^{3/2}\ln(\text{height})$	.523	.788	$-.324$
$10^{3/2}\ln(\text{width})$	.510	$-.594$	$-.622$
$10^{3/2}\ln(\text{length})$	.683	$-.159$	.713
ϕ	23.303	.598	.360

Recall that eigenvectors are not uniquely defined. Eigenvectors of a matrix A corresponding to ϕ along with the zero vector constitute the null space of $A - \phi I$. Often the null space has rank 1 so every eigenvector is a multiple of every other eigenvector. If we standardize the eigenvectors of S so that each has a maximum element of 1, we get the following eigenvectors.

	a_1	a_2	a_3
$10^{3/2}\ln(\text{height})$	.764	1	$-.451$
$10^{3/2}\ln(\text{width})$	.747	$-.748$	$-.876$
$10^{3/2}\ln(\text{length})$	1	$-.205$	1
ϕ	23.30	.60	.36

The first principal component accounts for $100(23.30)/(23.30 + .60 + .36) = 96\%$ of the predictive capability (variance) of the variables. The first two components account for $100(23.30 + .60)/(24.26) = 98.5\%$ of the predictive capability (variance) of the variables. The elements of a_1 are all positive and approximately equal so $a_1'y$ can be interpreted as a measure of overall size. The elements of a_2 are a large positive value for $10^{3/2}\ln(\text{height})$, a large negative value for $10^{3/2}\ln(\text{width})$, and a small value for $10^{3/2}\ln(\text{length})$. The component $a_2'y$ can be interpreted as a comparison of the $\ln(\text{height})$ and the $\ln(\text{width})$. Finally, if one considers the value $a_{31} = -.451$ small relative to $a_{32} = -.876$ and $a_{33} = 1$, one can interpret $a_3'y$ as a comparison of width versus length.

Interpretations such as these necessarily involve rounding values to make them more interpretable. The interpretations just given are actually appropriate for the three linear combinations of y, $b_1'y$, $b_2'y$, and $b_3'y$ given below.

$$
\begin{array}{cccc}
 & b_1 & b_2 & b_3 \\
10^{3/2}\ln(\text{height}) & 1 & 1 & 0 \\
10^{3/2}\ln(\text{width}) & 1 & -1 & -1 \\
10^{3/2}\ln(\text{length}) & 1 & 0 & 1
\end{array}
$$

The first interpreted component is

$$
\begin{aligned}
b_1'y &= 10^{3/2}\ln\left[(\text{height})(\text{width})(\text{length})\right] \\
&= 10^{3/2}\ln[\text{volume}]
\end{aligned}
$$

where the volume is that of a box. It is interesting to note that in this particular example, the first principal component can be interpreted without changing the coefficients of a_1.

$$
\begin{aligned}
a_1'y &= 10^{3/2}[.764\ln(\text{height}) + .747\ln(\text{width}) + \ln(\text{length})] \\
&= 10^{3/2}\ln[(\text{height})^{.764}(\text{width})^{.747}(\text{length})].
\end{aligned}
$$

The component $a_1'y$ can be thought of as measuring the log volume with adjustments made for the fact that painted turtle shells are somewhat curved and thus not a perfect box. Because the first principal component accounts for 96% of the predictive capability, to a very large extent, if you know this pseudo-volume measurement, you know the height, length, and width.

In this example, we have sought to interpret the elements of the vectors a_i. Alternatively, one could base interpretations on estimates of the correlations $\text{Corr}(y_h, a_i'y)$ that were discussed in Section 2. The estimates of $\text{Corr}(y_h, a_1'y)$ are very uniform so they also suggest that a_1 is an overall size factor.

Linear combinations $b_i'y$ that are determined by the effort to interpret principal components will be called *interpreted components*. Although it does not seem to be common practice, it is interesting to examine how well interpreted components predict the original data and compare that to how well the corresponding principal components predict the original data.

As long as the interpreted components are linearly independent, a full set of q components will predict the original data perfectly. *Any* nonsingular transformation of y will predict y perfectly because it amounts to simply changing the coordinate system. If we restrict attention to r components, we know from the theoretical results on joint prediction that the interpreted components can predict no better than the actual principal components. In general, to evaluate the predictive capability of r interpreted components, write $B_r = [b_1, \ldots, b_r]$ and compute

$$\sum_{i=1}^{n} [y_i - \hat{E}(y|B_r'y_i)]'[y_i - \hat{E}(y|B_r'y_i)] = (n-1)\mathrm{tr}\{S - SB_r(B_r'SB_r)^{-1}B_r'S\}.$$

One hundred times this value divided by $(n-1)\sum_{i=1}^{q}\phi_i = (n-1)\mathrm{tr}(S)$ gives the percent of the predictive error unaccounted for by the r interpreted components. If this is not much greater than the corresponding percentage for the first r principal components, the interpretations are to some extent validated.

EXAMPLE 3.3.2.   Using the first two interpreted components from Example 3.3.1

$$\mathrm{tr}[SB_2(B_2'SB_2)^{-1}B_2'S] = 23.88$$

and

$$\frac{100\,\mathrm{tr}[S - SB_2(B_2'SB_2)^{-1}B_2'S]}{\mathrm{tr}[S]} = \frac{100(24.26 - 23.88)}{24.26}$$

$$= \frac{100(.38)}{24.26}$$

$$= 1.6$$

Using the first two principal components

$$\frac{100\,\mathrm{tr}[S - SA_2(A_2'SA_2)^{-1}A_2'S]}{\mathrm{tr}[S]} = \frac{100\sum_{j=1}^{3}\phi_j - \sum_{j=1}^{2}\phi_j}{\sum_{j=1}^{3}\phi_j}$$

$$= \frac{100(.36)}{24.26}$$

$$= 1.5\,.$$

Thus, in this example, there is almost no loss of predictive capability by using the two interpreted components rather than the first two principal components.

There is one aspect of principal component analysis that is often overlooked. It is possible that the most interesting components are those that have the least predictive power. Such components are taking on very similar values for all cases in the sample. It may be that these components can be used to characterize the population. Jolliffe (1986) has a fairly extensive discussion of uses for the *last few* principal components.

EXAMPLE 3.3.3.   The smallest eigenvalue of S is .36 and corresponds to the linear combination

$$a_3'y = 10^{3/2}\left[-.451\ln(\text{height}) - .876\ln(\text{width}) + \ln(\text{length})\right].$$

This linear combination accounts for only 1.5% of the variability in the data. It is essentially a constant. All male painted turtles in the sample have about the same value for this combination. The linear combination is a comparison of the ln-length with the ln-width and ln-height. This might be considered as a measurement of the general shape of the carapace. One would certainly be very suspicious of any new data that were supposedly the shell dimensions of a male painted turtle but which had a substantially different value of $a_3'y$. On the other hand, this should not be thought of as a discrimination tool except in the sense of identifying whether data is or is not consistent with the male painted turtle data. We have no evidence that other species of turtles will produce substantially different values of $a_3'y$.

III.4   Factor Analysis

Principal components are often used in an attempt to identify factors underlying the observed data. There is also a formal modeling procedure called *factor analysis* that is used to address this issue. The model is similar to a multivariate linear model, but several of the assumptions are changed. It is assumed that each observation vector y_i has $\text{E}(y_i) = \mu$ and $\text{Cov}(y_i) = \Sigma$. For n observations, the factor analysis model is

$$Y = J\mu' + XB + e \tag{1}$$

where Y is $n \times q$ and X is $n \times r$. Most of the usual multivariate linear model assumptions about the rows of e are made,

$$\begin{aligned}
\text{E}(\varepsilon_i) &= 0 \\
\text{Cov}(\varepsilon_i, \varepsilon_j) &= 0 \qquad i \neq j,
\end{aligned}$$

and for Ψ nonnegative definite

$$\text{Cov}(\varepsilon_i) = \Psi,$$

but Ψ is assumed to be *diagonal*. The primary change in assumptions relates to X. If we have assumed that $\text{E}(y_i') = \mu'$, the corresponding row $x_i'B$

better be random with mean zero. In fact, each row of X is assumed to be an unobservable random vector with

$$\begin{aligned} \mathrm{E}(x_i) &= 0, \\ \mathrm{Cov}(x_i, x_j) &= 0 \qquad i \neq j, \\ \mathrm{Cov}(x_i) &= I_r, \end{aligned}$$

and

$$\mathrm{Cov}(x_i, \varepsilon_j) = 0 \qquad \text{any } i, j.$$

The matrix B remains a fixed, but unknown, matrix of parameters.

The idea behind the model is that there is a vector x consisting of r underlying factors. Each element of the observation vector y is some linear combination of the factors plus an error term, i.e., $y = \mu + B'x + \varepsilon$. The errors for each element of the observation vector are allowed to have different variances, but they are assumed to be uncorrelated. For the model to be of interest, the number of factors r should be less than the number of dependent variables q. Based on this model, $y_i = \mu + B'x_i + \varepsilon_i$, $i = 1, \ldots, n$ and

$$\begin{aligned} \mathrm{Cov}(y_i) &= \mathrm{Cov}(B'x_i + \varepsilon_i) \\ &= B'B + \Psi. \end{aligned}$$

In most of the discussion below, we will work directly with the matrix $B'B$. It is convenient to have a notation for this matrix. Write

$$\Lambda = B'B.$$

The matrix Λ is characterized by two properties: (1) Λ is nonnegative definite and (2) $r(\Lambda) = r$. Recalling our initial assumption that $\mathrm{Cov}(y_i) = \Sigma$, the factor analysis model has imposed the restriction that

$$\Sigma = \Lambda + \Psi. \tag{2}$$

Clearly, one cannot have $r(\Lambda) = r > q$. It is equally clear that if $r = q$, one can always find matrices Λ and Ψ that satisfy (2). Just choose $\Lambda = \Sigma$ and $\Psi = 0$. If $r < q$, equation (2) places a real restriction on Σ.

In practice, Σ is unknown so one seeks matrices $\hat{B}$ and $\hat{\Psi}$ such that

$$S \doteq \hat{B}'\hat{B} + \hat{\Psi}.$$

The interesting questions now become, how many factors r are needed to get a good approximation and which matrices $\hat{B}$ and $\hat{\Psi}$ give good approximations. The first question is certainly amenable to analysis. Clearly, $r = q$ will always work so there must be ways to decide when $r < q$ is doing an adequate job. The second question ends up being tricky. The problem is that if U is an orthogonal matrix, $U\hat{B}$ works just as well as $\hat{B}$ because

$$\hat{B}'\hat{B} = (U\hat{B})'(U\hat{B}).$$

TERMINOLOGY AND APPLICATIONS

Factor analysis uses some distinctive terminology for various components of the model. Elements of a row of X are called *common factors*. These are factors that apply to all of the columns of the corresponding row of Y. While the common factors all affect each column of Y, different linear combinations of the common factors apply to different columns. Note that a column of X consists of n observations from a common factor. The elements of the matrix B are called the *factor loadings*. The columns of B are the coefficients applied to the common factors in order to generate the dependent variables. The j'th row of B consists of all the factor loadings that apply to the j'th common factor. The elements of a row of e are called *unique* or *specific factors*. These are uncorrelated random variables that are added to the linear combinations of common factors to generate the observations. They are distinct random variables for each distinct observation, i.e., they are specific to the observation.

The diagonal elements of Λ are called communalities. Writing $\Lambda = [\lambda_{ij}]$, the *communality* of the i'th variable is generally denoted as

$$h_i^2 = \lambda_{ii} .$$

Note that if $B = [\beta_{ij}]$,

$$h_i^2 = \sum_{k=1}^{r} \beta_{ki}^2 .$$

The i'th diagonal element of Ψ is called the *uniqueness*, *specificity*, or *specific variance* of the i'th variable. The total variance is

$$\mathrm{tr}[\Sigma] = \mathrm{tr}[\Lambda] + \mathrm{tr}[\Psi] .$$

The *total communality* is

$$v \equiv \mathrm{tr}[\Lambda] = \sum_{i=1}^{q} h_i^2 = \sum_{i=1}^{q} \sum_{k=1}^{r} \beta_{ki}^2 .$$

The matrix

$$\Lambda = \Sigma - \Psi$$

is called the *reduced covariance matrix*, for obvious reasons. Often the observations are standardized so that Σ is actually a correlation matrix. If this has been done, Λ is the reduced correlation matrix.

In practice, factor analysis is used primarily to obtain estimates of B. One then tries to interpret the estimated factor loadings in some way that makes sense relative to the subject matter of the data. As is discussed below, this is a fairly controversial procedure. One of the reasons for the controversy is that B is not uniquely defined. Given any orthogonal $r \times r$ matrix U, write $X_0 = XU'$ and $B_0 = UB$, then

$$XB = XU'UB = X_0B_0$$

where X_0 again satisfies the assumptions made about X. Unlike standard linear models, X is not observed so there is no way to tell X and X_0 apart. There is also no way to tell B and B_0 apart. Actually, this indeterminacy is used in factor analysis to increase the interpretability of B. This will be discussed again below. At the moment, we examine ways in which the matrix B is interpreted.

One of the key points in interpreting B is recognizing that it is the rows of B that are important and not the columns. A column of B is used to explain one dependent variable. A row of B consists of all of the coefficients that affect a single common factor. The q elements in the j'th row of B represent the contributions made by the j'th common factor to the q dependent variables. Traditionally, if a factor has all of its large loadings with the same sign, the subject matter specialist tries to identify some common attribute of the dependent variables that correspond to the high loadings. This common attribute is then considered to be the underlying factor. A *bipolar* factor involves high loadings that are both positive and negative, the user identifies common attributes for both the group of dependent variables with positive signs and the group with negative signs. The underlying factor is taken to be one that causes individuals who are high on some scores to be low on other scores. The following example involves estimated factor loadings. Estimation is discussed in the following two subsections.

EXAMPLE 3.4.1.   Lawley and Maxwell (1971) and Johnson and Wichern (1988) examine data on the examination scores of 220 male students. The dependent variable vector consists of test scores on (Gaelic, English, history, arithmetic, algebra, geometry). The correlation matrix is

$$R = \begin{bmatrix} 1.000 & .439 & .410 & .288 & .329 & .248 \\ .439 & 1.000 & .351 & .354 & .320 & .329 \\ .410 & .351 & 1.000 & .164 & .190 & .181 \\ .288 & .354 & .164 & 1.000 & .595 & .470 \\ .329 & .320 & .190 & .595 & 1.000 & .464 \\ .248 & .329 & .181 & .470 & .464 & 1.000 \end{bmatrix}.$$

For $r = 2$, maximum likelihood estimation gives one choice of estimates

$$\hat{B} = \begin{bmatrix} .553 & .568 & .392 & .740 & .724 & .595 \\ -.429 & -.288 & -.450 & .273 & .211 & .132 \end{bmatrix}$$

and

$$(\hat{\psi}_1, \ldots, \hat{\psi}_6) = (.510, .594, .644, .377, .431, .628).$$

Factor interpretation involves looking at the rows of $\hat{B}$ and trying to interpret them. Write

$$B = \begin{bmatrix} b'_1 \\ b'_2 \end{bmatrix}.$$

All of the elements of b_1 are large and fairly substantial. This suggests that the first factor is a factor that indicates general intelligence. The second factor is bipolar with positive scores on math subjects and negative scores on nonmath subjects. The second factor might be classified as some sort of math-nonmath factor. This example will be examined again later with a slightly different slant.

Rather than taking the factor analysis model as a serious model for the behavior of data, it may be more appropriate to view factor analysis as a data analytic procedure that seeks to discover structure in the covariance matrix and may *suggest* the presence of underlying factors.

MAXIMUM LIKELIHOOD THEORY

Maximum likelihood theory can be used for both estimation and testing. Maximum likelihood factor analysis is based on assuming that the random vectors in the factor model have a joint multivariate normal distribution and rewriting the factor analysis model as a standard multivariate linear model. To do this, the random terms are pooled together as, say

$$\xi_i = B'x_i + \varepsilon_i$$

and

$$\xi = XB + e.$$

Write

$$\Lambda = B'B.$$

The factor analysis model implies that

$$Y = J\mu' + \xi \tag{3}$$

where

$$\begin{aligned} E(\xi_i) &= 0, \\ \mathrm{Cov}(\xi_i, \xi_j) &= 0 \qquad i \neq j, \end{aligned}$$

and

$$\mathrm{Cov}(\xi_i) = \Lambda + \Psi$$

with Ψ diagonal, Λ nonnegative definite, and $r(\Lambda) = r$.

The only difference between this model and the multivariate linear model of Chapter I is the imposition of additional structure on the covariance matrix. In Chapter I, the assumption was simply that

$$\mathrm{Cov}(\xi_i) = \Sigma.$$

The new model places the restriction on Σ that

$$\Sigma = \Lambda + \Psi \tag{4}$$

where $r(\Lambda) = r$ and Ψ is diagonal. For ξ_i's with a joint multivariate normal distribution, the likelihood function for an arbitrary Σ was discussed in Chapter I. Clearly, the likelihood can be maximized subject to the restrictions that $\Sigma = \Lambda + \Psi$, Λ is nonnegative definite, $r(\Lambda) = r$, and Ψ is diagonal.

Because $\Lambda + \Psi$ is just a particular choice of Σ, as in Chapter I the maximum likelihood estimate of μ is always the least squares estimate, $\hat{\mu} = \bar{y}_{..}$. This simplifies the maximization problem. Unfortunately, with the additional restrictions on Σ, closed form estimates of the covariance matrix are no longer available. Computational methods for finding MLE's are discussed in Lawley and Maxwell (1971) and Jöreskog (1975).

EXERCISE 3.6.   Show that the maximization problem reduces to finding a rank r matrix $\hat{\Lambda}$ and a diagonal matrix $\hat{\Psi}$ that minimize

$$\log(|\Lambda + \Psi|) + \text{tr}\{(\Lambda + \Psi)^{-1}\hat{\Sigma}_q\}$$

where $\hat{\Sigma}_q = \dfrac{n-1}{n}S$ and $\hat{\Lambda}$ is nonnegative definite.

One advantage of the maximum likelihood method is that standard asymptotic results apply. Maximum likelihood estimates are asymptotically normal. Minus two times the likelihood ratio test statistic is asymptotically chi-squared under the null hypothesis. See Geweke and Singleton (1980) for a discussion of sample size requirements for the asymptotic test.

Of specific interest are tests for examining the rank of Λ. If $r < s$, the restriction $r(\Lambda) = r$ is more stringent than the restriction $r(\Lambda) = s$. To test $H_0 : r(\Lambda) = r$ versus $H_A : r(\Lambda) = s$, one can use the likelihood ratio test statistic. This is just the maximum value of the likelihood under $r(\Lambda) = r$ divided by the maximum value of the likelihood under $r(\Lambda) = s$. Under H_0, -2 times the log of this ratio has an asymptotic chi-squared distribution. The degrees of freedom are the difference in the number of independent parameters for the models with $r(\Lambda) = s$ and $r(\Lambda) = r$. If we denote

$$\hat{\Sigma}_r = \hat{\Lambda} + \hat{\Psi}$$

when $r(\Lambda) = r$ with a similar notation for $r(\Lambda) = s$, -2 times the log of the likelihood ratio test statistic is easily shown to be

$$n\left[\ln\left(\frac{|\hat{\Sigma}_r|}{|\hat{\Sigma}_s|}\right) + \text{tr}\{(\hat{\Sigma}_r^{-1} - \hat{\Sigma}_s^{-1})\hat{\Sigma}_q\}\right]$$

where again

$$\hat{\Sigma}_q = \frac{n-1}{n}S.$$

As will be seen below, the degrees of freedom for the test are

if $s = q$ $\qquad df = q(q+1)/2 - q - [qr - r(r-1)/2]$
if $s < q$ $\qquad df = [qs - s(s-1)/2] - [qr - r(r-1)/2].$

The formula for degrees of freedom are easily derived given the number of independent parameters in each model. If $r(\Lambda) \equiv r = q$, the covariance matrix Σ is unrestricted. The independent parameters are the q elements of μ and the $q(q+1)/2$ distinct elements of Σ. Recall that because Σ is symmetric not all of its elements are distinct. Thus for $r = q$, the model has

$$q + q(q+1)/2$$

degrees of freedom.

Counting the degrees of freedom when $r(\Lambda) = r < q$ is a bit more complicated. The model involves the restriction

$$\Sigma = \Lambda + \Psi$$

where Ψ is diagonal and Λ is nonnegative definite with rank r. Clearly, Ψ has q independent parameters, the diagonal elements. Because Λ is of rank r, the last $q - r$ columns of the $q \times q$ matrix are linear combinations of the first r columns. Thus, the independent parameters are at most the elements of the first r columns. There are qr of these parameters. However, Λ is also symmetric. All of the parameters above the diagonal are redundant. In the first r columns there are $1 + 2 + \cdots + (r-1) = r(r-1)/2$ of these redundant values. Thus, Λ has $qr - r(r-1)/2$ parameters. Finally, μ again involves q independent parameters. Adding the number of independent parameters in μ, Ψ, and Λ gives the model degrees of freedom as

$$q + q + [qr - r(r-1)/2].$$

Taking differences in model degrees of freedom gives the test degrees of freedom indicated above.

Thus far in the discussion, we have ignored B in favor of $\Lambda = B'B$. Given a function of Λ, say $B = f(\Lambda)$ and the maximum likelihood estimate $\hat{\Lambda}$, the MLE of B is $\hat{B} = f(\hat{\Lambda})$. The problem is in defining the function f. There are an uncountably infinite number of ways to define f. If f defines B and U is an orthogonal matrix, then

$$f_1(\Lambda) = U f(\Lambda)$$

is just as good a definition of B because $\Lambda = B'B = B'U'UB$. As mentioned, this indeterminacy is used to make the results more interpretable. The matrix B is redefined until the user gets a pleasing $\hat{B}$. The procedure starts with any $\hat{B}$ and then $\hat{B}$ is rotated (multiplied by on orthogonal matrix) until $\hat{B}$ seems to be interpretable to the user. In fact, there are some standard rotations, e.g., varimax and quartimax, that are often used to increase interpretability. For a more complete discussion of rotations see

Williams (1979). Often, in an effort to make B well defined, it is taken to be $D\left(\sqrt{\phi_1},\ldots,\sqrt{\phi_r}\right) A_r'$ where $A_r = [a_1,\ldots,a_r]$ with a_i an eigenvector of Λ corresponding to ϕ_i that has length one. If the ϕ_i's are distinct, this accomplishes its purpose but at the price of introducing a considerable arbitrary element into the definition of B. With this definition, a parameterization based on B is strictly equivalent to that based on Λ.

EXAMPLE 3.4.2.   For $r = 2$ the orthogonal matrices U used in rotations are 2×2 matrices. Thus the effects of orthogonal rotations can be plotted. The plots consist of q points, one for each dependent variable. Each point consists of the two values in each column of $\hat{B}$. Figure III.1 gives a plot of the unrotated factor loadings presented in Example 3.4.1 for the examination score data. The points labeled 1 through 6 indicate the corresponding dependent variable $h = 1,\ldots,6$. Two commonly used rotations are the varimax rotation and the quartimax rotation, cf. Exercise 3.5.10. The *varimax* rotation for this data is

$$\hat{B}_V = \begin{bmatrix} 0.232 & 0.321 & 0.085 & 0.770 & 0.723 & 0.572 \\ 0.660 & 0.551 & 0.591 & 0.173 & 0.215 & 0.213 \end{bmatrix}$$

and the *quartimax* rotation is

$$\hat{B}_Q = \begin{bmatrix} 0.260 & 0.344 & 0.111 & 0.777 & 0.731 & 0.580 \\ 0.650 & 0.536 & 0.587 & 0.139 & 0.184 & 0.188 \end{bmatrix}.$$

Plots of these factor loadings are presented in Figures III.2 and III.3. Comparison of Figure III.1 with these plots indicates the rotations about the origin $(0,0)$ that were used to obtain $\hat{B}_V$ and $\hat{B}_Q$. In this example, the rotations are quite similar. Rather than isolating a general intelligence factor and a bipolar factor as seen in the unrotated factors, these both identify factors that can be interpreted as one for mathematics ability and one for nonmathematics ability.

The factor analysis model for maximum likelihood assumes that the matrix of common factors X has rows consisting of independent observations from a multivariate normal distribution with mean zero and covariance matrix I_r. While X is not observable, it is possible to predict the rows of X. In Exercise 3.7, it will be seen that $\hat{E}(x_i|Y) = B(\Lambda + \Psi)^{-1}(y_i - \mu)$. Thus, estimated best linear predictors of the x_i's can be obtained. These *factor scores* are frequently used to check the assumption of multivariate normality. Bivariate plots can be examined for elliptical shapes and outliers. Univariate plots can be checked for normality.

PRINCIPAL FACTOR ESTIMATION

It would be nice to have a method for estimating the parameters of model (3) that did not depend on the assumption of normality. Thurstone (1931)

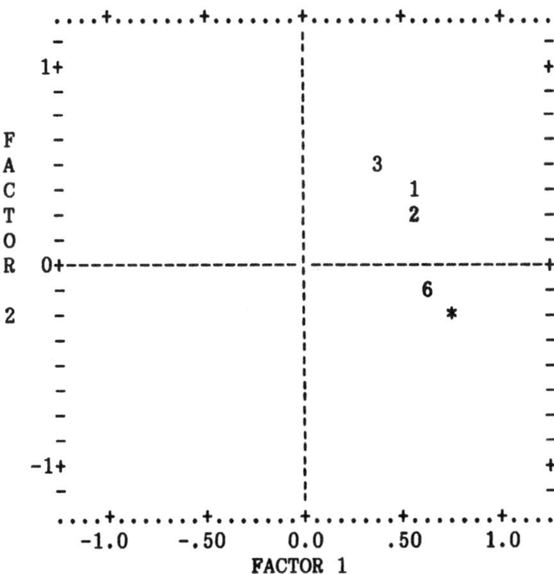

FIGURE III.1. Unrotated Factor Loadings

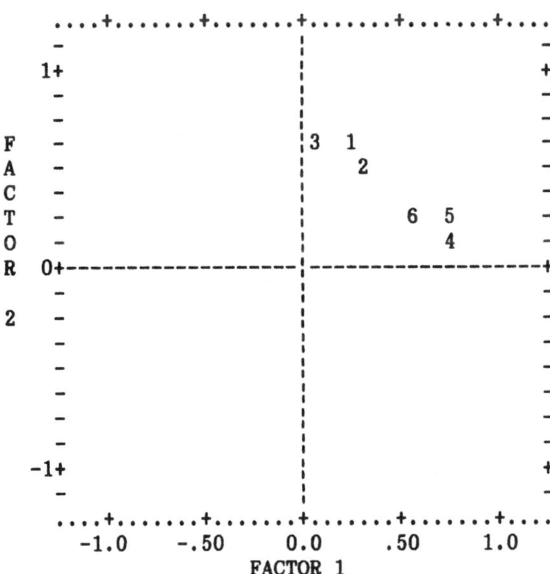

FIGURE III.2. Varimax Factor Loadings

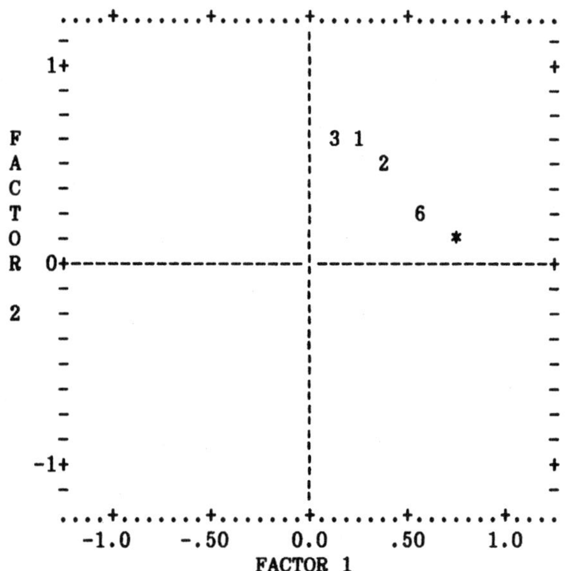

FIGURE III.3. Quartimax Factor Loadings

and Thompson (1934) have proposed principal factor estimation as such a method. The parameters to be estimated are μ, Λ, and Ψ. As mentioned earlier, model (3) is just a standard multivariate linear model with a peculiar choice for Σ. The results of Section I.2 imply that $\bar{y}.$ is the best linear unbiased estimate of μ.

It remains to estimate Λ and Ψ. If Ψ is known, estimation of Λ is easy. Using equation (4)

$$\Lambda = \Sigma - \Psi$$

where Λ is assumed to be nonnegative definite of rank r. If it were not for the rank condition, a natural estimate would be

$$\tilde{\Lambda} = S - \Psi .$$

Incorporating the rank condition, one natural way to proceed is to choose a nonnegative definite matrix of rank r, say $\hat{\Lambda}$, that minimizes, say

$$\operatorname{tr}\{(S - \Psi) - \Lambda\} .$$

Although other functions of $(S - \Psi) - \Lambda$ might be reasonable, the trace is a convenient choice because we have already solved a version of this problem.

Let $\phi_1 \geq \cdots \geq \phi_q$ be the eigenvalues of S, let $a_1, \ldots, a_q$ be the corresponding eigenvectors, let $A_r = [a_1, \ldots, a_r]$, and let B be a $q \times r$ matrix of rank r. In our discussion of principal components, we established that

$$\operatorname{tr}[S - SA_r(A'_r SA_r)^{-1}A'_r S] = \min_{B} \operatorname{tr}[S - SB(B'SB)^{-1}B'S] .$$

Clearly, $SB(B'SB)^{-1}B'S$ is nonnegative definite of rank r. If we consider the problem of estimating Σ when $r(\Sigma) = r$ and restrict ourselves to the class of estimates $SB(B'SB)^{-1}B'S$ then the matrix $SA_r(A_r'SA_r)^{-1}A_r'S$ is an optimal rank r estimate of S.

Applying this result in the factor analysis problem gives an optimal estimate

$$\hat{\Lambda} = \tilde{\Lambda}A_r(A_r'\tilde{\Lambda}A_r)^{-1}A_r'\tilde{\Lambda}$$

where A_r consists of eigenvectors of $\tilde{\Lambda} = S - \Psi$. If we choose the eigenvectors so that $A_r'A_r = I_r$, $\hat{\Lambda}$ simplifies to

$$\hat{\Lambda} = A_r D(\phi_1, \ldots, \phi_r)A_r'$$

where $\phi_1, \ldots, \phi_r$ are the r largest eigenvalues of $\tilde{\Lambda}$. An obvious estimate of B is

$$\hat{B} = D(\sqrt{\phi_1}, \ldots, \sqrt{\phi_r})A_r' .$$

Of course, any rotation of $\hat{B}$ is an equally appropriate estimate.

All of this assumes that Ψ is known. In practice, one makes an initial guess Ψ_0 that leads to initial estimates $\tilde{\Lambda}_0 = S - \Psi_0$ and $\hat{\Lambda}_0$. Having computed $\hat{\Lambda}_0$, compute Ψ_1 from the diagonal elements of $S - \hat{\Lambda}_0$ and repeat the process to obtain $\hat{\Lambda}_1$. This iterative procedure can be repeated until convergence. A common choice for $\Psi_0 = D(\psi_{i0})$ is

$$\psi_{i0} = 1/s^{ii}$$

where s^{ii} is the i'th diagonal element of S^{-1}.

Another common choice for Ψ_0 is taking $\psi_{i0} = 0$ for all i. This choice yields $\tilde{\Lambda} = S$ and the rows of $\hat{B} = D(\sqrt{\phi_i})A_r'$ are eigenvectors of S. These are the same vectors as used to determine principal components. In fact, principal components are often used to address questions about underlying factors. The difference is that in a principal component analysis the elements of the eigenvector determine a linear combination of the dependent variable y. In the factor analysis model, the elements of an eigenvector, say a_1, are the q coefficients applied to the first hypothetical factor. Although factor interpretations are based on these q values, in the factor analysis model data is generated using the r values $a_{1h}, \ldots, a_{rh}$ taken *across* eigenvectors.

Some of the problems with principal factor estimation are that r is assumed to be known, there are no tests available for the value of r, and the matrix $S - \Psi$ may not be nonnegative definite.

In our examination of principal components, we found that the eigenvectors of Σ provided solutions to several different problems: sequential prediction, joint prediction, sequential variance maximization, and geometrical interpretation. The principal factor estimation method can also be motivated by a sequential optimization problem, cf. Gnanadesikan (1977).

DISCUSSION

There is no question that model (3) is a reasonable model. There is considerable controversy about whether the factor analysis model (1) has any meaning beyond that of model (3). Factor analysis is a very frequently used methodology. Obviously, its users like it. Users like to rotate the estimated factor loadings $\hat{B}$ and interpret their results. On the other hand, many people, often of a more theoretical bent, are deeply disturbed by the indeterminacy of the factors and the factor loadings. Many people claim it is impossible to understand the nature of the underlying factors and the basis of their interpretation. Personally, I have always tried to straddle this particular fence. There are people on both sides that I respect.

An important criterion for evaluating models is that if a model is useful it should be useful for making predictions about future observables. The maximum likelihood model (3), like all linear models, satisfies this criterion. The prediction of a new case would be $\bar{y}_{..}$. The peculiar covariance matrix of model (3) plays a key role in predicting the unobserved elements of a new case when some of the elements have been observed.

The factor analysis model (1) looks like it is more than the corresponding linear model. The interpretation of factor loadings depends on (1) being more than the linear model. If the factor analysis model really is more than the linear model, it should provide predictions that are distinct from the linear model. When the factor analysis model is correct, these predictions should be better than the linear model predictions.

Unfortunately, the factor analysis model does not seem to lend itself to prediction except through the corresponding linear model. One can predict the factor vectors x_i (assuming that μ, B, and Ψ are known), but this does not affect prediction of y_i.

EXERCISE 3.7.   Show that
a) $\hat{E}(x_i|Y) = \hat{E}(x_i|y_i) = B(\Lambda + \Psi)^{-1}(y_i - \mu)$
b) $\hat{E}(\mu + B'x_i + \varepsilon_i|y_i) = y_i$
In (b), ε_i is the i'th row of e. Do not use the fact that $\hat{E}(y_i|y_i) = y_i$.

Though the factor analysis model may not hold up to careful scrutiny, it does not follow that the data analytic method known as factor analysis is a worthless endeavor. Rather than thinking of factor analysis as a theoretical method of estimating the loadings on some unspecified factors, it may be better to think of it as a data analytic method for identifying structure in the covariance matrix. As a data analytic method, it is neither surprising nor disconcerting that different people (using different rotations) obtain different results. It is more important whether, in practice, users working on similar problems often obtain similar results.

The factor analysis model is one motivation for this method of data analysis. We now will present a slightly different view. We begin by decomposing the covariance matrix into the sum of r different covariance matrices plus Ψ. In other words, write

$$B = \begin{bmatrix} b_1' \\ \vdots \\ b_r' \end{bmatrix}$$

and

$$\Lambda_i = b_i b_i' .$$

Thus

$$\begin{aligned} \Sigma &= \Lambda + \Psi \\ &= B'B + \Psi \\ &= \sum_{i=1}^r b_i b_i' + \Psi \\ &= \sum_{i=1}^r \Lambda_i + \Psi . \end{aligned}$$

We can think of y as being a random observation vector and Λ_i as being the covariance matrix for some factor, say w_i, where $y = \mu + \sum_{i=1}^r w_i + \varepsilon$ with $\text{Cov}(w_i, w_j) = 0$, $\text{Cov}(\varepsilon) = \Psi$, and $\text{Cov}(w_i, \varepsilon) = 0$. In the usual factor analysis model with factors $x = (x_1, \ldots, x_r)'$ and $B' = [b_1, \ldots, b_r]$, we have $w_i = x_i b_i$. The question then becomes what kind of underlying factor w_i would generate a covariance matrix such as Λ_i. The advantage to this point of view is that attention is directed towards explaining the observable correlations. In traditional factor analysis, attention is directed towards estimating the ill-defined factor loadings. Of course, the end result is the same.

Just as the matrix B is not unique, neither is the decomposition

$$\Lambda = \sum_{i=1}^r \Lambda_i .$$

In practice, one would rotate B to make the Λ_i's more interpretable. Moreover, as will be seen below, one need not actually compute Λ_i to discover its important structure. The key features of Λ_i are obvious from examination of b_i.

EXAMPLE 3.4.3.   Using $\hat{B}$ from Example 3.4.1

$$\hat{\Lambda}_1 = \hat{b}_1 \hat{b}_1' = \begin{bmatrix} .31 & .31 & .22 & .41 & .40 & .33 \\ .31 & .32 & .22 & .42 & .41 & .34 \\ .22 & .22 & .15 & .29 & .28 & .23 \\ .41 & .42 & .29 & .55 & .54 & .44 \\ .40 & .41 & .28 & .44 & .52 & .43 \\ .33 & .34 & .23 & .44 & .43 & .35 \end{bmatrix}$$

All of the variances and covariances are uniformly high because all of the elements of b_1 are uniformly high. The factor w_1 must be some kind of overall measure – call it general intelligence.

The examination of the second covariance matrix

$$\hat{\Lambda}_2 = \begin{bmatrix} .18 & .12 & .19 & -.12 & -.09 & -.07 \\ .12 & .08 & .13 & -.08 & -.06 & -.04 \\ .19 & .13 & .20 & -.12 & -.09 & -.06 \\ -.12 & -.08 & -.12 & .07 & .06 & .04 \\ -.09 & -.06 & -.09 & .06 & .04 & .03 \\ -.07 & -.04 & -.06 & .04 & .03 & .02 \end{bmatrix}$$

is trickier. The factor w_2 has two parts; there is positive correlation among the first three variables: Gaelic, English, and history. There is positive correlation among the last three variables: arithmetic, algebra, and geometry. However, the first three variables are negatively correlated with the last three variables. Thus, w_2 can be interpreted as a math factor and a non-math factor that are negatively correlated.

A totally different approach to dealing with Λ_2 is to decide that any variable with a *variance* less than, say, .09 is essentially constant. This leads to

$$\tilde{\Lambda}_2 = \begin{bmatrix} .18 & 0 & .19 & 0 & 0 & 0 \\ 0 & 0 & 0 & 0 & 0 & 0 \\ .19 & 0 & .20 & 0 & 0 & 0 \\ 0 & 0 & 0 & 0 & 0 & 0 \\ 0 & 0 & 0 & 0 & 0 & 0 \\ 0 & 0 & 0 & 0 & 0 & 0 \end{bmatrix}$$

Thus, the second factor puts weight on only Gaelic and history. The second factor would then be interpreted as some attribute that only Gaelic and history have in common.

Either analysis of Λ_2 can be arrived at by direct examination of

$$b_2' = (-.429, -.288, -.450, .273, .211, .132).$$

The pattern of positives and negatives determines the corresponding pattern in Λ_2. Similarly, the requirement that a variance be greater than .09 to be considered nonzero corresponds to a variable having an absolute factor loading greater than .3. Only Gaelic and history have factor loadings with absolute values greater than .3. Both are negative so Λ_i will display the positive correlation between them.

The examination of underlying factors is, of necessity, a very slippery enterprise. The parts of factor analysis that are consistent with traditional ideas of modeling are estimation of Λ and Ψ and the determination of the rank of Λ. The rest is pure data analysis. It is impossible to prove that underlying factors actually exist. The argument in factor analysis is that if these factors existed they could help explain the data.

Factor analysis can only suggest that certain factors might exist. The appropriate question is not whether they really exist, but whether their existence is a useful idea. For example, does the idea of a factor for general intelligence help people to understand the nature of test scores. A more stringent test of usefulness is whether the idea of a general intelligence factor leads to accurate predictions about future observable events. Recall from Exercise 3.7 that one can predict factor scores, so those predictions can be used as a tool in making predictions about future observables for the individuals in the study.

An interesting if unrelated example of these criteria for usefulness involves the force of gravity. For most of us, it is impossible to prove that such a force exists. However, the idea of this force allows one to both explain and predict the behavior of physical objects. The fact that accurate predictions can be made does not prove that gravity exists. If an idea explains current data in an intelligible manner and/or allows accurate prediction, it is a useful idea. For example, the usefulness of Newton's laws of motion cannot be disregarded just because they break down for speeds approaching that of light.

III.5   Additional Exercises

EXERCISE 3.5.1.   a) Find the vector b that minimizes

$$\sum_{i=1}^{q} \left[y_i - \mu_i - b'(x - \mu_x) \right]^2 .$$

b) For given weights w_i, $i = 1, \ldots, q$, find the vector b that minimizes

$$\sum_{i=1}^{q} w_i^2 \left[y_i - \mu_i - b'(x - \mu_x) \right]^2 .$$

c) Find the vectors b_i that minimize

$$\sum_{i=1}^{q} w_i^2 \left[y_i - \mu_i - b_i'(x - \mu_x) \right]^2 .$$

EXERCISE 3.5.2.   In a population of large industrial corporations, the covariance matrix for $y_1 = assets/10^6$ and $y_2 = net\ income/10^6$ is

$$\Sigma = \begin{bmatrix} 75 & 5 \\ 5 & 1 \end{bmatrix} .$$

a) Determine the principal components.
b) What proportion of the total prediction variance is explained by $a_1'y$?
c) Interpret $a_1'y$.
d) Repeat (a), (b), and (c) for principal components based on the correlation matrix.

EXERCISE 3.5.3.   What are the principal components associated with

$$\Sigma = \begin{bmatrix} 5 & 0 & 0 & 0 \\ 0 & 3 & 0 & 0 \\ 0 & 0 & 3 & 0 \\ 0 & 0 & 0 & 2 \end{bmatrix}.$$

Discuss the problem of reducing the variables to a two-dimensional space.

EXERCISE 3.5.4.   Let $v_1 = (2, 1, 1, 0)'$, $v_2 = (0, 1, -1, 0)'$, $v_3 = (0, 0, 0, 2)'$ and

$$\Sigma = \sum_{i=1}^{3} v_i v_i'.$$

a) Find the principal components of Σ.
b) What is the predictive variance of each principal component? What percentage of the maximum prediction error is accounted for by the first two principal components?
c) Interpret the principal components.
d) What are the correlations between the principal components and the original variables?

EXERCISE 3.5.5.   Do a principal components analysis of the female turtle carapace data of Exercise 1.8.1.

EXERCISE 3.5.6.   The data in Table III.1 is a subset of the Chapman data reported by Dixon and Massey (1983). It contains the age, systolic blood pressure, diastolic blood pressure, cholesterol, height, and weight for a group of men in the Los Angeles Heart Study. Do a principal components analysis of the data.

EXERCISE 3.5.7.   Assume a two-factor model with

$$\Sigma = \begin{bmatrix} .15 & 0 & .05 \\ 0 & .20 & -.01 \\ .05 & -.01 & .05 \end{bmatrix},$$

and

$$B = \begin{bmatrix} .3 & .2 & .1 \\ .2 & -.3 & .1 \end{bmatrix}.$$

What is Ψ? What are the communalities?

TABLE III.1. Chapman Data

age	sbp	dbp	chol	ht	wt	age	sbp	dbp	chol	ht	wt
44	124	80	254	70	190	37	110	70	312	71	170
35	110	70	240	73	216	33	132	90	302	69	161
41	114	80	279	68	178	41	112	80	394	69	167
31	100	80	284	68	149	38	114	70	358	69	198
61	190	110	315	68	182	52	100	78	336	70	162
61	130	88	250	70	185	31	114	80	251	71	150
44	130	94	298	68	161	44	110	80	322	68	196
58	110	74	384	67	175	31	108	70	281	67	130
52	120	80	310	66	144	40	110	74	336	68	166
52	120	80	337	67	130	36	110	80	314	73	178
52	130	80	367	69	162	42	136	82	383	69	187
40	120	90	273	68	175	28	124	82	360	67	148
49	130	75	273	66	155	40	120	85	369	71	180
34	120	80	314	74	156	40	150	100	333	70	172
37	115	70	243	65	151	35	100	70	253	68	141
63	140	90	341	74	168	32	120	80	268	68	176
28	138	80	245	70	185	31	110	80	257	71	154
40	115	82	302	69	225	52	130	90	474	69	145
51	148	110	302	69	247	45	110	80	391	69	159
33	120	70	386	66	146	39	106	80	248	67	181

EXERCISE 3.5.8.   Using the vectors v_1 and v_2 from Exercise 3.5.3, let

$$\Lambda = v_1 v_1' + v_2 v_2'.$$

Give the eigenvector solution for B and another set of loadings that generates Λ.

EXERCISE 3.5.9.   Given that

$$\Sigma = \begin{bmatrix} 1 & .3 & .09 \\ .3 & 1 & .3 \\ .09 & .3 & 1 \end{bmatrix}$$

and

$$\Psi = D(.1, .2, .3)$$

find Λ and two choices of B.

EXERCISE 3.5.10.   Find definitions for the well-known factor loading matrix rotations varimax, direct quartimin, quartimax, equamax, and orthoblique. What is each of the rotations specifically designed to accomplish? Apply each rotation to the covariance matrices of Exercise 3.5.9.

EXERCISE 3.5.11.   Do a factor analysis of the female turtle carapace data of Exercise 1.8.1. Include tests for the numbers of factors and examine various factor loading rotations.

EXERCISE 3.5.12.   Do a factor analysis of the Chapman data discussed in Exercise 3.5.6.

EXERCISE 3.5.13.   Show the following determinant equality.

$$|\Psi + BB'| = |I + B'\Psi^{-1}B|\,|\Psi|.$$

EXERCISE 3.5.14.   Find the likelihood ratio test for

$$H_0 : \Sigma = \sigma^2\left[(1-\rho)I + \rho JJ'\right]$$

against the general alternative.

EXERCISE 3.5.15.   Let $Y = (y_1, \ldots, y_n)'$ and $Y_k = (y_{n+1}, \ldots, y_{n+k})'$. Show that

$$\hat{E}\left[y_{n+k+1}\big|Y\right] = \hat{E}\left[\hat{E}(y_{n+k+1}|Y, Y_k)\big|Y\right].$$

Hint: Use the definition of $\hat{E}(\cdot|\cdot)$, Proposition 3.1.1, and results on the inverse of a partitioned matrix.

Chapter IV

Frequency Analysis of Time Series

Consider a sequence of observations $y_1, y_2, \ldots, y_n$ taken at equally spaced time intervals. Some of the many sources of such data are production from industrial units, national economic data, and regularly recorded biological data (e.g., blood pressures taken at regular intervals). The distinguishing characteristic of time series data is that because data are being taken on the same object over time, the individual observations are correlated.

A key feature of time series data is that they are often *cyclical* in nature. For example, retail sales go up at Christmas time every year. The number of people who vacation in Montana goes up every summer, down in the fall and spring, and up during the ski season. Time series analysis is designed to model cyclical elements.

There are two main schools of time series analysis: frequency domain analysis and time domain analysis. The frequency domain can be viewed as regression on independent variables that isolate the frequencies of the cyclic behavior. The regression variables are cosines and sines evaluated at known frequencies and times. Most properly, the regression coefficients are taken as random variables so the appropriate linear model is a mixed model as in Christensen (1987, Chapter XII). The justification for this approach is based on a very powerful result in probability theory called the Spectral Representation Theorem. Frequency domain analysis is the subject of this chapter.

Chapter V discusses time domain analysis. Time domain analysis involves the modeling of observed time series as processes generated by a series of random errors. An important idea in the time domain is that of an autoregressive model. For example, the autoregressive model of order 2 is

$$y_t = \beta_1 y_{t-1} + \beta_2 y_{t-2} + e_t \tag{1}$$

where the e_t's are uncorrelated errors. The current observation y_t is being regressed on the previous two observations. Because the design matrix for model (1) consists of random observations, it does not satisfy the standard linear model assumption that the design matrix is fixed. Recall from Christensen (1987, Chapter VI) that regression is closely related to best linear prediction. Prediction theory is based on having random predictors such as in model (1). Best linear prediction is an important tool in time domain analysis.

In addition to cyclical elements, there may also be *trends* in the data. For example, from year to year the retail sales at Christmas time may display a relatively steady increase. The blood pressure of an overweight person on a diet may display a steady tendency to decrease. This is an important aspect of the analysis of time series. One way to handle trend is to remove it prior to analyzing the cyclical behavior of the time series. Another way of viewing, say, an increasing trend is to consider it as a cycle for which the downturn is nowhere in sight. In this view, trend is just a very slowly oscillating cycle.

Time series analysis is a large and important subject and there is a huge literature available. The books by Shumway (1988), Brockwell and Davis (1987), and Fuller (1976) discuss both the frequency and time domains. Bloomfield (1976) and Koopmans (1974) are devoted to the frequency domain. At a more advanced level are the frequency domain books by Brillinger (1981) and Hannan (1970). Many other books are also available that examine frequency domain ideas.

We begin with an introduction to stationary processes. This is followed in Section 2 with some basic ideas on analyzing data to identify important frequencies. The methods and models of Section 2 are related to stochastic processes in Section 3. The relationship suggests that random effects models are more appropriate than fixed effects models for stationary time series. Two random effects models are examined in Sections 4 and 5: one with and one without uncorrelated residual errors.

A traditional data analytic tool for time series is linear filtering. In Section 6 various types of linear filters are defined and their frequency properties are examined. The linear filters discussed in Section 6 are the basis for the time domain models of Chapter V. The chapter closes with some ideas on the relationship between two different time series and a discussion of the Fourier series notation for data analysis.

IV.1   Stationary Processes

One fruitful approach to modeling time series data is through the use of stationary processes. A stationary random process is just a group of random variables that exhibit a property called stationarity. Typically, the groups of random variables are large: either countably or uncountably infinite. In this chapter and the next we consider sequences of random variables $y_1, y_2, y_3, \ldots$ or more generally $\ldots, y_{-1}, y_0, y_1, \ldots$. In Chapter VI we consider random variables $y(u)$ for vectors u in $\mathbf{R}^d$. Of course, in any application, only a finite number of random variables can be observed. From these observations, linear model theory will be used to draw inferences about the entire process.

The property of stationarity is simply that the mechanisms generating the process do not vary — they remain stationary. The sense in which this occurs is that no matter where you start to examine the process, the distribution of the process looks the same. A formal definition of a stationary sequence can be based on examining an arbitrary number of random variables, say k, both at the start of the process and at any other time t. Let $C_1, \ldots, C_k$ be arbitrary (Borel) sets, then $\ldots, y_1, y_2, \ldots$ is a *discrete time stationary process* if for any t,

$$\Pr[y_1 \in C_1, \ldots, y_k \in C_k] = \Pr[y_{t+1} \in C_1, \ldots, y_{t+k} \in C_k]. \tag{1}$$

Thus, the random vectors $(y_1, \ldots, y_k)'$ and $(y_{t+1}, \ldots, y_{t+k})'$ have the same distribution for any values t and k. In particular, if the expectation exists,

$$\mathrm{E}(y_t) = \mu \tag{2}$$

for any t and some scaler μ. If second moments exist then for any t and k

$$\mathrm{Cov}(y_t, y_{t+k}) = \sigma(k) \tag{3}$$

for some *covariance (autocovariance) function* $\sigma(\cdot)$ that does not depend on t. Note that $\sigma(0)$ is the variance of y_t.

A concept related to stationarity is that of second-order stationarity, also known as covariance stationarity, weak stationarity, and stationarity in the wide sense. A process is said to be *second-order stationary* if it satisfies conditions (2) and (3). The name derives from the fact that conditions (2) and (3) only involve second-order moments of the process. As mentioned in the previous paragraph, any process with second moments that satisfies the stationarity condition (1) also satisfies (2) and (3) so any stationary process with second moments is second-order stationary. The converse, of course, does not hold. That random variables have the same first and second moments does not imply that they have the same distributions.

Interestingly, there is an important subclass of stationary processes for which stationarity and second-order stationarity are equivalent. If $(y_{t+1}, \ldots, y_{t+k})$ has a multivariate normal distribution for all t and k, the process is called a *Gaussian* process. Because multivariate normal distributions are completely determined by their means and covariances, conditions (2) and (3) imply that a Gaussian process is stationary.

In applying linear models to observations from stochastic processes, the assumption of second-order stationarity will lead to BLUE's. To obtain tests and confidence intervals or maximum likelihood estimates, the data need to result from a stationary Gaussian process.

In this chapter and the next, we discuss the use of stationary sequences (either second order or Gaussian) to analyze time series data. In reality, time is a continuous variable so random variables observable in time should be characterized as $y(t)$ for $t \in \mathbf{R}$. We assume that this continuous time process will only be observed at equally spaced times $t_1, t_2, \ldots$. (In other words, $t_k - t_{k-1}$ is the same for all k.) The time series we are concerned with is $y_1 = y(t_1)$, $y_2 = y(t_2)$, $\ldots$. If the $y(t)$ process is stationary, as defined in Chapter VI, then the sequence $y_1, y_2, \ldots$ is also stationary.

IV.2 Basic Data Analysis

Frequency domain analysis of an observed time series is based on identifying the frequencies associated with cycles displayed by the data. The most familiar mathematical functions that display cyclical behavior are the sine and cosine functions. The frequency domain analysis of time series can be viewed as doing regression on sines and cosines. In particular, if n is an odd number, we can fit the regression model for $t = 1, \ldots, n$

$$y_t = \alpha_0 + \sum_{k=1}^{\frac{n-1}{2}} \left[\alpha_k \cos\left(2\pi \frac{k}{n} t\right) + \beta_k \sin\left(2\pi \frac{k}{n} t\right) \right]. \tag{1}$$

Here the α's and β's are unknown regression parameters. The independent variables are the sine and cosine functions evaluated as indicated. The independent variables are grouped by their frequency of oscillation. If $\cos\left(2\pi \frac{k}{n} t\right)$ is graphed on $[0, n]$, the function will complete k cycles. Thus the frequency, the number of cycles in one unit of time, is k/n. Similarly, the frequency of $\sin\left(2\pi \frac{k}{n} t\right)$ is k/n. Notice that the model has $1 + 2\left(\frac{n-1}{2}\right) = n$ independent variables so the model is saturated (fits the data perfectly). For this reason we have not included an error term e_t.

If the number of observations n is even, a slightly different model is used: for $t = 1, \ldots, n$

$$y_t = \alpha_0 + \sum_{k=1}^{\frac{n}{2}-1} \left[\alpha_k \cos\left(2\pi \frac{k}{n} t\right) + \beta_k \sin\left(2\pi \frac{k}{n} t\right) \right] + \alpha_{\frac{n}{2}}(-1)^t. \tag{2}$$

Again, there are n predictors in the model, one for each observation, so again the data are fit perfectly. Note that the upper limit of the sums in (1) and (2) can both be written as $\left[\frac{n-1}{2}\right]$ where $\left[\frac{n-1}{2}\right]$ is the greatest integer contained in $\frac{n-1}{2}$. Considering that a saturated model is always fitted to the data, it may be more appropriate to view these models as data analytic tools rather than as realistic models for the data.

It is convenient to write these models in matrix form. Let

$$Y = (y_1, \ldots, y_n)',$$

$$C_k = \left[\cos\left(2\pi \frac{k}{n} 1\right), \cos\left(2\pi \frac{k}{n} 2\right), \ldots, \cos\left(2\pi \frac{k}{n} n\right)\right]',$$

$$S_k = \left[\sin\left(2\pi \frac{k}{n} 1\right), \sin\left(2\pi \frac{k}{n} 2\right), \ldots, \sin\left(2\pi \frac{k}{n} n\right)\right]'$$

and

$$Z_k = [C_k, S_k].$$

Also, let $\gamma_k = [\alpha_k, \beta_k]'$. The model for both n even and n odd can be written

$$Y = J\alpha_0 + \sum_{k=1}^{\left[\frac{n-1}{2}\right]} Z_k \gamma_k + \delta_{\left[\frac{n}{2}\right]\frac{n}{2}} C_{\frac{n}{2}} \alpha_{\frac{n}{2}} \qquad (3)$$

where J is a column of 1's,

$$\delta_{ab} = \begin{cases} 1 & a = b \\ 0 & a \neq b \end{cases}$$

and $\left[\frac{n}{2}\right]$ is the greatest integer in $\frac{n}{2}$, i.e., for n even $\left[\frac{n}{2}\right] = \frac{n}{2}$ and for n odd $\left[\frac{n}{2}\right] = \frac{n-1}{2}$. Note that for n even, $C_{\frac{n}{2}} = (-1, 1, -1, 1, \ldots, -1, 1)'$. Also observe that $C_0 = J$.

One reason that these particular models are used is that the independent variable vectors in (3) are orthogonal. In particular,

$$C_i' C_j = \begin{cases} n & \text{if } i = j \in \{0, n/2\} \\ n/2 & \text{if } i = j \notin \{0, n/2\} \\ 0 & \text{if } i \neq j \end{cases} \qquad (4)$$

$$S_i' S_j = \begin{cases} n/2 & \text{if } i = j \notin \{0, n/2\} \\ 0 & \text{otherwise} \end{cases} \qquad (5)$$

$$C_i' S_j = 0 \qquad \text{any } i \text{ and } j. \qquad (6)$$

See Exercise 4.9.15 for a proof of these relationships. Because the vectors are orthogonal, the sum of squares associated with each independent variable does not depend on the other variables that may or may not be included in any submodel. The order of fitting of variables is also irrelevant. Denote the sum of squares associated with C_k as $SS(C_k)$ and the sum of squares for S_k as $SS(S_k)$. The total sum of squares associated with the frequency k/n is $SS(C_k) + SS(S_k)$.

The periodogram is a function $P(\nu)$ of the frequencies that indicates how important a particular frequency is in the time series. The periodogram is defined only for $\nu = k/n$, $k = 1, \ldots, \left[\frac{n-1}{2}\right]$. The periodogram is

$$P(k/n) = \{SS(C_k) + SS(S_k)\}/2.$$

This is precisely the mean square associated with the frequency k/n. Clearly, if the mean square for the frequency k/n is large relative to the mean squares for the other frequencies, then the frequency k/n is important in explaining the time series.

As always, our linear model is only an approximation to reality. The true frequencies associated with a time series are unlikely to be among the values k/n. If the true frequency is, say $\nu \in \left[\frac{k-1}{n}, \frac{k}{n}\right]$, then we could expect the effect of this frequency to show up in $SS(C_{k-1})$, $SS(C_k)$, $SS(S_{k-1})$ and $SS(S_k)$. If we want a measure of the importance of all the frequencies in a neighborhood of $\frac{k}{n}$, it makes sense to compute the mean square for a group of frequencies near $\frac{k}{n}$. This idea is used to define a smoothed version of the periodogram called an *estimate of the spectral density* or more simply a *spectral estimator*. For r odd, define

$$\hat{f}_r(k/n) = \frac{1}{r} \sum_{i=-(r-1)/2}^{(r-1)/2} P\left(\frac{k+i}{n}\right) \tag{7}$$

which is the mean square for the $2r$ variables $C_{k+\ell}, S_{k+\ell} : \ell = -(r-1)/2, \ldots, (r-1)/2$. The frequencies $(k+\ell)/n$ for $\ell = -(r-1)/2, \ldots, (r-1)/2$ will be called the r *neighborhood* of k/n. Picking an r neighborhood is equivalent to picking frequencies in a band with *bandwidth* r/n. Choosing $r = 3$, i.e., examining the mean square for the frequencies $\frac{k}{n} - 1$, $\frac{k}{n}$, $\frac{k}{n} + 1$ seems particularly appealing to the author, but may not be particularly common in applications. (For consistent estimation of the spectral density, r must be an increasing function of n.)

Rather than using the simple average of the periodograms in the r neighborhood, the spectral density estimate is often taken as a weighted average of the periodogram values. The weights can be defined by evaluating a weighting function at appropriate points. One common choice for a weighting function is the cosine. Within this context, the simple average corresponds to a rectangular weighting function.

Although the function $\hat{f}_r(\cdot)$ is certainly a reasonable thing to examine, the name of the function must seem totally bizarre to anyone without a previous knowledge of time series analysis. The genesis of this name will be discussed in the next section.

EXAMPLE 4.2.1.   Tukey (1977) reported data, extracted from *The World Almanac*, on the production of bituminous coal in the United States between 1920 and 1968. In 1969 the method of reporting the production figures changed. Bituminous coal was subdivided into bituminous, subbituminous, and lignite. These three figures were combined to yield the data in Table IV.1 and Figure IV.1. The data reported are for the years 1920 to 1980. Data for 1981 to 1987 are available in Figure IV.10 and Table V.2. They will not be involved in any time series examples except to evaluate the quality of forecasts.

Table IV.2 gives the periodogram and spectral density estimates based on $r = 3$ and 5. Note that the frequencies ν reported in Table IV.1 are not the Fourier frequencies $k/61$ but rather $k/63$. In the interest of computational efficiency, BMDP has extended the time series from 61 to 63 observations by including artificial values $y_{62} = y_{63} = \bar{y}_{..}$. Clearly, this procedure will not affect the estimate of the mean. All effects for positive frequencies are orthogonal to the mean so working with $y_i - \bar{y}_{.}$ is equivalent to working

TABLE IV.1. Coal Production Data

	1920	1930	1940	1950	1960	1970	1980
0	569	468	461	516	416	602.9	823.7
1	416	382	511	534	403	552.2	
2	422	310	583	467	422	595.3	
3	565	334	590	457	459	591.7	
4	484	359	620	392	467	603.4	
5	520	372	578	467	512	648.4	
6	573	439	534	500	534	678.7	
7	518	446	631	493	552	691.3	
8	501	349	600	410	545	665.2	
9	505	395	438	412	560.5	776.3	

Values are in millions of short tons.

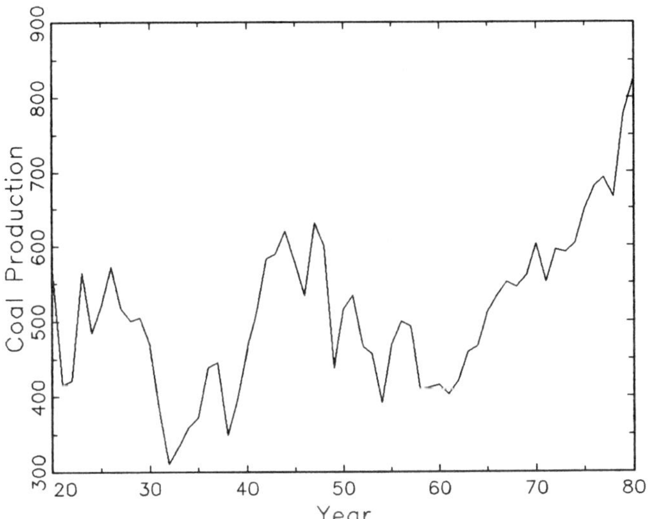

FIGURE IV.1. Coal Production Data 1920 - 1980

with the original data. After correcting for the mean, the artificial observations are zero; thus the artificial observations do not increase the sums of squares for positive frequencies. Nonetheless, it cannot be denied that we are obtaining the solution to a slightly different problem than we set out to solve. Figures IV.2, IV.3 and IV.4 contain plots, on a base 10 logarithmic scale, of the periodogram and the spectral density estimates. The horizontal bars in Figures IV.3 and IV.4 are explained in Example 4.5.1. Perhaps the two most noteworthy aspects of these plots are that there seem to be substantial effects for low frequencies and small effects for high frequencies.

TABLE IV.2. Periodogram and Spectral Density Estimates

ν	$P(k/n)$	$\hat{f}_3$	$\hat{f}_5$
.0000	—	0.408E+05	0.792E+05
.0159	0.612E+05	0.660E+05	0.536E+05
.0317	0.137E+06	0.689E+05	0.457E+05
.0476	0.878E+04	0.558E+05	0.494E+05
.0635	0.216E+05	0.162E+05	0.407E+05
.0794	0.183E+05	0.193E+05	0.141E+05
.0952	0.181E+05	0.134E+05	0.126E+05
.1111	0.373E+04	0.766E+04	0.935E+04
.1270	0.117E+04	0.344E+04	0.642E+04
.1429	0.541E+04	0.343E+04	0.313E+04
.1587	0.371E+04	0.359E+04	0.270E+04
.1746	0.164E+04	0.230E+04	0.299E+04
.1905	0.156E+04	0.194E+04	0.266E+04
.2063	0.262E+04	0.265E+04	0.198E+04
.2222	0.376E+04	0.224E+04	0.252E+04
.2381	0.340E+03	0.281E+04	0.221E+04
.2540	0.432E+04	0.156E+04	0.302E+04
.2698	0.255E+02	0.367E+04	0.405E+04
.2857	0.667E+04	0.520E+04	0.479E+04
.3016	0.890E+04	0.653E+04	0.454E+04
.3175	0.403E+04	0.533E+04	0.503E+04
.3333	0.306E+04	0.318E+04	0.472E+04
.3492	0.247E+04	0.356E+04	0.295E+04
.3651	0.516E+04	0.256E+04	0.243E+04
.3810	0.421E+02	0.221E+04	0.190E+04
.3968	0.141E+04	0.615E+03	0.141E+04
.4127	0.390E+03	0.619E+03	0.460E+03
.4286	0.521E+02	0.281E+03	0.467E+03
.4444	0.400E+03	0.177E+03	0.270E+03
.4603	0.806E+02	0.302E+03	0.239E+03
.4762	0.426E+03	0.249E+03	0.277E+03
.4921	0.239E+03	0.301E+03	0.282E+03

Low frequencies are consistent with trend in the data. To eliminate the effect of a trend, one can perform the frequency analysis on the residuals from a simple linear regression on time. Figure IV.5 gives $\hat{f}_5$ with the trend removed. The plot is not markedly different from the others.

IV.3 Spectral Approximation of Stationary Time Series

We have shown that model (4.2.3) is a useful tool for identifying important frequencies in the data. We now justify model (4.2.3) as an approximate

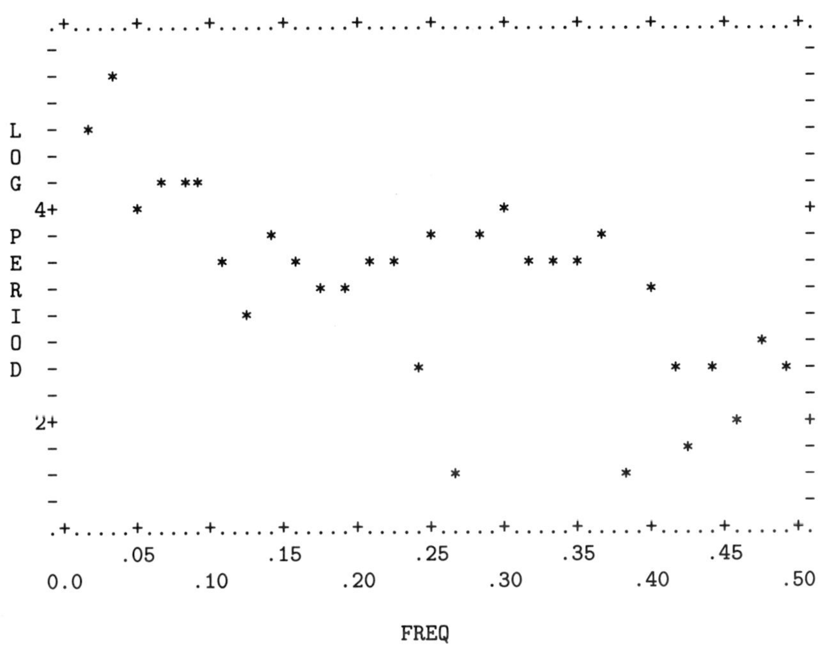

FIGURE IV.2. Log Periodogram versus Frequency

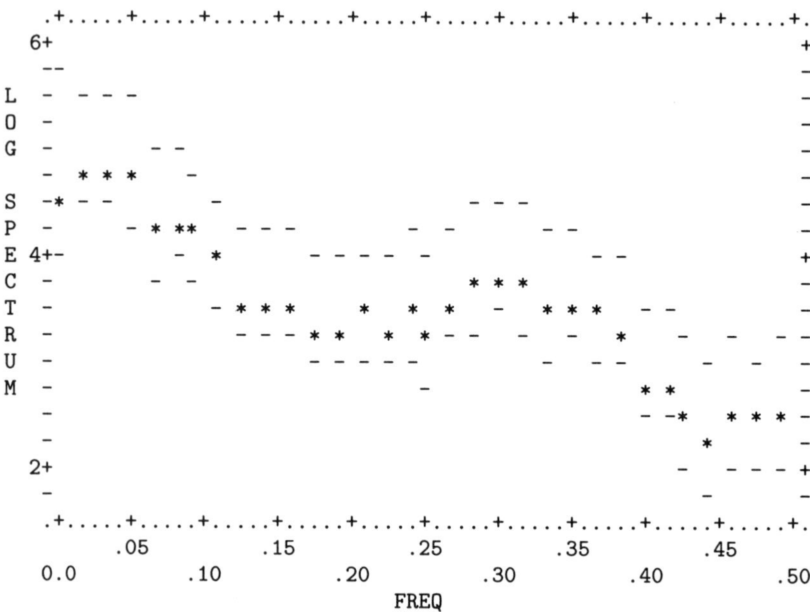

FIGURE IV.3. Log Spectrum versus Frequency for $r = 3$

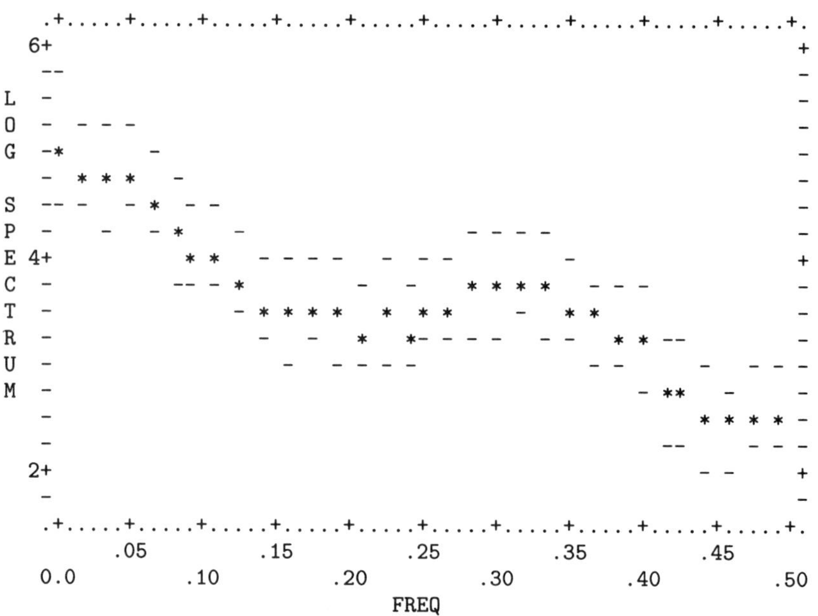

FIGURE IV.4. Log Spectrum versus Frequency for $r = 5$

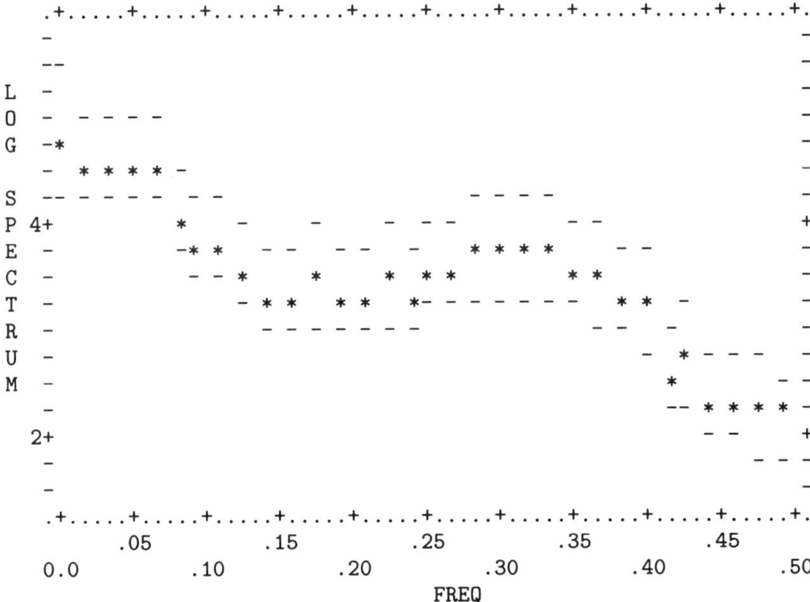

FIGURE IV.5. Log Spectrum versus Frequency, $r = 5$, detrended series

model for observations from a second-order stationary process. The argument is based on the *Spectral Representation Theorem*. This result can be found in a variety of advanced books on probability. The discussion in Doob (1953, Sections X.3, X.4) is particularly germane, but quite sophisticated mathematically. The mathematics in Breiman (1968, Section 11.6) are probably easier to follow, but the discussion is less clearly applicable to the problem at hand.

In this section, applications of various results in probability are discussed. The results themselves are given without proof. In particular, we will use the fact that any second-order stationary process can be approximated by model (4.2.3) with the regression coefficients taken as random variables. This defines a random effects model as in Christensen (1987, Chapter XII). The variance components are related to the autocovariance function through something called the spectral distribution function. Estimation of the variance components is via least squares, thus mimicking Henderson's method 3 which was discussed in Christensen (1987, Section XII.9).

Let $\ldots y_{-2}, y_{-1}, y_0, y_1, y_2, \ldots$ be a second-order stationary process with $E(y_t) = \mu$ and covariance (autocovariance) function $\sigma(k) = \mathrm{Cov}(y_t, y_{t+k})$. As Doob (1953, p. 486) points out, the Spectral Representation Theorem implies that the process $y_t - \mu$ can be approximated arbitrarily closely by a process based on sines and cosines. In particular, for n large and some α_k's and β_k's,

$$y_t - \mu \doteq z_t \tag{1}$$

where

$$z_t = \sum_{k=0}^{\left[\frac{n-1}{2}\right]} \left[\alpha_k \cos\left(2\pi \frac{k}{n}t\right) + \beta_k \sin\left(2\pi \frac{k}{n}t\right)\right] + \delta_{\left[\frac{n}{2}\right]\frac{n}{2}} \cos(\pi t)\alpha_{\frac{n}{2}}$$

with, for all k and k', $\mathrm{E}(\alpha_k) = \mathrm{E}(\beta_k) = 0$, $\mathrm{Var}(\alpha_k) = \mathrm{Var}(\beta_k) = \sigma_k^2$, $\mathrm{Cov}(\alpha_k, \beta_{k'}) = 0$, and for $k \neq k'$, $\mathrm{Cov}(\alpha_k, \alpha_{k'}) = \mathrm{Cov}(\beta_k, \beta_{k'}) = 0$. Although it has been suppressed in the notation, σ_k^2 also depends on n. Note that for $k = 0$, $\sin\left(2\pi \frac{k}{n}t\right) = 0$ and $\cos\left(2\pi \frac{k}{n}t\right) = 1$ for all t. Thus, the right-hand side of (1) is identical to the right-hand sides of models (4.2.1) and (4.2.2), except that we had not previously assumed random regression coefficients.

In our later data analysis, the random effects α_0 and β_0 will not be considered. This requires some justification. First, β_0 is multiplied by $0 \equiv \sin(2\pi 0/nt)$ so β_0 has no effect. The random effect α_0 cannot be analyzed because it is statistically indistinguishable from the parameter μ. Because $1 \equiv \cos(2\pi 0/nt)$, α_0 is added to *every* observation y_t. Similarly, μ is added to every y_t. On the basis of one realization of the series, the two effects are hopelessly confounded. Fortunately, our interest is often in predicting the future of this realization or in simply explaining the behavior of these observations. For either purpose, it is reasonable to consider α_0 as fixed. With α_0 fixed, the mean of y_t is $\mu + \alpha_0$. Because this involves two parameters for one object, we will suppress μ in the notation and use α_0 to denote the mean.

The spectral approximation (1) along with the discussion in the previous paragraph is the justification for using model (4.2.3) as an approximate model for stationary time series. Based on the approximation, in model (4.2.3) we assume that

$$\begin{aligned}
\mathrm{E}(\gamma_k) &= 0, \\
\mathrm{Cov}(\gamma_k) &= \sigma_k^2 I_2, \\
\mathrm{Cov}(\gamma_k, \gamma_{k'}) &= 0 \qquad k \neq k',
\end{aligned}$$

α_0 is a fixed effect and for n even $\alpha_{\frac{n}{2}}$ is a random effect with zero mean, variance $\sigma_{\frac{n}{2}}^2$, and zero covariance with the other random effects. Under these assumptions, model (4.2.3) is a mixed model as in Christensen (1987, Chapter XII).

The key parameters in model (4.2.3) are the variance components, the σ_k^2's. These are closely tied to additional aspects of the Spectral Representation Theorem. To illustrate these aspects, assume that $\mu = 0$ and α_0 is a random effect. The Spectral Representation implies that there is a *unique* right continuous function F called the *spectral distribution function* that is (a) defined on $[-1/2, 1/2]$, (b) symmetric about zero, i.e., $F(v-) - F(0) = F(0) - F(-v)$ where $F(v-) = \lim_{\eta \nearrow v} F(\eta)$, and (c) satisfies

$$\sigma(k) = \int_{-1/2}^{1/2} e^{2\pi i v k} dF(v) \tag{2}$$

$$= \int_{-1/2}^{1/2} \cos(2\pi v k) dF(v).$$

By definition, $e^{2\pi i v k} = \cos(2\pi v k) + i \sin(2\pi v k)$. The integrals are Riemann-Stieltjes integrals. They are similar to the standard Riemann integrals. If $\mathcal{P}_n = (v_{0n}, \ldots, v_{nn})$ defines a sequence of partitions of $[-1/2, 1/2]$ with $v_{in} - v_{i-1,n}$ approaching zero for all i and n, then

$$\int_{-1/2}^{1/2} \cos(2\pi v k) dF(v) = \lim_{n \to \infty} \sum_{i=0}^{n-1} \cos(2\pi v_{in} k)[F(v_{i+1,n}) - F(v_{in})].$$

The second equality in (2) follows from the symmetry of F and the fact that $\sin(2\pi v k)$ is an odd function in v.

EXAMPLE 4.3.1.   Consider the stochastic process defined by

$$z_t = \sum_{j=0}^{\frac{n-1}{2}} \left[\alpha_j \cos\left(2\pi \frac{j}{n} t \right) + \beta_j \sin\left(2\pi \frac{j}{n} t \right) \right]$$

where n is odd and the α's and β's are random effects that satisfy the assumptions made along with equation (1). Note that this process is just the right-hand side of (1). In this example, we examine properties of the process that is used to approximate second-order stationary time series. Using the assumptions about the random effects and applying the formula $\cos(a - b) = \cos(a)\cos(b) + \sin(a)\sin(b)$ to $\cos(2\pi \frac{j}{n} k) = \cos(-2\pi \frac{j}{n} k) = \cos(\{2\pi(j/n)(t+k)\} - \{2\pi(j/n)t\})$, we get

$$\sigma(k) = \mathrm{Cov}(z_t, z_{t+k}) = \mathrm{E}(z_t z_{t+k}) \tag{3}$$

$$= \sum_{j=0}^{\frac{n-1}{2}} \sigma_j^2 \cos\left(2\pi \frac{j}{n} t \right) \cos\left(2\pi \frac{j}{n} (t+k) \right)$$

$$+ \sum_{j=0}^{\frac{n-1}{2}} \sigma_j^2 \sin\left(2\pi \frac{j}{n} t \right) \sin\left(2\pi \frac{j}{n} (t+k) \right)$$

$$= \sum_{j=0}^{\frac{n-1}{2}} \sigma_j^2 \cos\left(2\pi \frac{j}{n} k \right).$$

Noticing that $\cos(2\pi(j/n)k) = \cos(-2\pi(j/n)k)$, the spectral distribution function F must be a step function that is zero at $v = -\frac{1}{2}$, has a jump of σ_0^2 at $v = 0$, and jumps of $\sigma_j^2/2$ at $v = \pm j/2n$. (Computing (2) for this function F is just like computing the "expected value" of $\cos(2\pi vk)$ where v is a random variable that takes on the values $\pm j/2n$ with "probability" $\sigma_j^2/2$ and the value 0 with "probability" σ_0^2. The only difference is that the "probabilities" do not add up to one.)

This random effects process is used to approximate an arbitrary time series process. The fact that it makes a good approximation to the series does not imply that it makes a realistic model for the time series. Realistic models for times series should probably have continuous spectral distributions. This process has a discrete spectral distribution. It may be reasonable to approximate a continuous distribution with a discrete distribution; it is less reasonable to *model* a continuous distribution as a discrete distribution. Most realistic models for time series do not "remember" forever what has happened in the past. More technically, they have a covariance function $\sigma(k)$ that converges to zero as $k \to \infty$. From equation (3), that does not occur with the approximation process. Nonetheless, the approximation process provides a good tool for introducing basic concepts of frequency domain analysis to people with a background in linear models.

A second-order stationary process y_t has a spectral distribution function F defined by (2). Typically F will not be a discrete distribution. The important point is that the distribution F determines the spectral distribution of the approximation process z_t and thus the variance components of the approximation model. In particular, the Spectral Representation Theorem implies that the approximation in (1) has

$$
\begin{aligned}
\sigma_0^2 &= F\left(\frac{1}{n}-\right) - F\left(-\frac{1}{n}\right) \\
\sigma_k^2 &= \left[F\left(\frac{k+1}{n}-\right) - F\left(\frac{k}{n}-\right)\right] + \left[F\left(-\frac{k}{n}\right) - F\left(-\frac{k+1}{n}\right)\right] \quad (4) \\
&= 2\left[F\left(\frac{k+1}{n}-\right) - F\left(\frac{k}{n}-\right)\right].
\end{aligned}
$$

Much of standard frequency domain time series relates to the spectral density function. Assuming that

$$
\sum_{k=0}^{\infty} |\sigma(k)| < \infty,
$$

the spectral distribution function $F(v)$ has a derivative $f(v)$ and

$$
\sigma(k) = \int_{-1/2}^{1/2} e^{2\pi ivk} f(v)dv \quad (5)
$$

$$= \int_{-1/2}^{1/2} \cos(2\pi v k) f(v)\, dv\,.$$

The function $f(v)$ is called the *spectral density*. From (4) we see that for $k \neq 0$

$$\sigma_k^2 \doteq 2f\left(\frac{k}{n}\right)\left[\frac{k+1}{n} - \frac{k}{n}\right]$$

and

$$\frac{n}{2}\sigma_k^2 \doteq f\left(\frac{k}{n}\right)\,. \tag{6}$$

We have assumed that the process has gone on in the infinite past. Thus, by second-order stationarity the random variables $\ldots, y_{-2}, y_{-1}, y_0, y_1, y_2, \ldots$ have the property that

$$\sigma(k) = \sigma(-k)\,.$$

Equation (5) leads to a well-known inverse relation

$$f(v) \;=\; \sum_{k=-\infty}^{\infty} \sigma(k) e^{-2\pi i v k} \tag{7}$$

$$\;=\; \sum_{k=-\infty}^{\infty} \sigma(k) \cos(2\pi v k)\,.$$

The last equality follows from the symmetry of $\sigma(\cdot)$.

White noise is a term used to describe one of the simplest yet most important examples of a second-order stationary process. A process e_t is said to be *white noise* if

$$E(e_t) = 0$$

and

$$\sigma(k) = \begin{cases} \sigma^2 & \text{if } k = 0 \\ 0 & \text{if } k \neq 0 \end{cases}\,.$$

Such processes are of particular importance in defining time domain models in Chapter V. Note that random sampling (with replacement) from any population with mean zero and finite variance generates white noise.

EXERCISE 4.1.   Let e_t be a white noise process. Show that the spectral density of e_t is

$$f_e(v) = \sigma^2\,.$$

IV.4   The Random Effects Model

Consider now the random effects version of model (4.2.3). In all its gory detail, the model is

$$Y = J\alpha_0 + \sum_{k=1}^{\left[\frac{n-1}{2}\right]} Z_k \gamma_k + \delta_{\left[\frac{n}{2}\right]\frac{n}{2}} C_{\frac{n}{2}} \alpha_{\frac{n}{2}} , \tag{1}$$

$$\begin{aligned}
\mathrm{Cov}(\gamma_k) &= \sigma_k^2 I_2 , \\
\mathrm{Var}(\alpha_{\frac{n}{2}}) &= \sigma_{\frac{n}{2}}^2 , \\
\mathrm{Cov}(\gamma_k, \gamma_{k'}) &= 0 \qquad k \neq k' , \\
\mathrm{Cov}(\gamma_k, \alpha_{\frac{n}{2}}) &= 0 .
\end{aligned}$$

Let $\mathrm{Cov}(Y) = V$, then

$$V = \sum_{j=1}^{\left[\frac{n-1}{2}\right]} \sigma_j^2 Z_j Z_j' + \delta_{\left[\frac{n}{2}\right]\frac{n}{2}} \sigma_{\frac{n}{2}}^2 C_{\frac{n}{2}} C_{\frac{n}{2}}' .$$

Model (4.2.3) can be rewritten as

$$Y = J\alpha_0 + \xi , \ \mathrm{E}(\xi) = 0 , \ \mathrm{Cov}(\xi) = V .$$

This random effects model is exactly correct for observations generated by the approximation process z_t of Section IV.3. It is approximately correct for other second-order stationary processes. As discussed in Example 4.3.1, the process z_t based on sines and cosines approximates the behavior of the true process but in broader terms may not be a very realistic model for the true process. In this section, we rely only on the quality of the approximation. We derive results assuming that model (1) is correct and mention the appropriate interpretation when model (1) is only an approximation.

For convenience, we will take n odd eliminating the need to bother with terms involving the subscript $n/2$. The case with n even is similar. Let P_k be the perpendicular projection operator onto the column space $C(C_k, S_k)$. Note that by the choice of the C_j's and S_j's,

$$P_k = \frac{2}{n}[C_k C_k' + S_k S_k'] = \frac{2}{n} Z_k Z_k' .$$

The periodogram is

$$P(k/n) = Y' P_k Y / 2 .$$

Proposition 4.4.1.

$$E\left[P(k/n)\right] = \frac{n}{2}\sigma_k^2 \doteq f\left(\frac{k}{n}\right) .$$

PROOF.   By Theorem 1.3.2 in Christensen (1987),

$$
\begin{aligned}
\mathrm{E}(Y'P_kY) &= \mathrm{tr}[P_kV] \\
&= \sum_j \sigma_j^2 \mathrm{tr}[P_kZ_jZ_j'] \\
&= \sigma_k^2 \mathrm{tr}[P_kZ_kZ_k'] \\
&= \sigma_k^2 \mathrm{tr}[Z_k'P_kZ_k] \\
&= \sigma_k^2 \mathrm{tr}[C_k'C_k + S_k'S_k] \\
&= \sigma_k^2 \left[\frac{n}{2} + \frac{n}{2}\right] \\
&= n\sigma_k^2 \,.
\end{aligned}
$$

Dividing by 2 and recalling (4.3.6) gives the result. $\square$

Proposition 4.4.2.   If the data have a multivariate normal distribution and model (1) is correct, then

$$
\frac{2P\left(\frac{k}{n}\right)}{\frac{n}{2}\sigma_k^2} \sim \chi^2(2) \,.
$$

PROOF.   This follows from checking the conditions of Christensen's (1987) Theorem 1.3.6. Note that

$$
2P\left(\frac{k}{n}\right) \bigg/ \frac{n}{2}\sigma_k^2 = Y'P_kY \bigg/ \frac{n}{2}\sigma_k^2 = Y'\left[\frac{2}{n\sigma_k^2}P_k\right]Y \,.
$$

Because the only fixed effect is $J\alpha_0$, and because $P_kJ = 0$, it suffices to show that $V\frac{2}{n\sigma_k^2}P_kV\frac{2}{n\sigma_k^2}P_kV = V\frac{2}{n\sigma_k^2}P_kV$. In fact, it suffices to show that

$$
\left(\frac{2}{n\sigma_k^2}\right)^2 P_kVP_kV = \left(\frac{2}{n\sigma_k^2}\right)P_kV.
$$

Note that

$$
\begin{aligned}
P_kV &= P_k\sum_j \sigma_j^2 Z_jZ_j' \\
&= \sigma_k^2 P_k[Z_kZ_k'] \\
&= \sigma_k^2 P_k\left[\frac{n}{2}P_k\right] \\
&= \frac{n}{2}\sigma_k^2 P_k \,.
\end{aligned}
$$

Clearly, $\left(\frac{2}{n\sigma_k^2}\right)^2 P_kVP_kV = \left(\frac{2}{n\sigma_k^2}\right)P_kV.$ $\square$

Of course, in practice, model (1) will not be true. However, if the y_t process is a stationary Gaussian process and n is large, then by (4.3.1), model (1) is approximately correct, so $P\left(\frac{k}{n}\right) \big/ \frac{n}{2}\sigma_k^2$ is approximately $\chi^2(2)$.

We can now derive confidence intervals for the $\frac{n}{2}\sigma_k^2$'s. A $(1-\alpha)100\%$ confidence interval for $\frac{n}{2}\sigma_k^2$ is based on

$$
\begin{aligned}
1 - \alpha &= \Pr\left[\chi^2\left(\frac{\alpha}{2},2\right) \le \frac{2P\left(\frac{k}{n}\right)}{\frac{n}{2}\sigma_k^2} \le \chi^2\left(1-\frac{\alpha}{2},2\right)\right] \\[2mm]
&= \Pr\left[\frac{2P\left(\frac{k}{n}\right)}{\chi^2\left(1-\frac{\alpha}{2},2\right)} \le \frac{n}{2}\sigma_k^2 \le \frac{2P\left(\frac{k}{n}\right)}{\chi\left(\frac{\alpha}{2},2\right)}\right].
\end{aligned}
$$

The confidence interval is

$$
\left(\frac{2P\left(\frac{k}{n}\right)}{\chi^2\left(1-\frac{\alpha}{2},2\right)}, \; \frac{2P\left(\frac{k}{n}\right)}{\chi^2\left(\frac{\alpha}{2},2\right)}\right).
$$

From (4.3.6), this is also an approximate confidence interval for $f\left(\frac{k}{n}\right)$.

Note that as n increases, the periodogram estimates the spectral density at more points, but the quality of the individual estimates does not improve. We are observing one realization of the process. To get improved periodogram estimates of the spectral density requires independent replication of the *series*, not additional observations on the realization at hand.

The spectral density estimate $\hat{f}\left(\frac{k}{n}\right)$ was defined in (4.2.7). Arguments similar to those just given establish that under model (1)

$$
E\left[\hat{f}_r\left(\frac{k}{n}\right)\right] = \frac{1}{r}\sum_{i=-(r-1)/2}^{(r-1)/2} \frac{n}{2}\sigma_{k+i}^2 \tag{2}
$$

$$
\doteq \frac{1}{r}\sum_{i=-(r-1)/2}^{(r-1)/2} f\left(\frac{k+i}{n}\right),
$$

which is the average of $f(\nu)$ in the r neighborhood of $\frac{k}{n}$. If $f(\nu)$ is continuous and n is large, all of these values should be similar so

$$
E\left[\hat{f}_r\left(\frac{k}{n}\right)\right] \doteq f\left(\frac{k}{n}\right).
$$

In fact, if we let r be an increasing function of n, under reasonable conditions on r and the process, it is possible to achieve consistent estimation of the spectral density from only one realization of the process.

If all the σ_{k+i}^2's are *equal* in the r neighborhood,

$$
\frac{2r\hat{f}_r\left(\frac{k}{n}\right)}{\frac{n}{2}\sigma_k^2} \sim \chi^2(2r). \tag{3}
$$

Again, if $f(\nu)$ is continuous and n is large, the distribution should hold approximately. This yields a confidence interval for $\frac{n}{2}\sigma_k^2 \doteq f(k/n)$ of

$$\left(\frac{2r\hat{f}_r\left(\frac{k}{n}\right)}{\chi^2\left(1-\frac{\alpha}{2},2r\right)}, \frac{2r\hat{f}_r\left(\frac{k}{n}\right)}{\chi^2\left(\frac{\alpha}{2},2r\right)}\right).$$

It is often convenient to have the length of the confidence intervals independent of the frequency. This can be accomplished by reporting the confidence intervals based on $\log \hat{f}_r\left(\frac{k}{n}\right)$ rather than $\hat{f}_r\left(\frac{k}{n}\right)$. The confidence interval for $\log\left(\frac{n}{2}\sigma_k^2\right)$ is

$$\left(\log \hat{f}_r\left(\frac{k}{n}\right) - \log\left[\frac{\chi^2\left(1-\frac{\alpha}{2},2r\right)}{2r}\right], \log \hat{f}_r\left(\frac{k}{n}\right) - \log\left[\frac{\chi^2\left(\frac{\alpha}{2},2r\right)}{2r}\right]\right).$$

As discussed earlier, $\hat{f}_r\left(\frac{k}{n}\right)$ is the mean square for the r neighborhood of $\frac{k}{n}$ and is a good indicator of the relative importance of frequencies near $\frac{k}{n}$. Confidence intervals allow more rigorous comparisons of the relative importance of the various frequencies.

EXERCISE 4.2.   Assuming that the random effects model (1) is Gaussian, prove that (2) and (3) hold.

IV.5   The Measurement Error Model

It would be nice if we could arrive at some justification for looking at reduced models. Predictions based on saturated models are notoriously poor. Unfortunately, reduced models are not possible in model (4.4.1) because the estimates for all of the $\frac{n}{2}\sigma_k^2$'s will be positive and there is no reason to conclude that any can be zero. In using model (4.4.1) we have overlooked the nearly ubiquitous fact that measurements on the same unit differ. It seems reasonable to model the observations as the sum of a simple stationary process and individual uncorrelated measurement errors. (Although the measurement errors can be incorporated into (4.4.1), it is convenient to isolate them.) This suggests the model

$$Y = J\alpha_0 + \sum_{k=1}^{\left[\frac{n-1}{2}\right]} Z_k\gamma_k + \delta_{\left[\frac{n}{2}\right]\frac{n}{2}}C_{\frac{n}{2}}\alpha_{\frac{n}{2}} + e \tag{1}$$

$$\begin{aligned} \mathrm{Cov}(e) &= \sigma^2 I_n \\ \mathrm{Cov}(\gamma_k) &= \sigma_k^2 I_2 \\ \mathrm{Var}(\alpha_{\frac{n}{2}}) &= \sigma_{\frac{n}{2}}^2 \end{aligned}$$

$$
\begin{aligned}
\mathrm{Cov}(e, \gamma_k) &= 0 \\
\mathrm{Cov}(e, \alpha_{\frac{n}{2}}) &= 0 \\
\mathrm{Cov}(\gamma_k, \gamma_{k'}) &= 0 \qquad k \neq k' \\
\mathrm{Cov}(\gamma_k, \alpha_{\frac{n}{2}}) &= 0.
\end{aligned}
$$

The analysis of this model is similar to that of model (4.4.1). When all variances in the r neighborhood of $\frac{k}{n}$ are equal

$$
\mathrm{E}\left[\hat{f}_r\left(\frac{k}{n}\right)\right] = \sigma^2 + \frac{n}{2}\sigma_k^2 \tag{2}
$$

and

$$
\frac{2r\,\hat{f}_r\left(\frac{k}{n}\right)}{\sigma^2 + \frac{n}{2}\sigma_k^2} \sim \chi^2(2r). \tag{3}
$$

This leads to confidence intervals for the values

$$
\sigma^2 + \frac{n}{2}\sigma_k^2 .
$$

If all of the σ_k^2's are zero, the various confidence intervals are estimates of the same thing, σ^2. Confidence intervals containing distinctly larger values suggest the existence of a nonzero variance σ_k^2.

Another way of identifying important frequencies is to modify the approach used in Christensen (1987, Example 13.2.4) for identifying important effects in saturated models for designed experiments. Let $s = \left[\frac{n-1}{2}\right]$ and let $w_1, \ldots, w_s$ be i.i.d. $\chi^2(2r)$. Construct the order statistics $w_{(1)} \leq \cdots \leq w_{(s)}$, and compute the expected order statistics $\mathrm{E}[w_{(1)}], \ldots, \mathrm{E}[w_{(s)}]$. If all the σ_k^2's are zero, $\hat{f}_r\left(\frac{1}{n}\right), \hat{f}_r\left(\frac{2}{n}\right), \ldots, \hat{f}_r\left(\frac{s}{n}\right)$ are each σ^2 times a $\chi^2(2r)$ random variable and the plot of $\left(\mathrm{E}[w_{(k)}], \hat{f}_r\left(\frac{k}{n}\right)\right)$ should form an approximate straight line. In any case, the unimportant frequencies should form a straight line. Values $\hat{f}_r\left(\frac{k}{n}\right)$ that are so large as to be inconsistent with the straight line indicate important frequencies.

There is one obvious problem with this method. The w_k's were assumed to be independent, but the $\hat{f}_r\left(\frac{k}{n}\right)$'s are only independent when $r = 1$. If $r \geq 3$ (recall that r is odd), then the neighborhoods overlap so, for example, $\hat{f}_r\left(\frac{k}{n}\right)$ and $\hat{f}_r\left(\frac{k-1}{n}\right)$ both involve $P\left(\frac{k}{n}\right)$ and $P\left(\frac{k-1}{n}\right)$ so the $\hat{f}_r\left(\frac{k}{n}\right)$'s are not independent. It is interesting to recall that normal plots are also plagued by a correlation problem when the residuals are used to check for normality. However, the order of magnitude of the correlation problem is very different in the two cases, cf. Example 4.5.1.

EXERCISE 4.3.   Assuming that model (1) is Gaussian with, for convenience, n odd, prove that (2) and (3) hold.

If a group of s frequencies, say $k_1/n, \ldots, k_s/n$ have been identified as important, the random effects model

$$Y = J\alpha_0 + \sum_{i=1}^{s} Z_{k_i}\gamma_{k_i} + e$$

together with the best linear unbiased prediction methods of Christensen (1987, Section XII.2) can be used to obtain predictions of both current and future observations. As always, our approach here has been based on linear models. Brockwell and Davis (1987, Section 5.6) discuss a prediction method for the frequency domain that is founded on Fourier analysis.

EXAMPLE 4.5.1.   Consider again the coal production data of Example 4.2.1. Tables IV.3, IV.4 and IV.5 give 95% confidence intervals for the spectral density based on $P(k)$, $\hat{f}_3$, and $\hat{f}_5$ respectively. The limits for $\hat{f}_3$, and $\hat{f}_5$ are displayed graphically in Figures IV.3 and IV.4 as horizontal bars. A similar graphical display is present in Figure IV.5 for the detrended version of $\hat{f}_5$. The intervals for $k/n = \nu = 0$ are constructed using half of the usual degrees of freedom. The confidence intervals confirm the impression that there are three levels to $f(\nu)$: an area of low frequencies with large contributions, an area of moderate frequencies with moderate contributions, and an area of high frequencies that account for little of the variability in the data.

Figures IV.6, IV.7 and IV.8 contain various chi-square probability plots for the values of $\hat{f}_5(\nu)$. The chi-square scores are $G^{-1}(i/(n+1))$ where $G(\cdot)$ is the cumulative distribution function for the $\chi^2(10)$ distribution. The value $\hat{f}_5(0)$ was dropped because the $\chi^2(10)$ distribution is not appropriate for it. Figure IV.6 contains the plot for the 31 frequencies other than $\nu = 0$. The four largest $\hat{f}_5$ values stand out as distinct from the rest. The frequencies associated with these $\hat{f}_5$ values are $1/63 = .0159$, $3/63$, $2/63$, and $4/63$, respectively. The fifth to eighth largest $\hat{f}_5$ values are also suspiciously large. To investigate this next set of frequencies, the four that are clearly disparate were dropped and a chi-square plot for the remaining 27 frequencies was created. This plot appears as Figure IV.7. Certainly the two largest order statistics are inconsistent with the line. The third largest $\hat{f}_5$ value also appears to be inconsistent. These three values correspond to the frequencies $5/63$, $6/63$, and $7/63$, respectively. Figure IV.8 contains the chi-square plot obtained by deleting the seven largest of the 31 $\hat{f}_5$ values. This plot is reasonably linear. The main problem is that it contains a group of six frequencies that are disturbingly low.

Figure IV.6 involves 31 $\hat{f}_5$ values, each with 10 degrees of freedom. Supposedly there are 310 degrees of freedom involved in the plot but there are only 61 observations. The extra degrees of freedom are generated by the averaging involved in computing $\hat{f}_5$. It is no accident that the seven frequencies identified form a contiguous group. This correlation problem

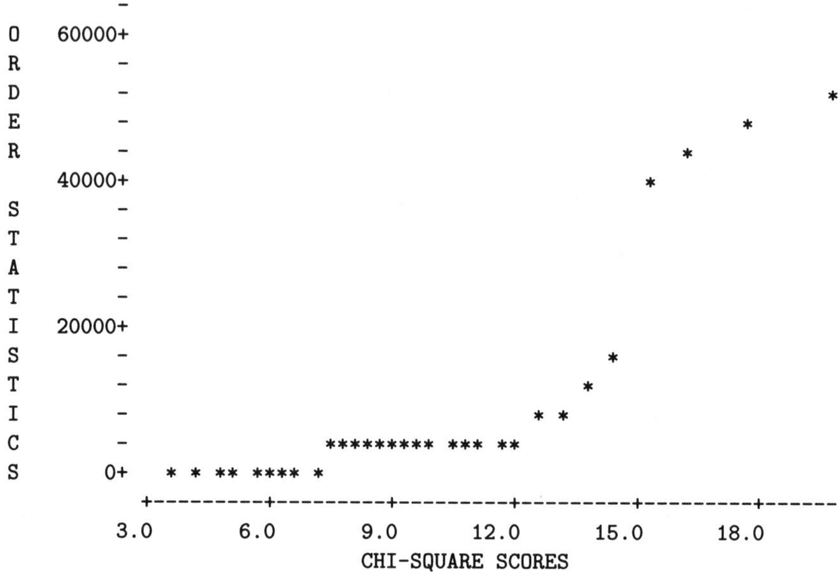

FIGURE IV.6. $\chi^2(10)$ Plot for 31 $\hat{f}_5$ Observations

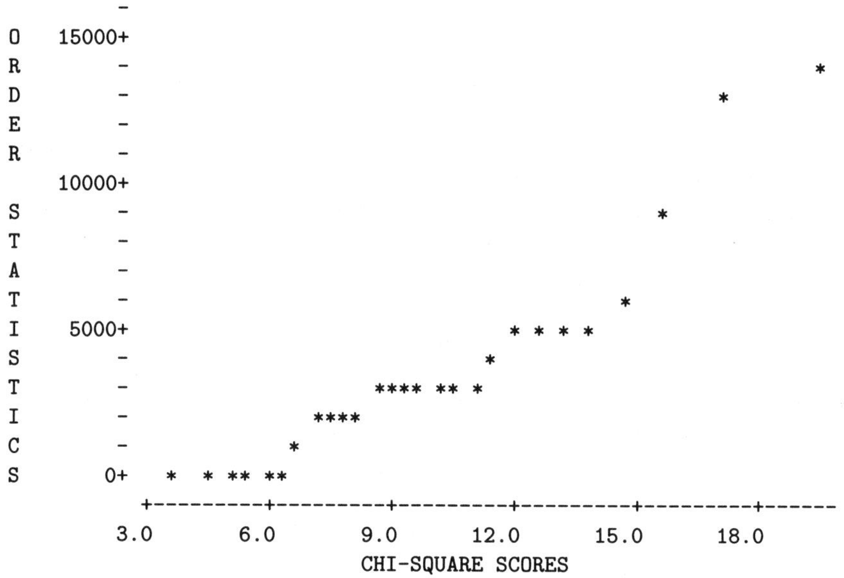

FIGURE IV.7. $\chi^2(10)$ Plot for 27 Smallest $\hat{f}_5$ Observations

TABLE IV.3. Confidence Intervals Based on the
Periodogram

ν	$P(k)$	Lower Limit	Upper Limit
0.0159	0.612E+05	0.166E+05	0.242E+07
0.0317	0.137E+06	0.371E+05	0.541E+07
0.0476	0.878E+04	0.238E+04	0.347E+06
0.0635	0.216E+05	0.586E+04	0.854E+06
0.0794	0.183E+05	0.497E+04	0.724E+06
0.0952	0.181E+05	0.490E+04	0.714E+06
0.1111	0.373E+04	0.101E+04	0.147E+06
0.1270	0.117E+04	0.318E+03	0.464E+05
0.1429	0.541E+04	0.147E+04	0.214E+06
0.1587	0.371E+04	0.101E+04	0.147E+06
0.1746	0.164E+04	0.444E+03	0.647E+05
0.1905	0.156E+04	0.423E+03	0.616E+05
0.2063	0.262E+04	0.712E+03	0.104E+06
0.2222	0.376E+04	0.102E+04	0.148E+06
0.2381	0.340E+03	0.921E+02	0.134E+05
0.2540	0.432E+04	0.117E+04	0.170E+06
0.2698	0.255E+02	0.690E+01	0.101E+04
0.2857	0.667E+04	0.181E+04	0.264E+06
0.3016	0.890E+04	0.241E+04	0.352E+06
0.3175	0.403E+04	0.109E+04	0.159E+06
0.3333	0.306E+04	0.829E+03	0.121E+06
0.3492	0.247E+04	0.669E+03	0.974E+05
0.3651	0.516E+04	0.140E+04	0.204E+06
0.3810	0.421E+02	0.114E+02	0.166E+04
0.3968	0.141E+04	0.383E+03	0.558E+05
0.4127	0.390E+03	0.106E+03	0.154E+05
0.4286	0.521E+02	0.141E+02	0.206E+04
0.4444	0.400E+03	0.108E+03	0.158E+05
0.4603	0.806E+02	0.218E+02	0.318E+04
0.4762	0.426E+03	0.116E+03	0.168E+05
0.4921	0.239E+03	0.647E+02	0.943E+04

can be eliminated by plotting the periodogram $\hat{f}_1$. Well, almost. A plot
corresponding to Figure IV.6 for the periodogram still has 62 rather than
61 degrees of freedom because the data were augmented with two artificial
observations but the frequency $k/n = 0$ with one degree of freedom was
dropped.

A series of periodogram chi-square plots isolated the important frequencies 2/63, 1/63, 4/63, 5/63, and 6/63 with the $\chi^2(2)$ plot for the remaining
26 frequencies given in Figure IV.9. Except for the fact that one of the
two largest observations seems strange (it isn't clear which), Figure IV.9 is
quite linear.

We now present a semi-quick, semi-dirty job of predicting the future of
the series based on the five frequencies that have been identified as im-

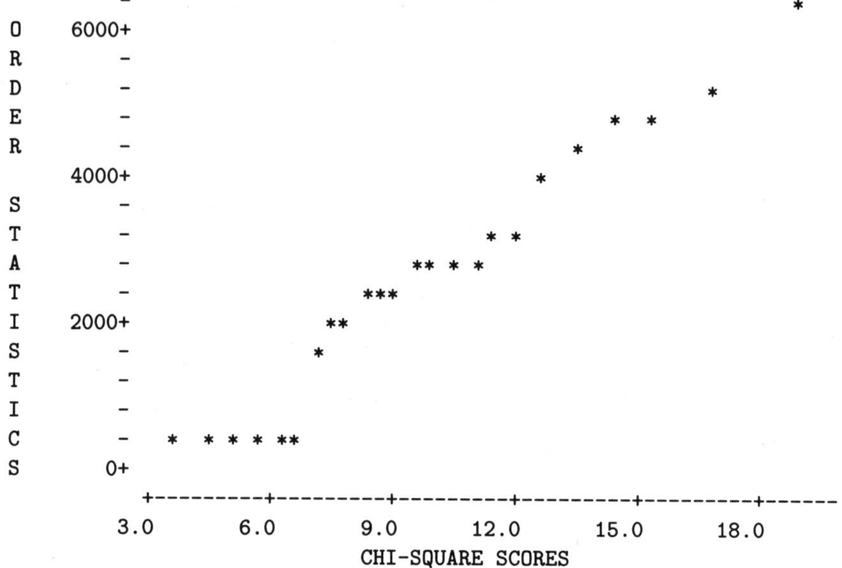

FIGURE IV.8. $\chi^2(10)$ Plot for 24 Smallest $\hat{f}_5$ Observations

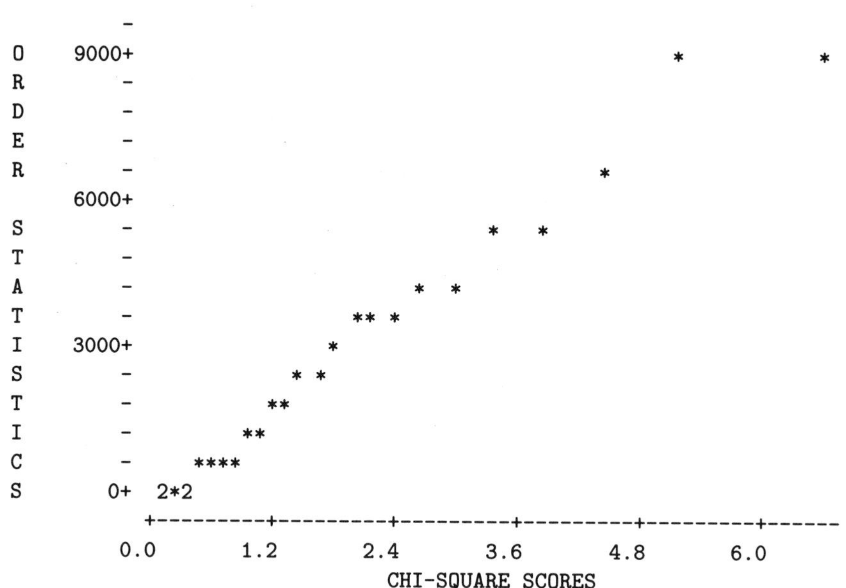

FIGURE IV.9. $\chi^2(2)$ Plot for 26 Smallest $P(k)$ Observations

TABLE IV.4. Confidence Intervals Based on $\hat{f}_3$

ν	$\hat{f}_3(\nu)$	Lower Limit	Upper Limit
0.0000	0.408E+05	0.131E+05	0.567E+06
0.0159	0.660E+05	0.274E+05	0.320E+06
0.0317	0.689E+05	0.286E+05	0.334E+06
0.0476	0.558E+05	0.232E+05	0.270E+06
0.0635	0.162E+05	0.674E+04	0.788E+05
0.0794	0.193E+05	0.803E+04	0.938E+05
0.0952	0.134E+05	0.556E+04	0.649E+05
0.1111	0.766E+04	0.318E+04	0.372E+05
0.1270	0.344E+04	0.143E+04	0.167E+05
0.1429	0.343E+04	0.143E+04	0.166E+05
0.1587	0.359E+04	0.149E+04	0.174E+05
0.1746	0.230E+04	0.956E+03	0.112E+05
0.1905	0.194E+04	0.806E+03	0.941E+04
0.2063	0.265E+04	0.110E+04	0.128E+05
0.2222	0.224E+04	0.931E+03	0.109E+05
0.2381	0.281E+04	0.116E+04	0.136E+05
0.2540	0.156E+04	0.648E+03	0.757E+04
0.2698	0.367E+04	0.152E+04	0.178E+05
0.2857	0.520E+04	0.216E+04	0.252E+05
0.3016	0.653E+04	0.271E+04	0.317E+05
0.3175	0.533E+04	0.221E+04	0.258E+05
0.3333	0.318E+04	0.132E+04	0.154E+05
0.3492	0.356E+04	0.148E+04	0.173E+05
0.3651	0.256E+04	0.106E+04	0.124E+05
0.3810	0.221E+04	0.916E+03	0.107E+05
0.3968	0.615E+03	0.256E+03	0.298E+04
0.4127	0.619E+03	0.257E+03	0.300E+04
0.4286	0.281E+03	0.116E+03	0.136E+04
0.4444	0.177E+03	0.737E+02	0.860E+03
0.4603	0.302E+03	0.125E+03	0.147E+04
0.4762	0.249E+03	0.103E+03	0.121E+04
0.4921	0.301E+03	0.125E+03	0.146E+04

portant. It won't really be quick and it certainly won't be clean. Before proceeding, a word of warning. We are modeling the data with cyclic functions. By inspection of Figure IV.1, the data appear to be overdue for a downturn in 1980. Our model comes to the same conclusion. Unfortunately, the actual data from 1981 to 1987 are unwilling to cooperate with our model.

The main problem in this analysis (other than the data themselves) is that the analysis presented thus far involves two artificial observations. The frequencies identified are based on $k/63$ rather than $k/61$. We begin with a reduced version of the measurement error random effects model (1),

$$Y = J\mu + Z_1\gamma_1 + Z_2\gamma_2 + Z_4\gamma_4 + Z_5\gamma_5 + Z_6\gamma_6 + e, \tag{4}$$

TABLE IV.5. 95% Confidence Intervals Based on $\hat{f}_5$

ν	$\hat{f}_5(\nu)$	Lower Limit	Upper Limit
0.0000	0.792E+05	0.309E+05	0.477E+06
0.0159	0.536E+05	0.262E+05	0.165E+06
0.0317	0.457E+05	0.223E+05	0.141E+06
0.0476	0.494E+05	0.241E+05	0.152E+06
0.0635	0.407E+05	0.199E+05	0.125E+06
0.0794	0.141E+05	0.689E+04	0.435E+05
0.0952	0.126E+05	0.615E+04	0.388E+05
0.1111	0.935E+04	0.456E+04	0.288E+05
0.1270	0.642E+04	0.314E+04	0.198E+05
0.1429	0.313E+04	0.153E+04	0.965E+04
0.1587	0.270E+04	0.132E+04	0.831E+04
0.1746	0.299E+04	0.146E+04	0.921E+04
0.1905	0.266E+04	0.130E+04	0.819E+04
0.2063	0.198E+04	0.969E+03	0.611E+04
0.2222	0.252E+04	0.123E+04	0.776E+04
0.2381	0.221E+04	0.108E+04	0.682E+04
0.2540	0.302E+04	0.148E+04	0.931E+04
0.2698	0.405E+04	0.198E+04	0.125E+05
0.2857	0.479E+04	0.234E+04	0.148E+05
0.3016	0.454E+04	0.222E+04	0.140E+05
0.3175	0.503E+04	0.245E+04	0.155E+05
0.3333	0.472E+04	0.231E+04	0.145E+05
0.3492	0.295E+04	0.144E+04	0.909E+04
0.3651	0.243E+04	0.119E+04	0.748E+04
0.3810	0.190E+04	0.925E+03	0.584E+04
0.3968	0.141E+04	0.690E+03	0.435E+04
0.4127	0.460E+03	0.224E+03	0.142E+04
0.4286	0.467E+03	0.228E+03	0.144E+04
0.4444	0.270E+03	0.132E+03	0.831E+03
0.4603	0.239E+03	0.117E+03	0.737E+03
0.4762	0.277E+03	0.135E+03	0.853E+03
0.4921	0.282E+03	0.138E+03	0.869E+03

$$\begin{aligned} \mathrm{E}(e) &= 0, \\ \mathrm{Cov}(e) &= \sigma^2 I \end{aligned}$$

where the rows of Z_k are

$$z'_{kt} = \left(\cos(2\pi \frac{k}{63} t), \sin(2\pi \frac{k}{63} t) \right)$$

$t = 1, \ldots, 61$. The cosines and sines are set up for a data set with 63 observations but we only have 61. Our analysis will not include the artificial observations so Y, J and every Z_k are defined with only 61 rows. A key reason for using the frequencies $k/63$ in the cosines and sines is because the vectors they determine are orthogonal in 63-dimensional Euclidean space. With only 61 rows, the vectors are not quite orthogonal.

Now to start getting dirty. After examining various plots and confidence intervals, not all of which have been presented here, it was decided to incorporate the following assumptions about the variance components.

$$\text{Cov}(\gamma_1) = \text{Cov}(\gamma_2) = \sigma_a^2 I$$

and

$$\text{Cov}(\gamma_4) = \text{Cov}(\gamma_5) = \text{Cov}(\gamma_6) = \sigma_b^2 I.$$

The disadvantage of these assumptions is that they may not be true. The advantage is that, if they approximate the truth, we should get better estimates of the variance components.

If we really had 63 observations, we could estimate the variance components directly from the periodogram values of Table IV.3. The measurement error variance σ^2 could be estimated by averaging all the periodogram values other than those for the five important frequencies. The average of the periodogram values for $k = 1, 2$ provides an estimate of $\sigma^2 + \frac{63}{2}\sigma_a^2$, so an estimate of σ_a^2 is

$$\hat{\sigma}_a^2 = \left(\frac{P(1) + P(2)}{2} - \hat{\sigma}^2\right)\left(\frac{2}{63}\right).$$

A similar result holds for estimating σ_b^2 using the average of $P(4)$, $P(5)$, and $P(6)$. All of this is a direct application of Henderson's method 3, see Christensen (1987, Section XII.9.) Generally, Henderson's method does not provide unique estimates of variance components; they depend on the order in which the components are estimated. However, due to the orthogonality of the columns involved, the estimates based on the periodogram are unique. With 61 instead of 63 observations the orthogonality breaks down. We will ignore this problem. It should be minor.

Fitting model (4) by least squares gives MSE $= 2327$. The mean square for frequencies $4/63$, $5/63$, and $6/63$ after fitting $1/63$ and $2/63$ is $MS(Z_4, Z_5, Z_6|Z_1, Z_2, J) = 18{,}546.5$ and the mean square for frequencies $1/63$ and $2/63$ ignoring the others is $MS(Z_1, Z_2|J) = 106{,}353.75$. The variance component estimates are

$$\hat{\sigma}_b^2 = (18{,}546.5 - 2327)\left(\frac{2}{61}\right) = 531.8$$

and

$$\hat{\sigma}_a^2 = (106{,}353.75 - 2327)\left(\frac{2}{61}\right) = 3411.$$

The multiplier $2/61$ is only an approximation to the appropriate multiplier which is found as in Christensen (1987, Section XII.9). (Alternatively, see the proof of Proposition 4.4.1.) The formula given above for $\hat{\sigma}_a^2$ is based on a not quite valid assumption of orthogonality. (I warned you that this was going to be dirty.)

We are now in a position to estimate the best linear unbiased predictors. We have to estimate the predictors because we do not know the variance components. Prediction is performed as in Christensen (1987, Section XII.2) and almost exactly as in his Example 12.2.6; see also Sections III.1 and VI.2. It is easily seen that the best linear unbiased predictor (BLUP) for an observation at time t is

$$\hat{y} = \hat{\mu} + x_0' \hat{\gamma}$$

where

$$\hat{\mu} = \bar{y}. = 511.8\,,$$

$$x_0 = \begin{bmatrix} \cos\left(2\pi(\frac{1}{63})t\right) \\ \sin\left(2\pi(\frac{1}{63})t\right) \\ \cos\left(2\pi(\frac{2}{63})t\right) \\ \sin\left(2\pi(\frac{2}{63})t\right) \\ \cos\left(2\pi(\frac{4}{63})t\right) \\ \sin\left(2\pi(\frac{4}{63})t\right) \\ \cos\left(2\pi(\frac{5}{63})t\right) \\ \sin\left(2\pi(\frac{5}{63})t\right) \\ \cos\left(2\pi(\frac{6}{63})t\right) \\ \sin\left(2\pi(\frac{6}{63})t\right) \end{bmatrix}$$

and $\hat{\gamma}$ is the BLUP of

$$\gamma = \begin{bmatrix} \gamma_1 \\ \gamma_2 \\ \gamma_4 \\ \gamma_5 \\ \gamma_6 \end{bmatrix}.$$

In finding $\hat{\gamma}$, the covariance matrix of the data is estimated as

$$2327\, I + 3410\, (Z_1 Z_1' + Z_2 Z_2') + 531.8\, (Z_4 Z_4' + Z_5 Z_5' + Z_6 Z_6')$$

and $\text{Cov}(Y, \gamma)$ is estimated by

$$[3411\, Z_1, 3411\, Z_2, 531.8\, Z_4, 531.8\, Z_5, 531.8\, Z_6]\,.$$

The estimated BLUP of γ is

$$\hat{\gamma} = \mathrm{Cov}(\gamma, Y)\left[\mathrm{Cov}(Y)\right]^{-1}(Y - J\hat{\mu}) = \begin{bmatrix} 52.4 \\ -41.9 \\ 62.4 \\ -73.8 \\ -17.2 \\ -24.0 \\ 5.9 \\ -32.0 \\ -12.3 \\ -25.8 \end{bmatrix}$$

where the estimated covariance matrices are used in place of the unknown true covariance matrices.

Figure IV.10 gives the data for 1920 to 1987 along with the predicted values. As suggested earlier, the model, which was fitted only to the data from 1920 to 1980, turns down after 1980. In point of fact, coal production continued to increase during the 1980s so the model and its predictions are not very good after 1980. Perhaps the data are simply not regular enough to allow predictions based on only 61 observations. Or perhaps the process generating the data changed around 1960. The nature of the plot certainly seems to have changed. Another possibility is that, given the discussion of Exercise 4.3.1, the very act of prediction may involve taking the random effects model more seriously than is appropriate.

For a time domain analysis of these data, see Example 5.6.1.

SUMMARY OF SECTIONS IV.2-IV.5

We began Section IV.2 by showing that the least squares fit of a model based on sines and cosines yields useful information about the frequencies that are important in describing the cyclic behavior of a time series. We then explained that any second-order stationary process can be approximated by a random effects model based on sines and cosines. The variances of the random effects were related to the spectral distribution function F, which is implicitly defined by the covariance function $\sigma(k)$. Estimates of $\frac{n}{2}$ times the variance components were obtained and related to the spectral density function $f(\nu) \equiv dF(\nu)/d\nu$. Confidence intervals and a χ^2 plot were suggested as methods for identifying important frequencies. Best linear unbiased prediction was suggested for obtaining forecasts based on important frequencies.

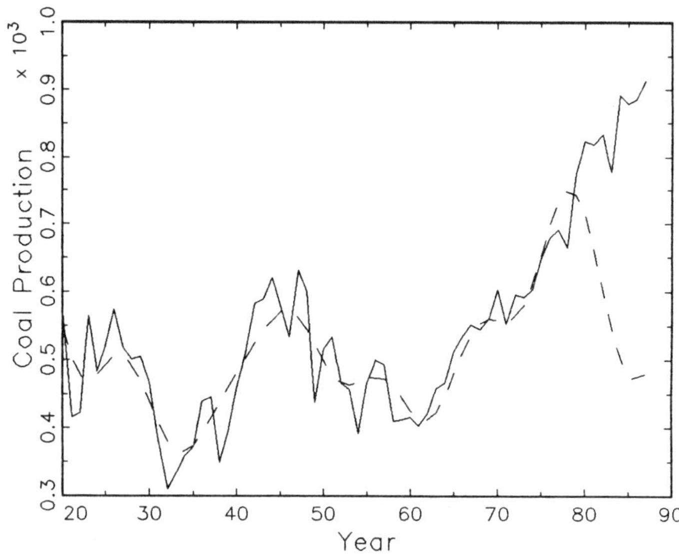

FIGURE IV.10. Coal Production (solid line) and Prediction (dashed line) for 1920 to 1987

IV.6 Linear Filtering

Many traditional ways of dealing with time series consist of computing simple linear functions of the observations to create a new series that is, in some sense, representative of the original series but is also better behaved than the original series. By "better behaved" we mean that the important structure of the series is clarified in the transformed series.

EXAMPLE 4.6.1.   A series that oscillates very quickly and erratically has very high frequencies that contribute substantially to the spectral approximation. (High frequencies are those near $1/2$.) To examine the structure of such a series, it might be wise to try to eliminate the high frequencies while retaining the relative importance of the low and moderate size frequencies. A traditional method of attenuating high frequencies is taking moving averages. For instance, a running average of order 5 (a centered 5-term moving average) is

$$w_t^{(5)} = \frac{1}{5}[y_{t-2} + y_{t-1} + y_t + y_{t+1} + y_{t+2}] .$$

A running average of order 6 is

$$w_t^{(6)} = \frac{1}{6}\left[\frac{1}{2}y_{t-3} + y_{t-2} + y_{t-1} + y_t + y_{t+1} + y_{t+2} + \frac{1}{2}y_{t+3}\right] .$$

EXAMPLE 4.6.2.   Consider a nonstationary process

$$y_t = \gamma_0 + \gamma_1 t + y_t^{(1)}$$

where γ_0 and γ_1 are fixed and $y_t^{(1)}$ is second-order stationary. This is a process with a linear time trend. The difference series

$$w_t = y_t - y_{t-1}$$

is second-order stationary. To see this, observe that

$$\begin{aligned}
E(w_t) &= E(y_i - y_{t-1}) \\
&= \gamma_0 + \gamma_1 t + E\big(y_t^{(1)}\big) \\
&\quad - \gamma_0 - \gamma_1(t-1) - E\big(y_{t-1}^{(1)}\big) \\
&= \gamma_1
\end{aligned}$$

and

$$\begin{aligned}
\text{Cov}(w_t, w_{t+k}) &= \text{Cov}\big(y_t^{(1)} - y_{t-1}^{(1)}, y_{t+k}^{(1)} - y_{t+k-1}^{(1)}\big) \\
&= \text{Cov}\big(y_t^{(1)}, y_{t+k}^{(1)}\big) - \text{Cov}\big(y_t^{(1)}, y_{t+k-1}^{(1)}\big) \\
&\quad - \text{Cov}\big(y_{t-1}^{(1)}, y_{t+k}^{(1)}\big) + \text{Cov}\big(y_{t-1}^{(1)}, y_{t+k-1}^{(1)}\big) \\
&= 2\sigma^{(1)}(k) - \sigma^{(1)}(k-1) - \sigma^{(1)}(k+1)
\end{aligned}$$

where $\sigma^{(1)}(\cdot)$ is the covariance function for $y_t^{(1)}$. Note that the covariance depends only on k. Often, the first difference operator is denoted ∇ so that w_t is

$$\nabla y_t \equiv y_t - y_{t-1}\,.$$

The idea that trends can be thought of as low-frequency cyclic effects was mentioned earlier. In this example, we can think of the frequency generated by $\gamma_0 + \gamma_1 t$ as being so low that the oscillation will never be observed. The first difference process eliminates low frequencies from the spectral approximation.

EXERCISE 4.4.   Consider the process $y_t = \gamma_0 + \gamma_1 t + \gamma_2 t^2 + y_t^{(1)}$ where $y_t^{(1)}$ is second-order stationary. Let $w_t = \nabla y_t$ and $z_t = \nabla w_t$. In other words,

$$\begin{aligned}
z_t &= \nabla(\nabla y_t) \\
&\equiv \nabla^2 y_t\,.
\end{aligned}$$

Show that z_t is second-order stationary.

These examples are special cases of the *general linear filter*

$$w_t = \sum_{s=-\infty}^{\infty} a_s y_{t-s} . \tag{1}$$

The special case

$$w_t = \sum_{s=0}^{\infty} a_s y_{t-s}$$

in which w_t depends only on the current and previous values of y_t will be referred to as a *causal* linear filter. The process y_t is viewed as causing w_t.

EXAMPLE 4.6.1., *continued.*   The process $w_t^{(5)}$ has $a_{-2} = a_{-1} = a_0 = a_1 = a_2 = \frac{1}{5}$ and $a_s = 0$ for all other s. For $w_t^{(6)}$, $a_{-3} = a_3 = \frac{1}{12}$, $a_{-2} = a_{-1} = a_0 = a_1 = a_2 = \frac{1}{6}$, and $a_s = 0$ for all others. Neither filter is causal.

EXAMPLE 4.6.2., *continued.*   The process ∇y_t has $a_0 = 1$, $a_1 = -1$, and $a_s = 0$ for all other s. The filter is causal.

Collectively, the series $\dots, a_{-1}, a_0, a_1, \dots$ is called the *impulse response function*; it is a function from the integers to the reals. If $\sum_{k=-\infty}^{\infty} |a_k| < \infty$, we can draw an analogy between a_k and $\sigma(k)$, thus defining a function similar to the spectral density as given in (4.3.7), say

$$A(\nu) = \sum_{k=-\infty}^{\infty} a_k e^{-2\pi i \nu k} . \tag{2}$$

(Because we need not have $a_k = a_{-k}$, we cannot, in general, reduce $A(\nu)$ to a sum involving only cosines.) The complex valued function $A(\nu)$ is called the *frequency response* function. As will be seen later, $A(\nu)$ identifies the effect that the filter has on different frequencies in the spectral approximation.

The behavior of the spectral approximation (4.3.1) is determined by the spectral density of y_t, say, $f_y(\nu)$. Similarly, the process w_t determined by (1) has a spectral density $f_w(\nu)$. We wish to show that

$$f_w(\nu) = |A(\nu)|^2 f_y(\nu) \tag{3}$$

where $|A(\nu)|^2 = A(\nu)\overline{A(\nu)}$ and $\overline{A(\nu)}$ is the complex conjugate of $A(\nu)$. (The conjugate of the complex number $a + ib$ with a and b real is $\overline{a + ib} = a - ib$.) If, for example,

$$|A(\nu)|^2 = \begin{cases} 1 & v \in [-.4, .4] \\ 0 & \text{otherwise} \end{cases} ,$$

then in the spectral approximation to w_t, the variance components σ_k^2 corresponding to frequencies $\frac{k}{n} \in [-.4, .4]$ are identical to the corresponding variance components in the spectral approximation to y_t. At the same time, the spectral approximation to w_t has $\sigma_k^2 = 0$ for $|\frac{k}{n}| > .4$ regardless of the size of the corresponding variance components in the approximation to y_t. A linear filter with $A(\nu)$ as given above perfectly eliminates high frequencies (greater than .4) while it leaves the contributions of the lower frequencies unchanged. In fact, a function

$$|A(\nu)|^2 = \begin{cases} 7 & v \in [-.4, .4] \\ 0 & \text{otherwise} \end{cases}$$

would be equally effective because it retains the same relative contributions of the frequencies below .4.

A filter that eliminates high frequencies but does little to low frequencies is called a *low-pass* filter. Symmetric moving averages such as those defined in Example 4.6.1 are low-pass filters, cf. Exercise 4.9.7d. Filters that eliminate low frequencies but have little effect on high frequencies are called *high-pass* filters. The first difference filter of Example 4.6.2 is such a filter, cf. Exercise 4.9.7a.

It is by no means clear that an impulse response function (i.e., a sequence $\ldots, a_{-1}, a_0, a_1, \ldots$) exists that generates either of the functions $|A(\nu)|^2$ given above but the examples illustrate the potential usefulness of the frequency response function in interpreting the result of applying a linear filter to a process.

Of course, to discuss $f_w(\nu)$ presupposes that w_t is second-order stationary. If $\sum_{s=-\infty}^{\infty} |a_s| < \infty$,

$$
\begin{aligned}
\mathrm{E}(w_t) &= \sum_{s=-\infty}^{\infty} a_s \mathrm{E}(y_{t-s}) \\
&= \sum_{s=-\infty}^{\infty} a_s \mu \\
&= \mu \sum_{s=-\infty}^{\infty} a_s ,
\end{aligned}
$$

which is a constant. Covariances for the w_t process depend only on k because

$$
\begin{aligned}
\mathrm{Cov}(w_t, w_{t+k}) &= \mathrm{Cov}\left(\sum_s a_s y_{t-s}, \sum_{s'} a_{s'} y_{t+k-s'} \right) \qquad (4) \\
&= \sum_s \sum_{s'} a_s a_{s'} \mathrm{Cov}(y_{t-s}, y_{t+k-s'}) \\
&= \sum_{s=-\infty}^{\infty} \sum_{s'=-\infty}^{\infty} a_s a_{s'} \sigma_y(k - s' + s) \\
&\equiv \sigma_w(k).
\end{aligned}
$$

Thus w_t is second-order stationary.

To establish (3), we use (4) and the representation of $\sigma_y(k)$ from (4.3.5).

$$
\begin{aligned}
\sigma_w(k) &= \sum_s \sum_{s'} a_s a_{s'} \sigma_y(k - s + s') \\
&= \sum_s \sum_{s'} a_s a_{s'} \int_{-1/2}^{1/2} f_y(\nu) e^{2\pi i \nu (k - s + s')} d\nu \\
&= \int_{-1/2}^{1/2} f_y(\nu) e^{2\pi i \nu k} \left[\sum_s \sum_{s'} a_s a_{s'} e^{2\pi i \nu (s' - s)} \right] d\nu \\
&= \int_{-1/2}^{1/2} f_y(\nu) e^{2\pi i \nu k} \sum_s a_s e^{-2\pi i \nu s} \sum_{s'} a_{s'} e^{2\pi i \nu s'} d\nu \qquad (5) \\
&= \int_{-1/2}^{1/2} f_y(\nu) e^{2\pi i \nu k} A(\nu) \overline{A(\nu)} d\nu \\
&= \int_{-1/2}^{1/2} |A(\nu)|^2 f_y(\nu) e^{2\pi i \nu k} d\nu .
\end{aligned}
$$

By the spectral representation theorem, the spectral distribution function and hence the spectral density are unique. Thus, $f_w(\nu)$ is the unique function with

$$
\sigma_w(k) = \int_{-1/2}^{1/2} f_w(\nu) e^{2\pi i \nu k} d\nu .
$$

By uniqueness and (5)

$$
f_w(\nu) = |A(\nu)|^2 f_y(\nu) .
$$

EXAMPLE 4.6.3.   In this example we simply present the results of applying low-pass and high-pass filters to the coal production data of Example 4.2.1. Figure IV.11 displays the results of a low-pass filter. The behavior of the filtered process is extremely regular. Figure IV.12 shows the results of a high-pass filter. The early filtered observations do not show much pattern. The behavior after observation 40 seems a bit more regular. The filters are simply the BMDP defaults and are described in Dixon et al. (1988).

RECURSIVE FILTERS

A *general recursive filter* defines a current value w_t using previous values of the w process along with a linear filter in the current and previous values of the y_t process. A general recursive filter is written

$$
w_t = \sum_{s=1}^{p} a_s w_{t-s} + y_t - \sum_{s=1}^{q} b_s y_{t-s} .
$$

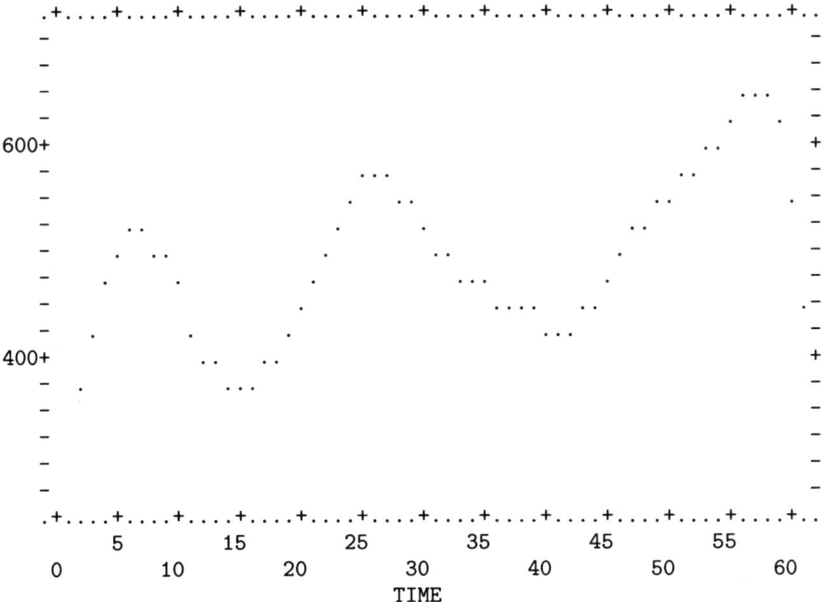

FIGURE IV.11. Low-Pass Filter of Coal Production Data

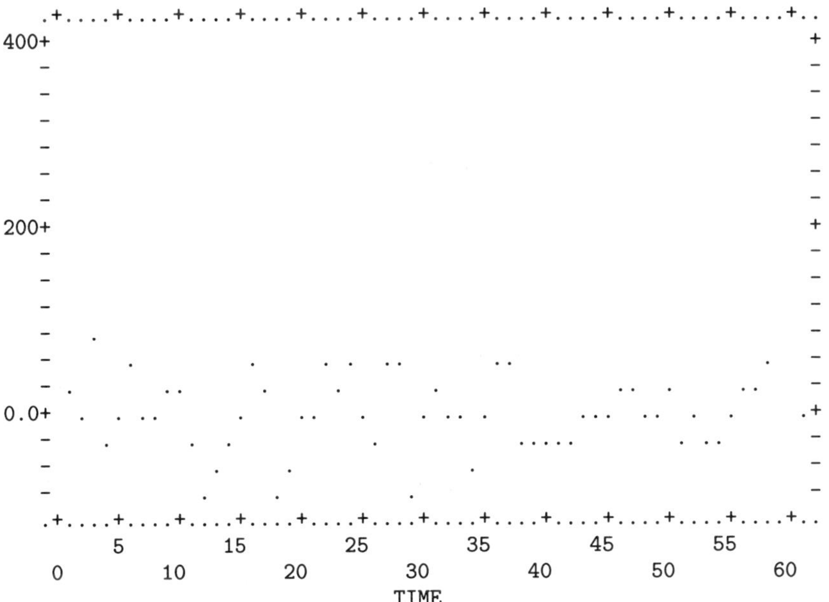

FIGURE IV.12. High-Pass Filter of Coal Production Data

If the w process is stationary, we can determine the frequency properties of the recursive filter. Let

$$w_t - \sum_{s=1}^{b} a_s w_{t-s} = z_t = y_t - \sum_{s=1}^{q} b_s y_{t-s}\,.$$

The process z_t is the result of a linear filter in the w process, so

$$f_z(\nu) = |A(\nu)|^2 f_w(\nu)\,.$$

It is also a linear filter in the y process so

$$f_z(\nu) = |B(\nu)|^2 f_y(\nu)\,.$$

Clearly,

$$f_w(\nu) = \frac{|B(\nu)|^2}{|A(\nu)|^2} f_y(\nu)\,,$$

thus $|B(\nu)|^2/|A(\nu)|^2$ is the frequency response function for the recursive filter.

A *simple recursive filter*, also called an *autoregressive filter*, is the special case

$$w_t = \sum_{s=1}^{p} a_s w_{t-s} + y_t\,.$$

EXERCISE 4.5.   Show that for a simple recursive filter

$$f_w(\nu) = \frac{1}{|A(\nu)|^2} f_y(\nu)$$

and thus that the frequency response function of the w process is $1/|A(\nu)|^2$.

SUMMARY

Properties of linear filters are of interest because they are commonly used in traditional data analysis to clarify the significant aspects of the time series. Recursive filters are also of interest in that much of time domain analysis is based on them.

IV.7   The Coherence of Two Time Series

Suppose we have two time series $y_1 = (y_{11}, y_{12}, \ldots)'$ and $y_2 = (y_{21}, y_{22}, \ldots)'$ with observations $Y_1 = (y_{11}, \ldots, y_{1n})'$ and $Y_2 = (y_{21}, \ldots, y_{2n})$. We wish to measure the correlation between y_1 and y_2 relative to the frequency k/n. One way to do that is to look at the sample partial correlation between Y_1 and Y_2 after eliminating all of the frequencies other than k/n, cf. Christensen (1987, Section VI.5) or Example 1.1.1. Because the vectors $C_0, \ldots, C_{[\frac{n}{2}]}, S_1, \ldots, S_{[\frac{n-1}{2}]}$ form an orthogonal basis for $\mathbf{R}^n$, eliminating all of the frequencies other than k/n amounts to projecting onto the column space $C(C_k, S_k)$. In particular, the squared sample partial correlation is

$$\frac{Y_1'[P_k Y_2(Y_2' P_k Y_2)^{-1} Y_2' P_k] Y_1}{Y_1' P_k Y_1}$$

or equivalently

$$(Y_1' P_k Y_2)^2 \big/ (Y_1' P_k Y_1)(Y_2' P_k Y_2).$$

Recall that $Y_1' P_k Y_1 = 2P_1(k/n)$, the periodogram for Y_1 evaluated at k/n, and $Y_2' P_k Y_2 = 2P_2(k/n)$. The numerator $Y_1' P_k Y_2$ is also closely related to the periodogram.

As in Section IV.2, it may be desired to pool results over an r neighborhood of the frequency k/n. Let

$$M_k = \sum_{\ell=-(r-1)/2}^{(r-1)/2} P_{k+\ell}.$$

The sample partial correlation between Y_1 and Y_2 eliminating all frequencies except those in the r neighborhood of k/n is called the *squared sample real coherence function*. It is

$$
\begin{aligned}
\hat{\gamma}_{12}^{2(R)}(k/n) &= \frac{Y_1'[M_k Y_2(Y_2' M_k Y_2)^{-1} Y_2' M_k] Y_1}{Y_1' M_k Y_1} \\
&= (Y_1' M_k Y_2)^2 \big/ (Y_1' M_k Y_1)(Y_2' M_k Y_2).
\end{aligned}
$$

The superscript (R) in $\hat{\gamma}_{12}^{2(R)}$ stands for "real" and will be explained in Section 8.

Consider the geometric interpretation of $\hat{\gamma}_{12}^{(R)}(k/n)$. M_k is the perpendicular projection operator onto the space spanned by $\{C_{k+\ell}, S_{k+\ell} : \ell = -(r-1)/2, \ldots, (r-1)/2\}$. This space has $2r$ degrees of freedom. The perpendicular projection operator $[M_k Y_2(Y_2' M_k Y_2)^{-1} Y_2' M_k]$ projects onto the one-dimensional subspace $C(M_k Y_2) \subset C(M_k)$. $M_k Y_2$ is the projection of Y_2 into the space associated with the r neighborhood of k/n. The value $\hat{\gamma}_{12}^{(R)}(k/n)$ is the squared length of the projection of Y_1 into $C(M_k Y_2)$ divided by the squared length of Y_1 projected into $C(M_k)$. $Y_1' M_k Y_1$ is the fraction of $Y_1' Y_1$ that is associated with the frequency k/n. $\hat{\gamma}_{12}^{2(R)}(k/n)$ is the proportion of the squared length of the vector Y_1 that can be attributed to the association between Y_1 and Y_2 (actually Y_1 and $M_k Y_2$). In particular, if $Y_1 = aY_2$ for some scaler a, it is easily seen that $\hat{\gamma}_{12}^{(R)}(k/n) = 1$ for any k.

It is interesting to consider the special case $Y_1 = C_{k+\ell}$ and $Y_2 = C_{k+j}$ for $j, \ell = -(r-1)/2, \ldots, (r-\ell)/2$, if $j = \ell$ as indicated above $\hat{\gamma}_{12}^{2(R)}(k/n) = 1$, but if $j \neq \ell$, $\hat{\gamma}_{12}^{2(R)}(k/n) = 0$. The same results hold for $Y_1 = S_{k+\ell}$ and $Y_2 = S_{k+j}$. If the two series have the same frequency and that frequency is in the r neighborhood, $\hat{\gamma}_{12}^{2(R)}(k/n) = 1$. However, if there are two different frequencies (even though both are in the r neighborhood) $\hat{\gamma}_{12}^{2(R)}(k/n)$ shows no relationship.

There is one problem with $\hat{\gamma}_{12}^{2(R)}(k/n)$ as a measure of the correlation between y_1 and y_2 relative to the frequency k/n. Suppose $Y_1 = C_k$ and $Y_2 = S_k$. Then both series are completely determined by the frequency k/n. More to the point, the relationship between these series is completely determined by k/n. However, it is easily seen that $\hat{\gamma}_{12}^{2(R)}(k/n) = 0$. Simply observe that $M_k Y_2 = S_k$ and thus $Y_1' M_k Y_2 = C_k' S_k = 0$. Obviously, we need another measure that can pick up relationships that are in the same frequency but orthogonal to one another. We begin by considering just the frequency k/n. The discussion is then extended to r neighborhoods.

Within $C(C_k, S_k)$ we began by looking at

$$Y_1'[P_k Y_2 (Y_2' P_k Y_2)^{-1} Y_2' P_k] Y_1 \Big/ Y_1' P_k Y_1 \, .$$

To detect an orthogonal relationship between Y_1 and Y_2 within $C(P_k)$, we should rotate $P_k Y_2$ by 90 degrees. This rotation is well defined because $C(P_k)$ is a two-dimensional space.

Let $c_k = C_k/\sqrt{n/2}$ and let $s_k = S_k/\sqrt{n/2}$. Thus, $c_k' c_k = 1$, $s_k' s_k = 1$, and

$$P_k = c_k c_k' + s_k s_k' \, .$$

A 90-degree rotation of $P_k Y_2$ is $F_k Y_2$ where

$$F_k = (s_k c_k' - c_k s_k') \, .$$

To see this we must establish that $(P_k Y_2)'(F_k Y_2) = 0$ and $(P_k Y_2)'(P_k Y_2) = (F_k Y_2)'(F_k Y_2)$. To see the first of these, note that

$$(P_k Y_2)'(F_k Y_2) = Y_2' P_k F_k Y_2 = Y_2' F_k Y_2$$

and that for any vector v

$$v' F_k v = (v' s_k)(c_k' v) - (v' c_k)(s_k' v) = 0 \, .$$

To see that $(P_k Y_2)'(P_k Y_2) = (F_k Y_2)'(F_k Y_2)$, observe that

$$\begin{aligned}
F_k' F_k &= (c_k s_k' - s_k c_k')(s_k c_k' - c_k s_k') \\
&= c_k s_k' s_k c_k' - s_k c_k' s_k c_k' - c_k s_k' c_k s_k' + s_k c_k' c_k s_k' \\
&= c_k c_k' - 0 - 0 + s_k s_k' \\
&= P_k \, .
\end{aligned}$$

We can now measure the orthogonal relationship between Y_1 and Y_2 in the space $C(C_k, S_k)$ by projecting Y_1 into $C(F_k Y_2)$ and comparing the squared length to $Y_1' P_k Y_1$, i.e.,

$$\frac{Y_1'[F_k Y_2 (Y_2' F_k' F_k Y_2)^{-1} Y_2' F_k] Y_1}{Y_1' P_k Y_1}.$$

Because $F_k' F_k = P_k$, this can be written as

$$(Y_1' F_k Y_2)^2 \big/ (Y_1' P_k Y_1)(Y_2' P_k Y_2).$$

We have projected Y_1 onto the space $C(P_k Y_2) \subset C(P_k)$ and onto $C(F_k Y_2) \subset C(P_k)$ where $C(P_k Y_2) \perp C(F_k Y_2)$. Because $r(P_k) = 2$,

$$Y_1' P_k Y_1 = Y_1'[P_k Y_2 (Y_2' P_k Y_2)^{-1} Y_2' P_k] Y_1 + Y_1'[F_k Y_2 (Y_2' P_k Y_2)^{-1} Y_2' F_k] Y_1.$$

Thus, all of the variability of $P_k Y_1$ can be accounted for in one of the two directions. In particular, the two correlation measures add up to 1, i.e.,

$$\frac{(Y_1' P_k Y_2)^2}{(Y_1' P_k Y_1)(Y_2' P_k Y_2)} + \frac{(Y_1' F_k Y_2)^2}{(Y_1' P_k Y_1)(Y_2' P_k Y_2)} = 1.$$

In a two-dimensional space, the partial correlation and the orthogonal partial correlation must add up to one.

The situation is not quite so degenerate when dealing with r neighborhoods. Let

$$G_k = \sum_{\ell=-(r-1)/2}^{(r-1)/2} F_{k+\ell}.$$

The partial correlation involves projecting Y_1 into $C(M_k Y_2)$. The orthogonal partial correlation is obtained by projecting Y_1 into $C(G_k Y_2)$. It is easily seen that $F_{k+\ell}' F_{k+j} = 0$ for any $\ell \neq j$. It follows that

$$\begin{aligned} G_k' G_k &= \sum_\ell \sum_j F_{k+\ell}' F_{k+j} = \sum_\ell F_{k+\ell}' F_{k+\ell} \\ &= \sum_\ell P_k = M_k. \end{aligned}$$

Also for any vector v,

$$v' G_k v = \sum_\ell v' F_{k+\ell} v = 0.$$

From these facts it is easily established that $(M_k Y_2)'(G_k Y_2) = 0$ and $(M_k Y_2)'(M_k Y_2) = (G_k Y_2)'(G_k Y_2)$. Thus, $G_k Y_2$ is an orthogonal rotation of $M_k Y_2$. In particular, $G_k Y_2$ is the sum of the orthogonal rotations in the spaces $C(P_{k+\ell})$.

As a measure of the orthogonal correlation between Y_1 and Y_2 in the frequency k/n, use the *squared sample imaginary coherence function*

$$\hat{\gamma}_{12}^{2(I)}(k/n) = \frac{Y_1'[G_k Y_2 (Y_2' M_k Y_2)^{-1} Y_2' G_k] Y_1}{Y_1' M_k Y_1}$$

$$= (Y_1' G_k Y_2)^2 \Big/ (Y_1' M_k Y_1)(Y_2' M_k Y_2).$$

The superscript (I) in $\hat{\gamma}_{12}^{2(I)}$ stands for "imaginary" and will be discussed in Section 8. The value $\hat{\gamma}_{12}^{2(I)}(k/n)$ is the fraction of $Y_1' M_k Y_1$ that is associated with projecting Y_1 into $C(G_k Y_2)$.

To see how the imaginary sample coherence works, suppose $Y_1 = C_{k+\ell}$ and $Y_2 = S_{k+j}$. For $\ell \neq j$, $\hat{\gamma}_{12}^{2(I)}(k/n) = 0$ and for $\ell = j$, $\hat{\gamma}_{12}^{2(I)}(k/n) = 1$. Of course, the same results hold for $Y_1 = S_{k+\ell}$ and $Y_2 = C_{k+j}$. The squared sample imaginary coherence identifies a relationship only if the frequency is the same but one process is a cosine while the other is a sine. These are analogous to the results for $\hat{\gamma}_{12}^{2(R)}(k/n)$ with $Y_1 = C_{k+\ell}$, $Y_2 = C_{k+j}$, and $Y_1 = S_{k+\ell}$, $Y_2 = S_{k+j}$.

As a total measure of the correlation between Y_1 and Y_2 relative to the frequency k/n, define the *squared sample coherence function*

$$\hat{\gamma}_{12}^{2}(k/n) = \hat{\gamma}_{12}^{2(R)}(k/n) + \hat{\gamma}_{12}^{2(I)}(k/n).$$

EXERCISE 4.6.   Let $Y_1 = a_1 C_{k+\ell} + b_1 S_{k+\ell}$ and let $Y_2 = a_2 C_{k+j} + b_2 S_{k+j}$. Show that for $j \neq \ell$, $\hat{\gamma}_{12}^{2} = 0$ and for $j = \ell$, $\hat{\gamma}_{12}^{2}(k/n) = 1$. What is $\hat{\gamma}_{12}^{2(I)}(k/n)$ when $b_1 = b_2 = 0$? What is $\hat{\gamma}_{12}^{2(I)}(k/n)$ when $a_1 = a_2 = 0$?

We now provide a test for the coherence. Let

$$Q_k = M_k Y_2 (Y_2' M_k Y_2)^{-1} Y_2' M_k + G_k Y_2 (Y_2' M_k Y_2)^{-1} Y_2' G_k.$$

The matrix Q_k is a perpendicular projection operator with rank two and $C(Q_k) \subset C(M_k)$. Note that

$$\hat{\gamma}_{12}^{2}(k/n) = \frac{Y_1' Q_k Y_1}{Y_1' M_k Y_1},$$

$$\frac{\hat{\gamma}_{12}^{2}(k/n)}{1 - \hat{\gamma}_{12}^{2}(k/n)} = \frac{Y_1' Q_k Y_1}{Y_1'(M_k - Q_k) Y_1},$$

and

$$\frac{2(r-1)}{2} \frac{\hat{\gamma}_{12}(k/n)}{1 - \hat{\gamma}_{12}(k/n)} = \frac{Y_1' Q_k Y_1/2}{Y_1'(M_k - Q_k) Y_1/2(r-1)}. \tag{1}$$

Let X be the design matrix for a linear model including all frequencies except those in the r neighborhood of k/n, i.e., the column space of X

contains C_i, S_i for all i other than those near k. The test statistic (1) is appropriate for testing

$$Y_1 = X\beta + e$$

against

$$Y_1 = X\beta + (M_k Y_2)\gamma_1 + (G_k Y_2)\gamma_2 + e.$$

Under a variety of conditions, if there is no relationship between Y_1 and Y_2 in the k/n frequencies, then (1) has either an $F(2, 2(r-1), 0)$ distribution or has this approximate distribution for large samples.

IV.8   Fourier Analysis

Our discussion of the frequency domain has been based on linear models. The traditional way of developing this material is via Fourier analysis. We now present the basic terminology used in Fourier analysis.

   Consider observations on two time series $Y_1 = (y_{11}, \cdots, y_{1n})'$ and $Y_2 = (y_{21}, \cdots, y_{2n})'$. We continue to use the notation of Section 7. Define the discrete Fourier transform of Y_1 as

$$
\begin{aligned}
Y_1(k) &= \frac{1}{\sqrt{n}}[Y_1'C_k - iY_1'S_k] \\
&= \frac{1}{\sqrt{n}}\sum_{t=1}^{n} y_{1t} e^{-2\pi i \frac{k}{n} t}
\end{aligned}
$$

and define $Y_2(k)$ similarly. It is easily shown that the periodogram of Y_1 is

$$P_1(k/n) = Y_1(k)\overline{Y_1(k)}$$

where $\overline{Y_1(k)}$ is the complex conjugate of $Y_1(k)$. Define the *cross-periodogram* as

$$
\begin{aligned}
P_{12}(k/n) &= Y_1(k)\overline{Y_2(k)} \\
&= Y_1'P_k Y_2 + iY_1'F_k Y_2.
\end{aligned}
$$

This is, in general, a complex valued function. Define the smoothed cross-spectral estimator from the r neighborhood of k/n as

$$\hat{f}_{12}(k/n) = \frac{1}{r}\sum_{\ell=-(r-1)/2}^{(r-1)/2} P_{12}\left(\frac{k+\ell}{n}\right).$$

The sample squared coherence equals

$$\hat{\gamma}_{12}^2(k/n) = \frac{|\hat{f}_{12}(k/n)|^2}{\hat{f}_1(k/n)\hat{f}_2(k/n)},$$

where $\hat{f}_1(k/n) = \frac{1}{2r}Y_1'M_kY_1$, $\hat{f}_2(k/n) = \frac{1}{2r}Y_2'M_kY_2$, and

$$|\hat{f}_{12}(k/n)|^2 = \hat{f}_{12}(k/n)\overline{\hat{f}_{12}(k/n)}\,.$$

Write the complex function $\hat{f}_{12}(k/n)$ as

$$\hat{f}_{12}(k/n) = \hat{f}_{12}^{(R)}(k/n) + i\hat{f}_{12}^{(I)}(k/n)\,.$$

Then

$$\hat{\gamma}_{12}^{2(R)}(k/n) = \frac{(\hat{f}_{12}^{(R)}(k/n))^2}{\hat{f}_1(k/n)\hat{f}_2(k/n)}$$

and

$$\hat{\gamma}_{12}^{(I)}(k/n) = \frac{(\hat{f}_{12}^{(I)}(k/n))^2}{\hat{f}_1(k/n)\hat{f}_2(k/n)}\,.$$

As discussed earlier, $\hat{f}_1(k/n)$ and $\hat{f}_2(k/n)$ are estimates of the corresponding theoretical spectral densities $f_1(\nu)$ and $f_2(\nu)$ of the two processes. If the covariances between the two processes are stationary, i.e., depend only on the time difference between the observations, define the cross-covariance function

$$\sigma_{12}(k) = \mathrm{Cov}(y_{1,t+k}, y_{2,t})\,.$$

The bivariate process (y_1, y_2) is defined to be second-order stationary if each marginal process is second-order stationary and if the cross-covariances are stationary. The spectral representation of $\sigma_{12}(k)$ for a bivariate second-order stationary process is

$$\sigma_{12}(k) = \int_{-1/2}^{1/2} e^{2\pi i\nu k} f_{12}(\nu)d\nu$$

which can be inverted as

$$f_{12}(\nu) = \sum_{k=-\infty}^{\infty} \sigma_{12}(k)e^{-2\pi i\nu k}\,.$$

The statistic $\hat{f}_{12}(k/n)$ can be viewed as an estimate of $f_{12}(k/n)$ and $\hat{\gamma}_{12}^2(k/n)$ can be viewed as an estimate of

$$\gamma_{12}^2(\nu) = \frac{|f_{12}(\nu)|^2}{f_1(\nu)f_2(\nu)}\,.$$

IV.9   Additional Exercises

EXERCISE 4.9.1.   Show that the following functions are nonnegative definite.

a)

$$\sigma(k) = \begin{cases} 1 & \text{if } k = 0 \\ \frac{14}{27} & \text{if } k = \pm 1 \\ \frac{4}{27} & \text{if } k = \pm 2 \\ 0 & \text{other } k \end{cases}$$

b)

$$\sigma(k) = \begin{cases} 1 & \text{if } k = 0 \\ \rho & \text{if } k = \pm 1 \\ \rho^2 & \text{if } k = \pm 2 \\ 0 & \text{other } k \end{cases}$$

EXERCISE 4.9.2.   Let α and β be random variables with $E(\alpha) = E(\beta) = 0$, $\text{Var}(\alpha) = \text{Var}(\beta) = \sigma^2$ and $\text{Cov}(\alpha, \beta) = 0$. Show that

$$y_t = \alpha \cos(2\pi\nu t) + \beta \sin(2\pi\nu t)$$

is a second-order stationary process.

EXERCISE 4.9.3.   Suppose e_t is second-order stationary. Which of the following processes are second-order stationary?
a) $y_t = \exp[e_t]$.
b) $y_t = y_{t-1} + e_t$.
c) $y_t = x_t e_t$ where x_t is another second-order stationary process independent of e_t.

EXERCISE 4.9.4.   Consider the simple linear regression model $y_t = \alpha + \beta t + e_t$ where e_t is a white noise process. Let w_t be the symmetric moving average of order 5 introduced in Example 4.6.1 as applied to the y_t process. Find $E(w_t)$ and $\text{Cov}(w_t, w_{t+k})$. Is the w_t process stationary? Why or why not?

EXERCISE 4.9.5.   Shumway (1988) reports data from Waldmeier (1960-1978, 1961) on the number of sunspots from 1748 to 1978. The data are collected monthly and a symmetric moving average of length 12 has been applied. The data in Table IV.6 are the moving averages corresponding to June and December (read across rows). Do a frequency analysis of these data including appropriate confidence intervals and plots.

EXERCISE 4.9.6.   Box and Jenkins (1970) report data on international air travel. The values in Table IV.7 are the number of passengers in thousands. Do a frequency analysis of these data including appropriate transformations of the data, confidence intervals, and plots.

EXERCISE 4.9.7.   Let e_t be a white noise process. The spectral density of e_t is $f_e(\nu) = \sigma^2$, cf. Exercise 4.1. Let $|\phi_1| < 1$ and $|\theta_1| < 1$.

TABLE IV.6. Sunspot Data

	1	2	3	4	5	6	7	8	9	10
000	89	84	70	49	47	48	41	32	17	12
010	11	9	11	11	17	31	45	47	46	53
020	62	63	72	87	72	60	53	46	49	38
030	26	22	18	11	19	36	50	67	78	106
040	112	101	94	82	80	68	53	36	39	32
050	11	7	11	19	41	88	139	157	142	126
060	107	87	80	69	51	39	31	24	14	10
070	12	24	46	79	108	130	138	133	127	118
080	109	93	76	67	63	61	55	47	41	41
090	33	23	20	17	10	7	5	4	7	7
100	6	13	24	34	41	45	43	43	45	47
110	48	43	35	29	20	11	7	8	7	3
120	0	0	0	1	3	5	7	12	15	14
130	20	34	48	47	43	41	36	31	24	24
140	23	17	11	7	6	4	1	1	6	9
150	10	16	23	34	47	51	60	63	61	67
160	71	70	62	50	41	28	14	9	7	12
170	24	55	93	116	139	142	126	105	82	87
180	82	64	50	37	29	25	19	11	12	15
190	28	38	47	61	65	97	123	122	121	99
200	78	69	67	64	60	55	45	40	30	21
210	16	7	3	4	9	22	36	54	76	93
220	95	95	91	78	68	61	55	44	43	48
230	41	31	24	17	7	6	17	36	57	73
240	106	138	135	113	98	102	92	68	52	45
250	33	18	13	11	13	13	7	3	2	6
260	16	31	44	54	62	60	55	62	74	65
270	55	53	41	26	14	13	11	7	5	6
280	6	7	17	35	56	71	77	84	87	79
290	71	64	53	43	35	27	26	28	20	12
300	11	10	5	3	3	5	11	23	33	42
310	51	63	62	53	60	63	51	49	51	43
320	33	21	13	6	3	3	3	2	4	9
330	24	47	59	56	69	101	98	84	78	65
340	51	38	31	27	23	14	7	6	8	16
350	24	41	61	65	72	71	69	77	68	63
360	57	39	28	22	15	11	9	6	5	8
370	15	34	57	77	101	116	110	109	103	91
380	76	67	58	49	48	31	21	17	9	9
390	19	33	56	89	126	152	145	135	139	136
400	118	87	72	70	47	32	26	15	7	4
410	12	35	81	137	164	188	200	187	181	161
420	131	114	84	56	49	38	30	28	21	10
430	11	15	25	45	73	91	101	107	110	106
440	105	105	84	67	69	71	51	39	32	35
450	25	16	16	12	15	26	57	89	110	

a) Find the spectral density of

$$y_t = e_t - \theta_1 e_{t-1}.$$

TABLE IV.7. Air Passenger Data

Year	Jan.	Feb.	March	April	May	June	July	Aug.	Sept.	Oct.	Nov.	Dec.
1949	112	118	132	129	121	135	148	148	136	119	104	118
1950	115	126	141	135	125	149	170	170	158	133	114	140
1951	145	150	178	163	172	178	199	199	184	162	146	166
1952	171	180	193	181	183	218	230	242	209	191	172	194
1953	196	196	236	235	229	243	264	272	237	211	180	201
1954	204	188	235	227	234	264	302	293	259	229	203	229
1955	242	233	267	269	270	315	364	347	312	274	237	278
1956	284	277	317	313	318	374	413	405	355	306	271	306
1957	315	301	356	348	355	422	465	467	404	347	305	336
1958	340	318	362	348	363	435	491	505	404	359	310	337
1959	360	342	406	396	420	472	548	559	463	407	362	405
1960	417	391	419	461	472	535	622	606	508	461	390	432

Sketch the graph of the spectral density for $\theta_1 = .5$ and $\theta_1 = 1$. Take $\sigma^2 = 1$. Which frequencies are most important in y_t?

b) Show that the spectral density of

$$y_t = \phi_1 y_{t-1} + e_t \, .$$

is

$$f_y(\nu) = \frac{\sigma^2}{1 - 2\phi_1 \cos(2\pi\nu) + \phi_1^2} \, .$$

Sketch the graph of the spectral density for $\phi_1 = .5$ and $\sigma^2 = 1$. Which frequencies are most important in y_t?

c) Find the spectral density of

$$y_t = \phi_1 y_{t-1} + e_t - \theta_1 e_{t-1} \, .$$

Sketch the graph of the spectral density for $\phi_1 = .5$, $\theta_1 = .5$, and $\sigma^2 = 1$. Which frequencies are most important in y_t?

d) Find the spectral density of the symmetric moving average of order 5,

$$w_t = \frac{1}{5} \left(e_{t-2} + e_{t-1} + e_t + e_{t+1} + e_{t+2} \right) .$$

Sketch the graph of the spectral density for $\sigma^2 = 1$. Which frequencies are most important in w_t?

EXERCISE 4.9.8.   Let e_t and ε_t be uncorrelated white noise processes. Let

$$y_t = \phi_1 y_{t-1} + e_t$$

and
$$w_t = \phi_1 w_{t-1} + y_t + \varepsilon_t .$$

Find the spectral density of w_t.

EXERCISE 4.9.9.   Let e_t be second-order stationary and define

$$y_t = \sum_{s=-\infty}^{\infty} a_s e_{t-s}$$

and

$$w_t = \sum_{r=-\infty}^{\infty} b_r y_{t-r}$$

where $\sum_{s=-\infty}^{\infty} |a_s| < \infty$ and $\sum_{r=-\infty}^{\infty} |b_r| < \infty$.
a) Show that
$$f_w(\nu) = |A(\nu)|^2 |B(\nu)|^2 f_e(\nu).$$

b) Show that

$$w_t = \sum_{s=-\infty}^{\infty} c_s e_{t-s}$$

where

$$c_s = \sum_{r=-\infty}^{\infty} a_{s-r} b_r .$$

EXERCISE 4.9.10.   Let y_t and w_t be two second-order stationary processes and suppose that $f_w(\nu) \le f_y(\nu)$ for all $\nu \in [-\pi, \pi]$. Show that

$$\Sigma_{yy} - \Sigma_{ww}$$

is nonnegative definite where Σ_{yy} and Σ_{ww} are the covariance matrices of $(y_1, \cdots, y_n)'$ and $(w_1, \cdots, w_n)'$ respectively.

EXERCISE 4.9.11.   Show that the spectral density

$$f(\nu) = \frac{\pi - |\nu|}{\pi^2} .$$

determines the covariance function

$$\sigma(k) = \begin{cases} \sigma^2 & \text{if } k = 0 \\ 4\sigma^2/(\pi|k|)^2 & \text{if } k \text{ is odd} \\ 0 & \text{otherwise.} \end{cases}$$

EXERCISE 4.9.12.   If y_t and w_t are independent second-order stationary processes with spectral densities f_y and f_w find the spectral distribution of the stationary process

$$z_t = y_t + w_t.$$

EXERCISE 4.9.13.   Suppose y_t is second-order stationary and $f_y(\nu)$ is nonnegative, bounded and $f_y(1/2) \neq 0$. Let

$$w_t = y_t - y_{t-1}.$$

Find $f_w(\nu)$ in terms of $f_y(\nu)$.

EXERCISE 4.9.14.   For $\nu \in [0, \pi]$ define

$$f_y(\nu) = \begin{cases} 50 & \nu \in [\frac{1}{4} - .01, \frac{1}{4} + .01] \\ 0 & \text{otherwise} \end{cases}$$

and $f_y(-\nu) = f_y(\nu)$. Find $\sigma(0)$, $\sigma(1)$, $\sigma(2)$.

EXERCISE 4.9.15.
a) Use the relationships

$$\sin(a + b) = \sin(a)\cos(b) + \cos(a)\sin(b)$$

and

$$\cos(a + b) = \cos(a)\cos(b) - \sin(a)\sin(b)$$

to show the following.

$$\cos(a)\cos(b) = \frac{1}{2}\{\cos(a + b) + \cos(a - b)\}.$$

$$\cos(a)\sin(b) = \frac{1}{2}\{\sin(a + b) - \sin(a - b)\}.$$

$$\sin(a)\sin(b) = \frac{1}{2}\{\cos(a - b) - \cos(a + b)\}.$$

b) Recall that for complex numbers x with $|x| < 1$,

$$\sum_{t=1}^{n} x^t = \frac{x - x^{n+1}}{1 - x}.$$

Apply this fact to $\exp(2\pi i \frac{j}{n})$ to show that for any $j = 1, \ldots, n - 1$,

$$\sum_{t=1}^{n} \cos\left(2\pi \frac{j}{n} t\right) = 0 = \sum_{t=1}^{n} \sin\left(2\pi \frac{j}{n} t\right).$$

c) Prove equations (4.2.4), (4.2.5), and (4.2.6).

Chapter V

Time Domain Analysis

In the frequency domain discussed in the previous chapter, a spectral approximation is used for an arbitrary second-order stationary time series $\ldots, y_{-1}, y_0, y_1, \ldots$. In the traditional approach, the variance components in the spectral approximation will be nonzero, thus there is little chance of developing a parsimonious model. (The model with measurement error is more promising in this regard.)

The time domain approach to time series analysis does not apply to arbitrary stationary processes. The time domain assumes that the process can be modeled as a simple recursive filter, a causal linear filter or, a general recursive filter of an uncorrelated stationary error process. Linear filters are discussed in Section IV.6. The uncorrelated stationary error process is often referred to as *white noise*. It has the same properties as measurement error. When the filtered process is stationary, a general recursive filter and hence a simple recursive filter can also be modeled as a causal linear filter. The particular choice of a model is determined by selecting a filter that does a good job of explaining the observed series and is relatively parsimonious, i.e., has few parameters.

In time domain analysis, the covariance function is again very important. In addition, partial correlations between observations are important, cf. Christensen (1987, Section VI.5) and Example 1.1.1. In this chapter, correlations are discussed in Section V.1. Section V.2 introduces the various time domain models. These sections are followed by discussions of prediction (forecasting), estimation, model selection, and seasonal adjustment. The chapter closes with consideration of a very general model: the state-space model. State-space methodology is closely tied to the Kalman filter. It is assumed that the reader is familiar with Chapter IV, especially Sections IV.1 and IV.6. Also, the reader needs to be familiar with best linear prediction which is discussed in Christensen (1987, Sections VI.3-VI.5) and Section III.1.

The time domain analysis of time series is discussed in many books and research articles. The classic text on the subject is Box and Jenkins (1970). Shorter discussions can be found in general texts on time series, e.g., Shumway (1988), Brockwell and Davis (1987), and Fuller (1976).

V.1   Correlations

We have seen that the (auto)covariance function $\sigma(k)$ determines the spec-

tral distribution and the variance components of the spectral approximation. Although time domain analysis is very different in spirit from frequency analysis, the covariance function again plays an important role.
Along with the covariance function time domain analysis uses the correlation function, the partial covariance function, and the partial correlation
function. The purpose of this section is to define these three additional functions and to relate partial correlation to best linear prediction. Assume a
second-order stationary process $\cdots, y_{-1}, y_0, y_1, \cdots$ with mean μ and covariance function $\sigma(k)$.

The *correlation function* is simply

$$\rho(k) \;=\; \mathrm{Cov}(y_t, y_{t+k}) \Big/ \sqrt{\mathrm{Var}(y_t)\mathrm{Var}(y_{t+k})}$$

$$=\; \sigma(k) \Big/ \sqrt{\sigma(0)\sigma(0)}$$

$$=\; \sigma(k)/\sigma(0)\,.$$

The *partial correlation function* is defined to be the partial correlation
between y_t and y_{t+k} given $y_{t+1}, \ldots, y_{t+k-1}$. Of course, this only makes
sense for $k \geq 2$. Defining notation similar to that in Section III.1, let
$y = (y_t, y_{t+k})'$ and $x = (y_{t+1}, \ldots, y_{t+k-1})'$. As seen in Example 1.1.1, the
partial covariance is the off-diagonal element of

$$V_{yy} - V_{yx}V_{xx}^{-1}V_{xy}$$

where

$$V_{yy} \;=\; \begin{bmatrix} \sigma(0) & \sigma(k) \\ \sigma(k) & \sigma(0) \end{bmatrix},$$

$$V_{yx} \equiv V_{xy}' \;\equiv\; \begin{bmatrix} \sigma_1' \\ \sigma_2' \end{bmatrix} = \begin{bmatrix} \sigma(1) & \sigma(2) & \cdots & \sigma(k-1) \\ \sigma(k-1) & \sigma(k-2) & \cdots & \sigma(1) \end{bmatrix}, \quad (1)$$

and

$$V_{xx} = \begin{bmatrix} \sigma(0) & \sigma(1) & \sigma(2) & \cdots & \sigma(k-2) \\ \sigma(1) & \sigma(0) & \sigma(1) & \cdots & \sigma(k-3) \\ \sigma(2) & \sigma(1) & \sigma(0) & \cdots & \sigma(k-4) \\ \vdots & \vdots & \vdots & & \vdots \\ \sigma(k-2) & \sigma(k-3) & \sigma(k-4) & \cdots & \sigma(0) \end{bmatrix}. \quad (2)$$

Note that none of these matrices depend on t; all are functions of k alone.

The partial correlation function is denoted $\phi(k)$. For $k = 1$, define $\phi(1) = \rho(1)$. For $k \geq 2$, $\phi(k)$ is the ratio of the partial covariance of y_t and y_{t+k} to
the product of the partial standard deviations of y_t and y_{t+k}. In particular,
the partial covariance is

$$\sigma(k) - \sigma_1' V_{xx}^{-1} \sigma_2\,.$$

The partial variances for y_t and y_{t+k} are respectively

$$\sigma(0) - \sigma_1' V_{xx}^{-1} \sigma_1$$

and

$$\sigma(0) - \sigma_2' V_{xx}^{-1} \sigma_2 .$$

Thus

$$\phi(k) = \{\sigma(k) - \sigma_1' V_{xx}^{-1} \sigma_2\} \Big/ \sqrt{\{\sigma(0) - \sigma_1' V_{xx}^{-1} \sigma_1\}\{\sigma(0) - \sigma_2' V_{xx}^{-1} \sigma_2\}} .$$

EXERCISE 5.1.   Let $H = [h_{ij}]$ be a $(k-1) \times (k-1)$ matrix where

$$h_{ij} = \begin{cases} 1 & \text{if } i + j = k \\ 0 & \text{otherwise.} \end{cases}$$

H is of the form

$$H = \begin{bmatrix} 0 & \cdots & 0 & 0 & 1 \\ 0 & \cdots & 0 & 1 & 0 \\ 0 & \cdots & 1 & 0 & 0 \\ \vdots & & \vdots & \vdots & \vdots \\ 1 & \cdots & 0 & 0 & 0 \end{bmatrix} .$$

Show that $\sigma_1' H = \sigma_2'$, $\sigma_2' H = \sigma_1'$, $H V_{xx} H = V_{xx}$, and $HH = I$.

Using the results of Exercise 5.1, we can show that

$$\sigma(0) - \sigma_1' V_{xx}^{-1} \sigma_1 = \sigma(0) - \sigma_2' V_{xx}^{-1} \sigma_2$$

i.e., the partial variances are the same for y_t and y_{t+k}. To see this, observe that

$$\begin{aligned} \sigma_2' V_{xx}^{-1} \sigma_2 &= \sigma_2' HH V_{xx}^{-1} HH \sigma_2 \\ &= \sigma_1' H V_{xx}^{-1} H \sigma_1 \\ &= \sigma_1' (H^{-1} V_{xx} H^{-1})^{-1} \sigma_1 \\ &= \sigma_1' (H V_{xx} H)^{-1} \sigma_1 \\ &= \sigma_1' V_{xx}^{-1} \sigma_1 . \end{aligned}$$

It follows that

$$\phi(k) = \{\sigma(k) - \sigma_1' V_{xx}^{-1} \sigma_2\} \Big/ \{\sigma(0) - \sigma_1' V_{xx}^{-1} \sigma_1\} .$$

PARTIAL CORRELATION AND BEST LINEAR PREDICTION

We now investigate the relationship between the coefficient of y_t in the best linear predictor of y_{t+k} based on $y_t, y_{t+1}, \ldots, y_{t+k-1}$ and the partial correlation between y_{t+k} and y_t. In particular, we will show that these are identical. From Section III.1, the best linear predictor is

$$\hat{E}(y_{t+k}|y_t, x) = \mu + \delta' \left[\begin{pmatrix} y_t \\ x \end{pmatrix} - J\mu \right]$$

where

$$\delta' = [\sigma(k), \sigma_2'] \begin{bmatrix} \sigma(0) & \sigma_1' \\ \sigma_1 & V_{xx} \end{bmatrix}^{-1}. \tag{3}$$

The coefficient of y_t is δ_k in $\delta' = (\delta_k, \ldots, \delta_1)$. Let $K = [\sigma(0) - \sigma_1' V_{xx}^{-1} \sigma_1]^{-1}$, from the standard result on the inverse of a partitioned matrix, cf. Christensen (1987, Exercise B.21a),

$$\begin{bmatrix} \sigma(0) & \sigma_1' \\ \sigma_1 & V_{xx} \end{bmatrix}^{-1} = \begin{bmatrix} K & -K\sigma_1' V_{xx}^{-1} \\ -V_{xx}^{-1}\sigma_1 K & V_{xx}^{-1} + V_{xx}^{-1}\sigma_1\sigma_1' V_{xx}^{-1} K \end{bmatrix},$$

so

$$\begin{aligned}
\delta_k &= [\sigma(k), \sigma_2'] \begin{bmatrix} K \\ -V_{xx}\sigma_1 K \end{bmatrix} \\[2mm]
&= (\sigma(k) - \sigma_2' V_{xx}^{-1}\sigma_1)K \\[2mm]
&= \{\sigma(k) - \sigma_1' V_{xx}^{-1}\sigma_2\} \Big/ \{\sigma(0) - \sigma_1' V_{xx}^{-1}\sigma_1\} \\[2mm]
&= \phi(k).
\end{aligned}$$

Because this is a second-order property of the process y_t (i.e., involves only means, variances, and covariances), and because y_t is second-order stationary, it follows immediately that $\phi(k)$ is also the coefficient of y_{t-k} in the best linear predictor of y_t from $y_{t-1}, \ldots, y_{t-k}$.

V.2   Time Domain Models

We now consider the various models used for analyzing time series data in the time domain. Generally, these are stationary models so their frequency properties can be studied using the methods of Chapter IV. Time domain models are just linear filters of the white noise process. White noise is the name used for the uncorrelated error process e_t where

$$\begin{aligned}
\mathrm{E}(e_t) &= 0 \\
\mathrm{Var}(e_t) &= \sigma^2 \\
\mathrm{Cov}(e_t, e_{t'}) &= 0 \qquad t \neq t'.
\end{aligned}$$

In particular, because time domain models are linear filters, their frequency properties are easily derived from Section IV.6 together with the fact, established in Exercise 4.1, that the spectral density of white noise is $f_e(\nu) = \sigma^2$.

AUTOREGRESSIVE MODELS: $AR(p)$'S

Let $\ldots, e_{-1}, e_0, e_1, \ldots$ be a stationary process of uncorrelated errors, i.e., white noise. An autoregressive model of order p, denoted $AR(p)$, is a model that states that the observable time series is a simple recursive filter of the error process involving p terms, i.e.,

$$y_t = \sum_{s=1}^{p} \phi_s y_{t-s} + e_t . \tag{1}$$

Because the $AR(p)$ model is just a simple recursive filter, the results of Exercises 4.1 and 4.5 (in Sections IV.3 and IV.6 respectively) fully determine which frequencies are important in a stationary autoregressive process.

Whether an autoregressive process is stationary depends on another way of looking at the process. Let B be the backshift operator, i.e.,

$$By_t \equiv y_{t-1} .$$

This is a very useful tool in writing models. For example, the ∇ operator of Example 4.6.2 is

$$\nabla y_t = (1 - B)y_t$$

and

$$B^2 y_t = B(By_t) = By_{t-1} = y_{t-2} .$$

To examine the $AR(p)$ model, let

$$\Phi(B) = 1 - \sum_{s=1}^{p} \phi_s B^s .$$

The $AR(p)$ model (1) can be rewritten as

$$y_t - \sum_{s=1}^{p} \phi_s y_{t-s} = e_t$$

or

$$\Phi(B)y_t = e_t . \tag{2}$$

Note that $\Phi(B)$ is a polynomial in the backshift operator. If we substitute a scaler variable x for B, we get a standard polynomial $\Phi(x)$. The roots of $\Phi(x)$ (i.e., the solutions to $\Phi(x) = 0$) are generally complex numbers. Let x_0 be an arbitrary root. If all of the roots satisfy $|x_0|^2 > 1$, then the rational polynomial $1/\Phi(x)$ can be written as an infinite polynomial. This follows from doing a Taylor's expansion of the function $1/\Phi(x)$ about 0, cf. Exercise 5.9.14. In particular, there exists

$$\Psi(B) = 1 + \sum_{s=1}^{\infty} \psi_s B^s$$

such that

$$\sum_{s=1}^{\infty} |\psi_s| < \infty \tag{3}$$

and

$$\Psi(B)\Phi(B) = 1. \tag{4}$$

Equation (4) states that the product of the polynomials $\Psi(B)$ and $\Phi(B)$ is the constant polynomial that takes only the value 1.

Applying $\Psi(B)$ to (2) yields

$$y_t = \Psi(B)\Phi(B)y_t = \Psi(B)e_t = e_t + \sum_{s=1}^{\infty} \psi_s e_{t-s} \tag{5}$$

so the $AR(p)$ model (1) can be written as a causal linear filter of the error process. Condition (3) implies that y_t is a mean zero second order stationary process (cf. Section IV.6). The restriction that all roots x_0 have $|x_0|^2 > 1$ is needed to obtain stationarity.

The coefficients in $\Psi(B)$ can be identified by solving an infinite system of equations. Note that (4) will occur if and only if the polynomials in the scalar x satisfy $\Psi(x)\Phi(x) = 1$. Because the constant terms in both $\Psi(x)$ and $\Phi(x)$ are 1, $\Psi(x)\Phi(x) = 1$ if and only if the coefficient of x^k in $\Psi(x)\Phi(x)$ equals zero for every $k \geq 1$.

EXERCISE 5.2.   Show that in an $AR(1)$ model with $|\phi_1| < 1$,

$$\psi_k = \phi_1^k.$$

The forms (2) and (5) are useful in finding the covariance function $\sigma_y(\cdot)$ of the process y_t. Let k be nonnegative and note that

$$\text{Cov}(\Phi(B)y_t, y_{t-k}) = \text{Cov}(e_t, y_{t-k}). \tag{6}$$

Using the fact that $\sigma_y(k) = \sigma_y(-k)$, the left-hand side is

$$\begin{aligned}
\text{Cov}(\Phi(B)y_t, y_{t-k}) &= \text{Cov}\left(y_t - \sum_{s=1}^{p} \phi_s y_{t-s}, y_{t-k}\right) \\
&= \sigma_y(-k) - \sum_{s=1}^{p} \phi_s \sigma_y(-k+s) \\
&= \sigma_y(k) - \sum_{s=1}^{p} \phi_s \sigma_y(k-s).
\end{aligned}$$

The right-hand side of (6) is

$$\text{Cov}(e_t, y_{t-k}) \;=\; \text{Cov}(e_t, \Psi(B)e_{t-k})$$

$$= \;\text{Cov}(e_t, e_{t-k}) + \sum_{s=1}^{\infty} \psi_s \text{Cov}(e_t, e_{t-k-s})$$

$$= \; \begin{cases} \sigma^2 & \text{if } k = 0 \\ 0 & \text{if } k \neq 0 \end{cases}.$$

Setting equal the reexpressions of the left and right sides of (6) gives

$$\sigma_y(0) - \sum_{s=1}^{p} \phi_s \sigma_y(s) = \sigma^2$$

and for $k \geq 1$

$$\sigma_y(k) - \sum_{s=1}^{p} \phi_s \sigma_y(k - s) = 0.$$

These are known as the *Yule-Walker equations* and for given ϕ_s's are solved to find $\sigma_y(k)$ for $k = 0, 1, \ldots$. The symmetry of $\sigma_y(\cdot)$ about zero completes its characterization.

The covariance function can also be characterized in terms of the ψ's. Details are given in equation (12) for autoregressive moving average ($ARMA$) models. Because autoregressive models are also $ARMA$ models, the later discussion applies to the current case. In particular, the covariance function for an $AR(1)$ process is given below in Example 5.2.3.

By combining the Yule-Walker equations with properties of best linear predictors, we can establish that the partial correlation function has the properties that

$$\phi(p) = \phi_p$$

and

$$\phi(k) = 0 \qquad k = p + 1, p + 2, \ldots .$$

There is a danger of getting the partial correlation function $\phi(k)$ confused with the coefficients $\phi_1, \ldots, \phi_p$ in the $AR(p)$ model. However, the equality $\phi(p) = \phi_p$ is the reason ϕ is used to denote both quantities.

To see that $\phi(p) = \phi_p$, consider the best linear predictor of y_t from $y_{t-1}, \ldots, y_{t-m}$. Best linear prediction was discussed in detail in Section 1. Because $\text{E}(y_t) = 0$, the best linear predictor is $\sum_{s=1}^{m} \delta_s y_{t-s}$ where

$$\begin{bmatrix} \sigma_y(0) & \sigma_y(1) & \cdots & \sigma_y(m-1) \\ \sigma_y(1) & \sigma_y(0) & \cdots & \sigma_y(m-2) \\ \vdots & \vdots & & \vdots \\ \sigma_y(m-1) & \sigma_y(m-2) & \cdots & \sigma_y(0) \end{bmatrix} \begin{bmatrix} \delta_1 \\ \vdots \\ \delta_m \end{bmatrix} = \begin{bmatrix} \sigma_y(1) \\ \vdots \\ \sigma_y(m) \end{bmatrix}$$

or equivalently $\delta_1, \ldots, \delta_m$ satisfy

$$\sigma_y(k) - \sum_{s=1}^{m} \delta_s \sigma_y(k - s) = 0 \tag{7}$$

for $k = 1, \ldots, m$.

The first thing to note is that for $m = p$, these are very similar to the Yule-Walker equations. The difference is in how they are used. The Yule-Walker equations assume that the ϕ_s's are known and are used to find the $\sigma(k)$'s. They are defined for $k \geq 0$. These equations assume the $\sigma(k)$'s are known and are used to find the δ_s's. They are defined for $k = 1, \ldots, p$. Nevertheless, the Yule-Walker equations imply that the prediction equations (7) have the solution

$$\delta_s = \phi_s,$$

$s = 1, \ldots, p$. If you think about model (1), it is only reasonable that the best linear predictor of y_t based on $y_{t-1}, \ldots, y_{t-p}$ would be $\sum_{s=1}^{p} \phi_s y_{t-s}$. As discussed in Section 1, $\delta_p = \phi(p)$, thus for an $AR(p)$ model

$$\phi(p) = \phi_p.$$

In addition, for $m > p$, a solution to (7) is given by

$$\begin{aligned} \delta_k &= \phi_k & k &= 1, \ldots, p \\ \delta_k &= 0 & k &= p+1, \ldots, m \end{aligned}$$

because the Yule-Walker equations imply that

$$\sigma_y(k) - \sum_{s=1}^{p} \phi_s \sigma_y(k - s) = 0$$

for $k = 1, \ldots, m$. Thus, for an $AR(p)$ model, the partial correlation function must satisfy

$$\phi(m) = \delta_m = 0 \qquad m > p.$$

This is a key fact in fitting autoregressive models. In practice, one estimates the partial correlation function. If the function is near zero for all values past, say 5, then an $AR(5)$ model is suggested.

EXAMPLE 5.2.1.   For $AR(1)$ and $AR(2)$ processes, the partial correlation functions have very simple forms. For an $AR(1)$ process, $y_t = \phi_1 y_{t-1} + e_t$

$$\phi(1) = \phi_1 = \rho(1)$$

and

$$\phi(k) = 0 \qquad k > 1.$$

For an $AR(2)$ process, $y_t = \phi_1 y_{t-1} + \phi_2 y_{t-2} + e_t$

$$\begin{aligned}
\phi(1) &= \rho(1), \\
\phi(2) &= \phi_2
\end{aligned}$$

and

$$\phi(k) = 0 \qquad k > 2.$$

For an $AR(3)$ process, $\phi(1) = \rho(1)$, $\phi(3) = \phi_3$, and $\phi(k) = 0$ for $k > 3$, but $\phi(2)$ is a more complicated function of the ϕ_i's and $\sigma(i)$'s.

To deal with time series that have a nonzero mean, write $E(y_t) = \mu$ and use the $AR(p)$ model

$$\Phi(B)(y_t - \mu) = e_t .$$

Equivalently, we can write

$$\begin{aligned}
\Phi(B)y_t &= \Phi(1)\mu + e_t \\
&= \left(1 - \sum_{s=1}^{p} \phi_s\right)\mu + e_t \\
&= \alpha + e_t
\end{aligned}$$

where $\alpha \equiv \left(1 - \sum_{s=1}^{p} \phi_s\right)\mu$. Another equivalent expression is

$$y_t = \alpha + \sum_{s=1}^{p} \phi_s y_{t-s} + e_t .$$

Note that

$$E(y_t) = \mu = \alpha \left/ \left(1 - \sum_{s=1}^{p} \phi_s\right)\right. .$$

MOVING AVERAGE MODELS: $MA(q)$'S

A moving average model of order q, denoted $MA(q)$, is just a causal linear filter of the error (white noise) process. In particular,

$$y_t = e_t - \sum_{s=1}^{q} \theta_s e_{t-s} . \tag{8}$$

The covariance function and spectral densities for linear filters were discussed in Section IV.6. Applying (4.6.4) gives

$$\sigma_y(k) = \begin{cases}
\sigma^2 \left(1 + \sum_{s=1}^{q} \theta_s^2\right) & k = 0 \\
\sigma^2 \left(-\theta_k + \sum_{s=1}^{q-k} \theta_{s+k}\theta_s\right) & k = 1, \ldots, q-1 \\
-\sigma^2 \theta_q & k = q \\
0 & k > q
\end{cases}$$

For model selection, this result is analogous to the fact that an $AR(p)$ model has $\phi(k) = 0$ for $k > p$. If we estimate $\sigma_y(k)$ or equivalently the correlation $\rho_y(k)$ and if the estimated correlations are negligible for say $k > 5$, then an $MA(5)$ model is suggested.

Using the backshift operator, (8) can be written as

$$y_t = \Theta(B)e_t \tag{9}$$

where

$$\Theta(B) = 1 - \sum_{s=1}^{q} \theta_s B^s .$$

Note that with only a finite number of nonzero coefficients in the filter, the process y_t is a mean zero second-order stationary process. If the roots of $\Theta(x)$ are all outside the unit circle in the complex plane, the process defined by (9) is said to be *invertible*. In such a case, $1/\Theta(B)$ can be written as an infinite polynomial in B and e_t can be written as a causal linear filter of y_t.

Processes with nonzero mean are written

$$y_t - \mu = \Theta(B)e_t .$$

EXAMPLE 5.2.2.   The $MA(1)$ process, $y_t = \mu + e_t - \theta_1 e_{t-1}$ has

$$
\begin{aligned}
\mathrm{E}(y_t) &= \mu \\
\sigma(0) &= \sigma^2(1 + \theta_1^2) \\
\sigma(1) &= -\sigma^2\theta_1 \\
\sigma(k) &= 0, \qquad k > 1
\end{aligned}
$$

and

$$\rho(1) = \frac{-\theta_1}{1 + \theta_1^2} .$$

AUTOREGRESSIVE MOVING AVERAGE MODELS: $ARMA(p, q)$'S

In many ways, the most important time domain models are the autoregressive moving average models. These are general recursive filters of the error process, i.e.,

$$y_t = \sum_{s=1}^{p} \phi_s y_{t-s} + e_t - \sum_{s=1}^{q} \theta_s e_{t-s} . \tag{10}$$

This is an $AR(p)$ with the addition of q moving average terms. It is also an $MA(q)$ with the addition of p autoregressive terms. Note that both the

$AR(p)$ and $MA(q)$ models are special cases of $ARMA$ models. Model (10) can be rewritten as

$$y_t - \sum_{s=1}^{p} \phi_s y_{t-s} = e_t - \sum_{s=1}^{q} \theta_s e_{t-s} \,,$$

or using the backshift operator,

$$\Phi(B)y_t = \Theta(B)e_t \,. \tag{11}$$

Here we assume that the roots of both $\Phi(x)$ and $\Theta(x)$ are outside the unit circle in the complex plane. This ensures that the process is both stationary and invertible. We also assume that the two polynomials have no roots in common. If the polynomials have s roots in common, then s common terms can be factored out of both sides of (11), thus creating an $ARMA(p-s, q-s)$ model. To eliminate duplication, we impose the condition of no common roots. The spectral density of model (11) can be obtained using the results in Section IV.6.

To compute the covariance function for an $ARMA(p, q)$, use the model in the form (11). Assuming the roots of the polynomial are outside the unit circle, the rational polynomial $1/\Phi(B)$ is itself an infinite polynomial. Write

$$\Psi(B) = \Theta(B)/\Phi(B)$$

thus, dividing (11) by $\Phi(B)$ gives

$$y_t = \Psi(B)e_t \,.$$

Computing the covariance function directly for $k \geq 0$

$$
\begin{aligned}
\sigma(k) = \mathrm{Cov}(y_t, y_{t+k}) \;&=\; \mathrm{Cov}(\Psi(B)e_t, \Psi(B)e_{t+k}) \\[2mm]
&=\; \mathrm{Cov}\left(\sum_{s=0}^{\infty} \psi_s e_{t-s}, \sum_{s'=0}^{\infty} \psi_{s'} e_{t+k-s'} \right) \\[2mm]
&=\; \sum_{s=0}^{\infty} \sum_{s'=0}^{\infty} \psi_s \psi_{s'} \mathrm{Cov}(e_{t-s}, e_{t+k-s'}) \\[2mm]
&=\; \sum_{s=0}^{\infty} \psi_s \psi_{s+k} \mathrm{Cov}(e_{t-s}, e_{t-s}) \\[2mm]
&=\; \sigma^2 \sum_{s=0}^{\infty} \psi_s \psi_{s+k} \,.
\end{aligned}
\tag{12}
$$

Equation (12) is just a special case of (4.6.4). Note that it can also be used to find the covariance function for an $AR(p)$ process where we consider an $AR(p)$ as an $ARMA(p, 0)$ process.

As usual, to model a process y_t with $E(y_t) = \mu$ use

$$\Phi(B)(y_t - \mu) = \Theta(B)e_t \, .$$

An equivalent form is

$$y_t = \alpha + \sum_{s=1}^{p} \phi_s y_{t-s} + e_t - \sum_{s=1}^{q} \theta_s e_{t-s}$$

where $\alpha = \left(1 - \sum_{s=1}^{p} \phi_s\right)\mu = \Phi(1)\mu$. Covariance properties are not affected by this change.

It is interesting to note that $ARMA$ models can be used to approximate the covariance structure of any stationary process with $\sigma(k) \to 0$ as $k \to \infty$. In particular, for such a covariance function and any integer $K \geq 0$, there exists an $ARMA(p, q)$ process y_t with covariance function $\sigma_y(k)$ such that

$$\sigma_y(k) = \sigma(k) \text{ for } k = 0, 1, \ldots, K.$$

EXAMPLE 5.2.3.   *The Covariance Function for an $AR(1)$.*
   Identify the $AR(1)$ process with an $ARMA(1, 0)$, thus

$$\Theta(B) = 1 \, .$$

Writing

$$\Phi(B) = 1 - \phi_1 B \, ,$$

it is easily seen that if $|\phi_1| < 1$,

$$\Psi(B) = 1/\Phi(B) = \sum_{s=0}^{\infty} \phi_1^s B^s \, .$$

Applying (12) and using the fact that $\sum_{s=0}^{\infty} v^s = \frac{1}{1-v}$ for $v \in (0, 1)$ yields

$$\begin{aligned}
\sigma(k) &= \sigma^2 \sum_{s=0}^{\infty} \phi_1^{2s+k} \\
&= \sigma^2 \phi_1^k \sum_{s=0}^{\infty} \left(\phi_1^2\right)^s \\
&= \sigma^2 \phi_1^k \Big/ \left(1 - \phi_1^2\right)
\end{aligned}$$

for $k \geq 0$. From the symmetry of the covariance function, the correlation function is

$$\rho(k) = \phi_1^{|k|} \, .$$

EXAMPLE 5.2.4.   The $ARMA(1, 1)$ model is

$$[1 - \phi_1(B)](y_t - \mu) = e_t - \theta_1 e_{t-1}$$

or

$$y_t = (1 - \phi_1)\mu + \phi_1 y_{t-1} + e_t - \theta_1 e_{t-1}. \tag{13}$$

The process is stationary if $|\phi_1| < 1$,

$$\begin{aligned}
\mathrm{E}(y_t) &= (1 - \phi_1)\mu + \phi_1 \mathrm{E}(y_{t-1}) + 0 \\
&= (1 - \phi_1)\mu + \phi_1\mu \\
&= \mu.
\end{aligned}$$

The covariance function $\sigma_y(\cdot)$ can be computed via application of equality (12). The autoregressive transformation is

$$\Phi(B) = 1 - \phi_1 B$$

so

$$\begin{aligned}
1/\Phi(B) &= 1 + \phi_1 B + \phi_1^2 B^2 + \phi_1^3 B^3 + \cdots \; . \\
&= \sum_{s=0}^{\infty} \phi_1^s B^s \; .
\end{aligned}$$

The moving average polynomial is

$$\Theta(B) = 1 - \theta_1 B$$

thus

$$\begin{aligned}
\Psi(B) &= \Theta(B)/\Phi(B) \tag{14} \\
&= \sum_{s=0}^{\infty} \phi_1^s B^s - \sum_{s=0}^{\infty} \theta_1 \phi_1^s B^{s+1} \\
&= 1 + \sum_{s=1}^{\infty} (\phi_1^s - \theta_1 \phi_1^{s-1}) B^s \\
&= 1 + \sum_{s=1}^{\infty} \phi_1^{s-1} (\phi_1 - \theta_1) B^s \; .
\end{aligned}$$

Applying (12),

$$\begin{aligned}
\sigma(0) &= \sigma^2 \left[1 + \sum_{s=1}^{\infty} \phi_1^{2(s-1)} (\phi_1 - \theta_1)^2 \right] \\
&= \sigma^2 [1 + (\phi_1 - \theta_1)^2/(1 - \phi_1^2)]
\end{aligned}$$

and for $k > 0$

$$\begin{aligned}
\sigma(k) &= \sigma^2 \left[\phi_1^{k-1}(\phi_1 - \theta_1) + \sum_{s=1}^{\infty} \phi_1^{2s-2+k} (\phi_1 - \theta_1)^2 \right] \\
&= \sigma^2 \{ (\phi_1 - \theta_1)[\phi_1^{k-1} + \phi_1^k(\phi_1 - \theta_1)/(1 - \phi_1^2)] \} \\
&= \sigma^2 (\phi_1 - \theta_1)\phi_1^{k-1}[1 + \phi_1(\phi_1 - \theta_1)/(1 - \phi_1^2)] \; .
\end{aligned}$$

Note that

$$\sigma(k) = \phi_1^{k-1}\sigma(1).$$

AUTOREGRESSIVE INTEGRATED MOVING AVERAGE MODELS: $ARIMA(p, d, q)$'s

The $ARMA(p, q)$ model and its special cases, the $AR(p)$ and $MA(q)$ models are used to model second-order stationary time series. Using $ARIMA(p, d, q)$ models is the time domain method for dealing with non-stationarity. Recall that in Example 4.6.2 and Exercise 4.4, the difference operator was used to transform series that were nonstationary into stationary processes. The $ARIMA(p, d, q)$ model assumes that the d'th difference

$$\nabla^d y_t = (1 - B)^d y_t$$

is a stationary $ARMA(p, q)$ process. The $ARIMA(p, d, q)$ model is written

$$\Phi(B)\nabla^d y_t = \Theta(B)e_t .$$

EXAMPLE 5.2.5.   The $ARIMA(1, 1, 1)$ model can be rewritten as

$$[1 - \phi_1(B)](y_t - y_{t-1}) = e_t - \theta_1 e_{t-1}$$

or

$$(y_t - y_{t-1}) - \phi_1(y_{t-1} - y_{t-2}) = e_t - \theta_1 e_{t-1}$$

or

$$y_t = (1 + \phi_1)y_{t-1} - \phi_1 y_{t-2} + e_t - \theta_1 e_{t-1} .$$

V.3   Time Domain Prediction

One of the prime motivations in analyzing time series is to be able to predict (forecast) the future of the series. The spectral approximation has limited value as a forecasting tool. The spectral model fits the data perfectly so the prediction of the future is that the past will reoccur. In particular, if n is even, the predictions for the next n observation will be precisely $y_1, \ldots, y_n$. If measurement error is included, reduced models can be fitted and more interesting predictions result.

These problems do not occur in the time domain. Time domain models are particularly well suited for making predictions. We wish to examine the best linear predictor of, say y_{n+k}, based on the observations actually in hand $y_1, \ldots, y_n$. For processes that are Gaussian, the best linear predictor is also the best predictor.

Let $Y = (y_1, \ldots, y_n)'$. We begin our discussion of prediction with examples.

EXAMPLE 5.3.1.   *Prediction for an AR(1) Model*
Because the model

$$y_t = \alpha + \phi_1 y_{t-1} + e_t$$

is linear and $\hat{E}(\cdot|Y)$ is a linear operator, we can develop, recursively, a formula for the BLP. For $k = 1, 2, \ldots$

$$
\begin{aligned}
\hat{E}(y_{n+k}|Y) &= \hat{E}(\alpha + \phi_1 y_{n+k-1} + e_{n+k}|Y) \\
&= \alpha + \phi_1 \hat{E}(y_{n+k-1}|Y) + \hat{E}(e_{n+k}|Y) \\
&= \alpha + \phi_1 \hat{E}(y_{n+k-1}|Y).
\end{aligned}
$$

The last equality holds by Proposition 3.1.5 because $\mathrm{E}(e_{n+k}) = 0$ and $\mathrm{Cov}(e_{n+k}, Y) = 0$. Similarly, for $k \geq 2$,

$$\hat{E}(y_{n+k-1}|Y) = \alpha + \phi_1 \hat{E}(y_{n+k-2}|Y)$$

so

$$
\begin{aligned}
\hat{E}(y_{n+k}|Y) &= \alpha + \phi_1[\alpha + \phi_1 \hat{E}(y_{n+k-2}|Y)] \\
&= \alpha(1 + \phi_1) + \phi_1^2 \hat{E}(y_{n+k-2}|Y).
\end{aligned}
$$

Continuing this procedure gives

$$
\begin{aligned}
\hat{E}(y_{n+k}|Y) &= \alpha \sum_{s=0}^{k-1} \phi_1^s + \phi_1^k \hat{E}(y_n|Y) \\
&= \alpha \sum_{s=0}^{k-1} \phi_1^s + \phi_1^k y_n
\end{aligned}
$$

where the last equality holds by Proposition 3.1.2 because we are predicting y_n from a vector of observations that includes y_n.

EXAMPLE 5.3.2.   *Prediction for a MA(1) Model*
The model is

$$y_t = \mu + e_t - \theta_1 e_{t-1}.$$

The BLP for $k = 2, 3, \ldots$ is

$$
\begin{aligned}
\hat{E}(y_{n+k}|Y) &= \hat{E}(\mu + e_{n+k} - \theta_1 e_{n+k-1}|Y) \\
&= \mu + \hat{E}(e_{n+k}|Y) - \theta_1 \hat{E}(e_{n+k-1}|Y) \\
&= \mu
\end{aligned}
\tag{1}
$$

where the last equality is a result of

$$E(e_{n+k}) = E(e_{n+k-1}) = 0$$

$$\text{Cov}(e_{n+k}, Y) = \text{Cov}(e_{n+k-1}, Y) = 0$$

and Proposition 3.1.5.

For $k = 1$, $\text{Cov}(e_{n+k-1}, Y) = \text{Cov}(e_n, Y) \neq 0$, so (1) does not hold. For $k = 1$, note that $\text{Cov}(y_{n+1}, y_{n-s}) = \sigma(s+1) = 0$, $s = 1, \ldots, n$, so $\text{Cov}(Y, y_{n+1}) \equiv V_{Y,y} = (0, 0, \ldots, 0, \sigma(1))'$. The covariance matrix of Y is of the form (5.1.2) with $k - 1 = n$. This can also be written

$$\text{Cov}(Y) = V_{YY} = \begin{bmatrix} V_{22} & \sigma_1 \\ \sigma_1' & \sigma(0) \end{bmatrix}$$

where V_{22} is defined by (5.1.2) with $k = n$ and σ_1' is defined by (5.1.1) with $k = n$. Using Christensen (1987, Exercise B.21) on inverses of partitioned matrices, we can show that

$$V_{YY}^{-1} V_{Yy} = \begin{bmatrix} -\sigma(1) V_{22}^{-1} \sigma_1 / [\sigma(0) - \sigma_1' V_{22}^{-1} \sigma_1] \\ \sigma(1) / [\sigma(0) - \sigma_1' V_{22}^{-1} \sigma_1] \end{bmatrix}.$$

Thus

$$\hat{E}(y_{n+1}|Y) = \mu + [Y - \mu J]' V_{YY}^{-1} V_{Yy} \tag{2}$$

where $V_{YY}^{-1} V_{Yy}$ is characterized as above. This is quite complicated and likely to get more so for an $MA(q)$ with $q > 1$. It is also difficult to compute V_{22}^{-1} for large n.

In dealing with moving average processes, it is mathematically convenient to condition not only on Y, but on $Y_\infty = (y_n, \ldots, y_1, y_0, y_{-1}, y_{-2}, \ldots)'$. This allows us to develop a simple recursive prediction method similar to that used for the $AR(1)$ process in Example 5.3.1. For an invertible $MA(q)$ process

$$(y_t - \mu) = \Theta(B) e_t$$

with $\Psi(B)\Theta(B) = 1$, we get

$$\Psi(B)(y_t - \mu) = e_t$$

where

$$\Psi(B)(y_t - \mu) = \sum_{s=0}^{\infty} \psi_s(y_{t-s} - \mu).$$

Thus, there is an invertible linear transformation between $(y_t, y_{t-1}, \ldots)'$ and $(e_t, e_{t-1}, \ldots)'$. In other words, there exists a nonsingular transformation between the two vectors for any value of t. Let $e_\infty = (e_n, e_{n-1}, \ldots)'$. By Proposition 3.1.4, for any random variable w

$$\hat{E}(w|Y_\infty) = \hat{E}(w|e_\infty). \tag{3}$$

EXAMPLE 5.3.3.   *MA(1) Prediction Using Y_∞.*
For $k = 2, 3, \ldots$ just as before

$$\hat{E}(y_{n+k}|Y_\infty) = \mu.$$

However, for $k = 1$

$$\begin{aligned}
\hat{E}(y_{n+1}|Y_\infty) &= \mu + \hat{E}(e_{n+1}|Y_\infty) - \theta_1 \hat{E}(e_n|Y_\infty) \\
&= \mu - \theta_1 \hat{E}(e_n|Y_\infty) \\
&= \mu - \theta_1 e_n
\end{aligned}$$

where the last equality follows from either Corollary 3.1.3 or Propositions 3.1.4 and 3.1.2.

Comparing the result above with (2), it becomes clear why predicting with Y_∞ is mathematically more convenient than predicting with Y. In general, (3) and Proposition 3.1.2 imply that

$$\hat{E}(e_{n+k}|Y_\infty) = \begin{cases} 0 & k \geq 1 \\ e_{n+k} & k \leq 0 \end{cases}. \tag{4}$$

This characterization is very useful in dealing with moving averages.

EXERCISE 5.3.   Consider an $AR(p)$ model with $p < n$.
a)   Show for $k = 0, 1, 2, \ldots$ that $\hat{E}(y_{n+k}|Y)$ depends only on $y_n, y_{n-1}, \ldots, y_{n-p+1}$.
b)   Show for $k = 0, 1, \ldots$ that $\hat{E}(y_{n+k}|Y) = \hat{E}(y_{n+k}|Y_\infty)$.
Hint: For (a), use induction on k.

We now consider prediction for an $ARMA(p, q)$ model. Predictions for $ARIMA$ models are found in a similar fashion as will be illustrated later. The forecasting procedure consists of building the predictions recursively. We wish to find $\hat{E}(y_{n+k}|Y_\infty)$. The value depends on all of p, q, and k. Moreover, to illustrate the recursive nature, assume that $p \geq 2$ and $q \geq 2$. The $ARMA(p, q)$ model is

$$y_t = \alpha + \sum_{s=1}^{p} \phi_s y_{t-s} + e_t - \sum_{s=1}^{q} \theta_s e_{t-s}$$

so

$$\begin{aligned}
\hat{E}(y_{n+1}|Y_\infty) &= \alpha + \sum_{s=1}^{p} \phi_s \hat{E}(y_{n+1-s}|Y_\infty) \\
&\qquad + \hat{E}(e_{n+1}|Y_\infty) - \sum_{s=1}^{q} \theta_s \hat{E}(e_{n+1-s}|Y_\infty) \\
&= \alpha + \sum_{s=0}^{p-1} \phi_{s+1} y_{n-s} + 0 - \sum_{s=0}^{q-1} \theta_s e_{n-s}.
\end{aligned}$$

For $k = 2$,

$$\hat{E}(y_{n+2}|Y_\infty) \;=\; \alpha + \sum_{s=1}^{p} \phi_s \hat{E}(y_{n+2-s}|Y_\infty)$$

$$+ \hat{E}(e_{n+2}|Y_\infty) - \sum_{s=1}^{q} \theta_s \hat{E}(e_{n+2-s}|Y_\infty)$$

$$=\; \alpha + \phi_1 \hat{E}(y_{n+1}|Y_\infty)$$

$$+ \sum_{s=0}^{p-2} \phi_{s+2} y_{n-s} + 0 - 0 - \sum_{s=0}^{q-2} \theta_{s+2} e_{n-s}$$

where $\hat{E}(y_{n+1}|Y_\infty)$ was found above.

The general recursive pattern is clear, for $k = 1, 2, \ldots$

$$\hat{E}(y_{n+k}|Y_\infty) = \alpha + \sum_{s=1}^{p} \phi_s \hat{E}(y_{n+k-s}|Y_\infty) + 0 - \sum_{s=1}^{q} \theta_s \hat{E}(e_{n+k-s}|Y_\infty) \quad (5)$$

where $\hat{E}(y_t|Y_\infty) = y_t$ if $t \le n$ and is found recursively for $t > n$ and where $\hat{E}(e_t|Y_\infty) = e_t$ for $t \le n$ and $\hat{E}(e_t|Y_\infty) = 0$ for $t > n$.

There are two problems in using this result. First, we do not know the values of α, the ϕ_i's, and the θ_i's. Second, we have not actually observed Y_∞. In particular, because we do not know Y_∞, we also do not know $e_n, e_{n-1}, \ldots, e_{n-q}$ and these error terms are explicitly involved in $\hat{E}(y_{n+k}|Y_\infty)$.

The first problem is generally handled by substituting estimates of α, the ϕ_i's, and the θ_i's for the actual parameters. Parameter estimation is discussed in Section 5.

One approach to the second problem is to assume that $e_t = 0$ for $t = 0, -1, -2, \ldots$. Note that by the invertibility of $\Phi(\cdot)$, this also implies that $y_t = \mu = \alpha / \left(1 - \sum_{s=1}^{p} \phi_s\right)$ for $t = 0, -1, -2, \ldots$. With this assumption, we can simply solve for $e_1, \ldots, e_n$. The $ARMA(p, q)$ model can be rearranged as

$$e_t = \sum_{s=1}^{q} \theta_s e_{t-s} + (y_t - \mu) - \sum_{s=1}^{p} \phi_s (y_{t-s} - \mu). \quad (6)$$

For $t = 1$, the right-hand side involves only "observed" random variables so

$$e_1 \;=\; 0 + (y_1 - \mu) - 0$$

$$=\; y_1 - \mu$$

$$=\; y_1 - \left\{ \alpha \Big/ \left(1 - \sum_{s=1}^{p} \phi_s \right) \right\}.$$

Similarly,

$$e_2 = \theta_1 e_1 + (y_2 - \mu) - \phi_1(y_1 - \mu).$$

Equation (6) can be used recursively to obtain $e_3, e_4, \ldots, e_n$. In practice, the parameters α, ϕ_i, $i = 1, \ldots, p$, and ϕ_i, $i = 1, \ldots, q$ must be estimated so estimated errors $\hat{e}_t$, $t = n - q, \ldots, n$ are used in the BLP, equation (5).

Another, less appealing, approach to dealing with the problem that Y_∞ is not observed is to assume that $e_1 = \cdots = e_q = 0$ and to otherwise ignore the problem. With this assumption about the errors, $e_n, \ldots, e_{n-q}$ can be computed using (6) and then substituted into (5) to give predictions. If n is much greater than p and q, the two approaches should give similar results. Yet another approach to dealing with the "prehistoric" values in Y_∞ is a method called backcasting, which is discussed in Section V.5.

EXAMPLE 5.3.4.   *Prediction for an ARMA(1, 1)*
Using equation (5),

$$\hat{E}(y_{n+1}|Y_\infty) \;=\; \alpha + \phi_1 y_n - \theta_1 e_n,$$

$$\begin{aligned}
\hat{E}(y_{n+2}|Y_\infty) &= \alpha + \phi_1 \hat{E}(y_{n+1}|Y_\infty) \\
&= \alpha + \phi_1[\alpha + \phi_1 y_n - \theta_1 e_n],
\end{aligned}$$

$$\begin{aligned}
\hat{E}(y_{n+3}|Y_\infty) &= \alpha + \phi_1 \hat{E}(y_{n+2}|Y_\infty) \\
&= \alpha + \phi_1\{\alpha + \phi_1[\alpha + \phi_1 y_n - \theta_1 e_n]\} \\
&= \alpha + \phi_1\alpha + \phi_1^2\alpha + \phi_1^3 y_n - \phi_1^2\theta_1 e_n,
\end{aligned}$$

and in general for $k \geq 1$

$$\hat{E}(y_{n+k}|Y_\infty) = \alpha \sum_{s=0}^{k-1} \phi_1^s + \phi_1^k y_n - \phi_1^{k-1}\theta_1 e_n.$$

EXERCISE 5.4.   a) Show that with $e_t = 0$ for $t = 0, -1, -2, \ldots$, an *ARMA(1, 1)* model has

$$e_t = (y_t - \mu) + \sum_{s=1}^{t-1}(\theta_1 - \phi_1)\theta_1^{t-1-s}(y_s - \mu)$$

for $t = 1, \ldots, n$.

b) Show that the assumption $e_1 = 0$ leads to

$$e_t = (y_t - \mu) + \left[\sum_{s=1}^{t-1}(\theta_1 - \phi_1)\theta_1^{t-1-s}(y_s - \mu)\right] - \theta_1^{t-1}(y_1 - \mu).$$

We complete our discussion of Y_∞ prediction in an $ARMA(p,q)$ model by finding the mean squared error of prediction,

$$\mathrm{E}[y_{n+k} - \hat{E}(y_{n+k}|Y_\infty)]^2 .$$

Because the BLP is an unbiased estimate of y_{n+k}, this is also called the *prediction variance*.

By definition

$$\Phi(B)[y_{n+k} - \mu] = \Theta(B)e_{n+k} . \tag{7}$$

Also, (5) can be restated as

$$\Phi(B)[\hat{E}(y_{n+k}|Y_\infty) - \mu] = \Theta(B)\hat{E}(e_{n+k}|Y_\infty) . \tag{8}$$

Subtracting (8) from (7) gives

$$\Phi(B)\{y_{n+k} - \hat{E}(y_{n+k}|Y_\infty)\} = \Theta(B)\{e_{n+k} - \hat{E}(e_{n+k}|Y_\infty)\} . \tag{9}$$

By assumption, we can write

$$\Psi(B) = \Theta(B)/\Phi(B) .$$

Multiplying (9) by $1/\Phi(B)$ gives

$$y_{n+k} - \hat{E}(y_{n+k}|Y_\infty) = \Psi(B)\{e_{n+k} - \hat{E}(e_{n+k}|Y_\infty)\} . \tag{10}$$

Recall that

$$\hat{E}(e_t|Y_\infty) = \begin{cases} e_t & t \leq n \\ 0 & t > n \end{cases}$$

so

$$e_t - \hat{E}(e_t|Y_\infty) = \begin{cases} 0 & t \leq n \\ e_t & t > n \end{cases} .$$

Substituting into (10) gives

$$y_{n+k} - \hat{E}(y_{n+k}|Y_\infty) = \sum_{s=0}^{k-1} \psi_s e_{n+k-s} . \tag{11}$$

From (11), the prediction variance is easily computed

$$\begin{aligned} \mathrm{E}[y_{n+k} - \hat{E}(y_{n+k}|Y_\infty)]^2 &= \mathrm{Var}\left(\sum_{s=0}^{k-1} \psi_s e_{n+k-s}\right) \\ &= \sum_{s=0}^{k-1} \psi_s^2 \mathrm{Var}(e_{n+k-s}) \\ &= \sigma^2 \sum_{s=0}^{k-1} \psi_s^2 . \end{aligned}$$

With estimated parameters, this probably underestimates the true prediction error. For a Gaussian process, we get a $(1-\alpha)100\%$ prediction interval for y_{n+k} with endpoints

$$\hat{E}(y_{n+k}|Y_\infty) \pm z\left(1 - \frac{\alpha}{2}\right)\sqrt{\sigma^2 \sum_{s=0}^{k-1} \psi_s^2}\,.$$

EXAMPLE 5.3.5.   *Prediction Variance in an $ARMA(1,1)$*

In (5.2.14), the polynomial transformation $\Psi(B)$ was given for an $ARMA(1,1)$. Applying this gives

$$\mathrm{E}[y_{n+1} - \hat{E}(y_{n+1}|Y_\infty)]^2 = \sigma^2$$

and for $k \geq 2$

$$\begin{aligned}
\mathrm{E}[y_{n+k} &- \hat{E}(y_{n+k}|Y_\infty)]^2 \\
&= \sigma^2 \left[1 + \sum_{s=1}^{k-1} \phi_1^{2(s-1)}(\phi_1 - \theta_1)^2\right] \\
&= \sigma^2 \left[1 + (\phi_1 - \theta_1)^2 \left\{\sum_{s=0}^{\infty} \phi_1^{2s} - \phi_1^{2(k-1)} \sum_{s=0}^{\infty} \phi_1^{2s}\right\}\right] \\
&= \sigma^2 \left[1 + (\phi_1 - \theta_1)^2 \{(1 - \phi_1^{2(k-1)})/(1 - \phi_1^2)\}\right] \\
&= \sigma^2 \left[1 + (\phi_1 - \theta_1)^2 (1 - \phi_1^{2(k-1)})/(1 - \phi_1^2)\right].
\end{aligned}$$

Recursive methods exist for finding predictions that depend only on Y and not on Y_∞. The difficulty of computing the predictors directly is that finding best linear predictors requires solving a system of equations based on an $n \times n$ matrix. The obvious method of solution involves taking the inverse of the $n \times n$ matrix. To be computationally practical, more sophisticated methods of finding a solution are needed. We begin by discussing two methods for finding $\hat{E}(y_{n+1}|Y)$. A more detailed discussion of prediction methods based on Y is given by Brockwell and Davis (1987, Section 5.2).

The *Durbin-Levinson algorithm* is based on repeated application of the partitioning results presented in the final subsection of Section 1. With $t = 1$ and $k = n$ the results apply to

$$\hat{E}(y_{n+1}|Y) = \mu + \delta'(Y - \mu J)$$

where $Y = (y_1, \ldots, y_n)'$. The Durbin-Levinson algorithm is based on finding the vector $\delta = (\delta_n, \ldots, \delta_1)$ in terms of a prediction problem that involves $n - 1$ rather than n predictors. If the order of the prediction problem can be reduced by one, successive application of the method reduces the prediction problem to one involving only a single predictor. The corresponding computational problem involves finding the inverse of a 1×1 matrix.

Section 1 shows that

$$\delta_n = \left(\sigma(k) - \sigma_2' V_{xx}^{-1} \sigma_1\right) K$$

where

$$K = [\sigma(0) - \sigma_1' V_{xx}^{-1} \sigma_1]^{-1}.$$

Write the other elements of δ as $\delta_* = (\delta_{n-1}, \ldots, \delta_1)$. Using the results of Section 1, it is easily seen that

$$
\begin{aligned}
\delta_* &= \sigma_2' V_{xx}^{-1} - \delta_n \sigma_1' V_{xx}^{-1} \\
&= \sigma_2' V_{xx}^{-1} - \delta_n \sigma_2' V_{xx}^{-1} H
\end{aligned}
$$

The vectors σ_1 and σ_2 have the same entries in reverse order, cf. Section 1. The vector $\sigma_2' V_{xx}^{-1}$ plays the role of δ in the solution to a reduced prediction problem,

$$\hat{E}(y_{n+1}|y_2, \ldots, y_n) = \mu + \sigma_2' V_{xx}^{-1} \begin{bmatrix} y_2 - \mu \\ \vdots \\ y_n - \mu \end{bmatrix}.$$

Thus solving the reduced problem leads to simple formula for K, δ_n, and δ_*.

The *innovations algorithm* is based on repeated application of Proposition 3.1.8 to obtain $\hat{E}(y_{n+1}|Y)$ as a linear combination of the prediction errors say, $\epsilon_1 \equiv y_1$ and, for $t > 1$, $\epsilon_t \equiv e(y_t|y_{t-1}, \ldots, y_1) \equiv y_t - \hat{E}(y_t|y_{t-1}, \ldots, y_1)$, cf. Exercise 5.5. Write the best linear predictor as

$$\hat{E}(y_{n+1}|Y) = \sum_{t=1}^{n} \eta_t \epsilon_t. \tag{12}$$

To use this equation one needs to know the η_t's and the ϵ_t's. The η_t's are determined by the repeated application of Proposition 3.1.8. Note that the η_t's depend on n. The ϵ_t's are found recursively. The value of $\epsilon_1 = y_1$ is known. For $t = 2, \ldots, n$, $\hat{E}(y_t|y_{t-1}, \ldots y_1)$ and thus ϵ_t can be found using exactly the same procedure as used for $\hat{E}(y_{n+1}|Y)$.

EXERCISE 5.5.   a) Using Proposition 3.1.8, find $\hat{E}(y_4|y_3, y_2, y_1)$ in terms of $\sigma(\cdot)$, $y_1, e(y_2|y_1)$, and $e(y_3|y_2, y_1)$.
b) Use induction to show that $y_1, e(y_2|y_1), \cdots, e(y_n|y_{n-1}, \cdots, y_1)$ are uncorrelated.

To obtain $\hat{E}(y_{n+k}|Y)$, first obtain $\hat{E}(y_{n+k}|Y, y_{n+1}, \ldots, y_{n+k-1})$ as a linear combination of prediction error terms. From Exercise 3.5.15 and the equivalent of equation (12) for predicting y_{n+k}

$$
\begin{aligned}
\hat{E}(y_{n+k}|Y) &= \hat{E}\big[\hat{E}(y_{n+k}|Y, y_{n+1}, \ldots, y_{n+k-1})\big|Y\big] \\
&= \hat{E}\left[\sum_{t=1}^{n+k} \eta_t \epsilon_t \Big| Y\right].
\end{aligned}
$$

By Proposition 3.1.4, Exercise 5.5 and Proposition 3.1.5, and Corollary 3.1.3

$$
\begin{aligned}
\hat{E}\left[\sum_{t=1}^{n+k} \eta_t \epsilon_t \Big| Y\right] &= \hat{E}\left[\sum_{t=1}^{n+k} \eta_t \epsilon_t \Big| \epsilon_n, \ldots, \epsilon_1\right] \\
&= \hat{E}\left[\sum_{t=1}^{n} \eta_t \epsilon_t \Big| \epsilon_n, \ldots, \epsilon_1\right] \\
&= \sum_{t=1}^{n} \eta_t \epsilon_t
\end{aligned}
$$

Note that the η_t's depend on $n + k$.

Methods related to the Durbin-Levinson and innovations algorithms can also be used to obtain the exact prediction variance.

To predict for an $ARIMA(p, d, q)$ model, let

$$
\Phi(B)\nabla^d y_t = \Theta(B)e_t
$$

and define

$$
z_t = \nabla^d y_t
$$

so that z_t is an $ARMA(p, q)$. A new problem in prediction is that $z_1, \ldots, z_d$ are at most partially observed. We have to assume "prehistoric" (i.e., $t = 0, -1, -2, \ldots$) values not only for the e_t's but also for the y_t's. With some reasonable assumption about the prehistoric y_t's (e.g., $y_t = \alpha + \beta t$) and again assuming that $e_t = 0$, $t = 0, -1, -2, \ldots$, predictions

$$
\hat{E}(z_{n+k}|Z_\infty)
$$

can be made for any $k > 0$. Here $Z_\infty = (z_n, z_{n-1}, \ldots)'$.

To predict y_{n+k}, write

$$
\begin{aligned}
\hat{E}(z_{n+k}|Z_\infty) &= \hat{E}(\nabla^d y_{n+k}|Z_\infty) & (13)\\
&= \hat{E}[(1 - B)^d y_{n+k}|Z_\infty] \\
&= \hat{E}\left[\sum_{s=0}^{d} \binom{d}{s}(-1)^d y_{n+k-s}\Big| Z_\infty\right] \\
&= \sum_{s=0}^{d} \binom{d}{s}(-1)^d \hat{E}(y_{n+k-s}|Z_\infty).
\end{aligned}
$$

Using the approximation

$$\hat{E}(y_t|Z_\infty) \doteq y_t \tag{14}$$

for $t \leq n$, equation (13) can be solved recursively to obtain $\hat{E}(y_{n+1}|Z_\infty), \hat{E}(y_{n+2}|Z_\infty), \ldots$. If n is much greater than p, d, and q, the assumptions about the prehistory should not have much effect on the predictions.

EXAMPLE 5.3.6. *Prediction for an $ARIMA(p, 1, q)$ model.*
  From equation (12),

$$\hat{E}(z_{n+k}|Z_\infty) = \hat{E}(y_{n+k}|Z_\infty) - \hat{E}(y_{n+k-1}|Z_\infty)$$

so

$$\hat{E}(y_{n+k}|Z_\infty) = \hat{E}(z_{n+k}|Z_\infty) + \hat{E}(y_{n+k-1}|Z_\infty).$$

In particular,

$$\hat{E}(y_{n+1}|Z_\infty) = \hat{E}(z_{n+1}|Z_\infty) + y_n.$$

The values $\hat{E}(z_{n+k}|Z_\infty)$ are found by applying the $ARMA(p, q)$ prediction results to the first difference process z_t. Values of $\hat{E}(y_{n+k}|Z_\infty)$ for $k > 1$ are found in a straightforward recursive manner.

V.4   Nonlinear Least Squares

As was seen in Christensen (1987, Chapter II), for linear models least squares estimates are not only intuitively appealing, but are also BLUE's, MLE's, and minimum variance unbiased estimates under appropriate assumptions. The idea of using estimates that minimize the sum of squared errors is a data analytic idea not a statistical idea; it does not depend on the statistical properties of the observations. The other properties of the estimates do depend on the statistical model.

  The idea of least squares estimation can be extended to very general nonlinear situations. In particular, the idea can be used to obtain estimates for time domain models. In many such extensions, least squares estimates are also maximum likelihood estimates for normal data. In this section we discuss least squares in general and discuss the Gauss-Newton algorithm for computing the estimates. The section closes with a discussion of nonlinear regression. Although nonlinear regression is not directly applicable to time series analysis, it is important in its own right and illustrates one application of nonlinear least squares methodology. Estimation for time domain models is discussed in Section V.5.

  Consider a vector of observations $v = (v_1, v_2, \ldots, v_n)$ and a parameter $\xi \in \mathbf{R}^p$. For known functions $f_i(\cdot)$ we can write a model

$$v_i = f_i(\xi) + e_i.$$

Here the e_i's are errors. The least squares estimates are values of ξ that minimize the sum of squared errors

$$\mathrm{SSE}(\xi) = \sum_{i=1}^{n} [v_i - f_i(\xi)]^2 \, .$$

In matrix notation write

$$F(\xi) = \begin{bmatrix} f_1(\xi) \\ \vdots \\ f_n(\xi) \end{bmatrix}$$

and $e = (e_1, \ldots, e_n)'$ so

$$v = F(\xi) + e \, .$$

Note that F is a function from $\mathbf{R}^p$ to $\mathbf{R}^n$. The sum of squared errors can be rewritten as

$$\mathrm{SSE}(\xi) = [v - F(\xi)]'[v - F(\xi)] \, . \tag{1}$$

The functions f_i can be very general. They can even depend on the v_j's, a situation that occurs in the time domain models discussed in Section 5. We now illustrate the application of this model to standard univariate linear models.

EXAMPLE 5.4.1.   Consider the linear function of ξ, $F(\xi) = Z\xi$ for a fixed $n \times p$ matrix Z of rank p. The model is $v = Z\xi + e$ and the criterion to be minimized is

$$[v - Z\xi]'[v - Z\xi] \, .$$

The minimizing value gives least squares estimates of ξ in the linear model.

THE GAUSS-NEWTON ALGORITHM

The *Gauss-Newton algorithm* is a method for finding least squares estimates in nonlinear problems. The algorithm consists of obtaining a sequence of linear least squares estimates that converge to the least squares estimate in the nonlinear problem.

The Gauss-Newton algorithm requires multivariate calculus. We use the notation set in Christensen (1987, Section XV.1). The procedure also requires an initial guess (estimate) for ξ, say ξ_0, and defines a series of estimates ξ_r that converge to the least squares estimate $\hat{\xi}$.

Given ξ_r, we define ξ_{r+1}. By the Mean Value Theorem, for ξ in a neighborhood of ξ_r

$$F(\xi) \doteq F(\xi_r) + dF(\xi_r)(\xi - \xi_r) \, , \tag{2}$$

where, because ξ_r is known, $F(\xi_r)$ and $dF(\xi_r)$ are known. The derivative $dF(\xi_r)$ is the $n \times p$ matrix of partial derivatives evaluated at ξ_r. We assume that $dF(\xi_r)$ has full column rank.

Define

$$\mathrm{SSE}_r(\xi) \equiv [v - F(\xi_r) - dF(\xi_r)(\xi - \xi_r)]'[v - F(\xi_r) - dF(\xi_r)(\xi - \xi_r)] \,. \quad (3)$$

Substituting the approximation (2) into equation (1), we see that $\mathrm{SSE}_r(\xi) \doteq \mathrm{SSE}(\xi)$ when ξ is near ξ_r. If ξ_r is near the least squares estimate $\hat{\xi}$, the minimum of $\mathrm{SSE}_r(\xi)$ should be close to the minimum of $\mathrm{SSE}(\xi)$. Now make the following identifications,

$$\begin{aligned} Y &= v - F(\xi_r) \\ X &= dF(\xi_r) \\ \beta &= (\xi - \xi_r) \,. \end{aligned}$$

With these identifications, minimizing $\mathrm{SSE}_r(\xi)$ is equivalent to minimizing

$$[Y - X\beta]'[Y - X\beta] \,.$$

From standard linear model theory, this is minimized by

$$\begin{aligned} \beta_{r+1} &= (X'X)^{-1}X'Y \\ &= ([dF(\xi_r)]'[dF(\xi_r)])^{-1}[dF(\xi_r)]'[v - F(\xi_r)] \,. \end{aligned}$$

Now $\beta = \xi - \xi_r$ so

$$\xi_{r+1} \equiv \xi_r + \beta_{r+1}$$

minimizes $\mathrm{SSE}_r(\xi)$. Although ξ_{r+1} minimizes $\mathrm{SSE}_r(\xi)$ exactly, ξ_{r+1} is only an approximation to the value $\hat{\xi}$ that minimizes $\mathrm{SSE}(\xi)$. However, as ξ_r converges to $\hat{\xi}$, the approximation (2) becomes increasingly better.

EXAMPLE 5.4.2.   Again consider the linear function $F(\xi) = Z\xi$. From standard linear model theory we know that $\hat{\xi} = (Z'Z)^{-1}Z'v$. Because $dF(\xi) = Z$, given any ξ_0

$$\begin{aligned} \beta_1 &= (Z'Z)^{-1}Z'(v - Z\xi_0) \\ &= (Z'Z)^{-1}Z'v - \xi_0 \end{aligned}$$

and

$$\xi_1 = \xi_0 + \beta_1 = (Z'Z)^{-1}Z'v = \hat{\xi}$$

Thus, for a linear least squares problem, the Gauss-Newton method converges to $\hat{\xi}$ in only one iteration.

An alternative to the Gauss-Newton method of finding least squares estimates is the method of steepest descent, cf. Draper and Smith (1981). Marquardt (1963) has presented a method that is a compromise between Gauss-Newton and the method of steepest descent. Marquardt's compromise is often used to improve the convergence properties of Gauss-Newton.

NONLINEAR REGRESSION

We present a very short account of the *nonlinear regression model* and the corresponding least squares estimates. Draper and Smith (1981, Chapter 10) have a more extensive introduction. Seber and Wild (1989) give a complete account.

Consider a situation in which there are observations $y_1, \ldots, y_n$ taken on a dependent variable and corresponding observations on independent variables denoted by the row vectors $x'_1, \ldots, x'_n$. As illustrated above, a linear model is

$$y_i = x'_i \beta + e_i .$$

Nonlinear regression is the model

$$y_i = f(x_i, \beta) + e_i \tag{4}$$

where $f(\cdot, \cdot)$ is a known function. This is a special case of the general model discussed above. Write $Y = (y_1, \ldots, y_n)$ and $e = (e_1, \ldots, e_n)$. In the notation of our discussion of the Gauss-Newton algorithm

$$\begin{aligned} v &= Y, \\ \xi &= \beta, \\ f_i(\cdot) &= f(x_i, \cdot) \end{aligned}$$

and

$$F(\beta) = \begin{bmatrix} f(x_1, \beta) \\ \vdots \\ f(x_n, \beta) \end{bmatrix}.$$

In this problem only one derivative vector needs to be determined, $(\partial f(x, \beta)/\partial \beta_1, \cdots, \partial f(x, \beta)/\partial \beta_p)$. However, the vector is evaluated at n different x_i values. Note that x need not be a p vector. We can now apply Gauss-Newton to find least squares estimates.

Nonlinear regression is a problem in which least squares estimates are MLE's under suitable conditions. Assume

$$e \sim N(0, \sigma^2 I),$$

then because $F(\beta)$ is a fixed vector

$$Y \sim N(F(\beta), \sigma^2 I)$$

with density

$$f(Y|\beta, \sigma^2) = (2\pi)^{-n/2} \sigma^{-n/2} \exp[-(Y - F(\beta))'(Y - F(\beta))/2\sigma^2].$$

The likelihood is the density taken as a function of β and σ^2 for fixed Y. The MLE's maximize the likelihood. For any σ^2, the likelihood is maximized by minimizing $[Y - F(\beta)]'[Y - F(\beta)]$. Thus, least squares estimates $\hat{\beta}$ are also MLE's. The MLE of σ^2 can be found by maximizing

$$L(\sigma^2) = (2\pi)^{-n/2}(\sigma^2)^{-n} \exp[-\left(Y - F(\hat{\beta})\right)'\left(Y - F(\hat{\beta})\right)/2\sigma^2].$$

Differentiation leads to

$$\hat{\sigma}^2 = [Y - F(\hat{\beta})]'[Y - F(\hat{\beta})]/n.$$

Asymptotic results for maximum likelihood estimation allow statistical inferences to be made for nonlinear regression models.

Some nonlinear relationships can be transformed into linear relationships. The nonlinear regression model (4) indicates that

$$y_i \doteq f(x_i, \beta).$$

If $f(\cdot, \cdot)$ can be written as

$$f(x_i, \beta) = f(x_i'\beta)$$

and f is invertible the relationship is said to be linearizable. In that case,

$$f^{-1}(y_i) \doteq x_i'\beta.$$

An alternative to fitting model (4) is to fit

$$f^{-1}(y_i) = x_i'\beta + \varepsilon_i. \tag{5}$$

The choice between analyzing the nonlinear model (4) and the linear model (5) is typically based on which model better approximates the assumption of independent identically distributed normal errors. Just as in linear regression, the nonlinear least squares fit of β in model (4) generates residuals

$$\hat{e}_i = y_i - \hat{y}_i = y_i - f(x_i, \hat{\beta}).$$

These can be plotted to examine normality and heteroscedasticity.

V.5   Estimation

In this section we discuss empirical estimates of correlations and partial correlations. These are empirical in the sense that they do not depend on any particular time domain model for the second-order stationary time series. We also discuss least squares and maximum likelihood estimates for the parameters of time domain models.

CORRELATIONS

The empirical estimate of the variance of a stationary process is similar to the sample variance of a simple random sample:

$$s(0) \equiv \frac{1}{n} \sum_{i=1}^{n} (y_i - \bar{y}.)^2 \, .$$

This differs from the unbiased estimate for a random sample in that it uses the multiplier $1/n$ instead of $1/(n-1)$. An empirical estimate of $\sigma(k) = \mathrm{Cov}(y_t, y_{t+k})$ can be based on the sample covariance of $(y_1, \ldots, y_{n-k})$ and $(y_{k+1}, \ldots, y_n)$. An estimate of $\sigma(k)$ is

$$s(k) = \frac{1}{n-k} \sum_{i=1}^{n-k} (y_i - \bar{y}.)(y_{i+k} - \bar{y}.) \, .$$

This differs from the estimate for simple random samples in that (a) the multiplier is $1/(n-k)$ rather than $1/(n-k-1)$ and (b) $\bar{y}.$ is used to estimate the mean of both samples rather than using $(y_1 + \cdots + y_{n-k})/(n-k)$ and $(y_{k+1} + \cdots + y_n)/(n-k)$, respectively.

Sometimes the alternative estimates

$$\hat{\sigma}(k) = \frac{n-k}{n} s(k) = \frac{1}{n} \sum_{i=1}^{n-k} (y_i - \bar{y}.)(y_{i+k} - \bar{y}.)$$

are used. The fact that the $\hat{\sigma}(k)$'s all use the same multiplier is convenient for some purposes. In particular, it ensures that estimates of covariance matrices such as (5.1.2) are nonnegative definite, cf. Brockwell and Davis (1987, Section 7.2). Note that $\hat{\sigma}(0) = s(0)$.

The correlation function $\rho(k) = \sigma(k)/\sigma(0)$ is estimated with either

$$r(k) = s(k)/s(0)$$

or

$$\hat{\rho}(k) = \hat{\sigma}(k)/\hat{\sigma}(0) \, .$$

Partial covariances and partial correlations as defined in Section 1 are functions of $\sigma(k)$. Either estimate of $\sigma(k)$ can be used to define estimated partial covariances and correlations. For example, substituting $\hat{\sigma}(k)$ into (5.1.1), (5.1.2), and (5.1.3) leads to an estimated partial correlation

$$\hat{\phi}(k) = \hat{\delta}_k$$

where $\hat{\delta}_k$ is the first component of $\hat{\delta} = (\hat{\delta}_k, \ldots, \hat{\delta}_1)$ and $\hat{\delta}$ is defined by replacing $\sigma(\cdot)$ with $\hat{\sigma}(\cdot)$ in (5.1.4). An alternative to using matrix operations for finding $\hat{\phi}(k)$ is to use the results of Exercise 6.8.9 iteratively.

Box and Jenkins (1970, p. 33) suggest that estimated covariances are only useful when $n \geq 50$ and $k \leq \frac{n}{4}$. Inferences for correlations can be based on asymptotic normality.

CONDITIONAL ESTIMATION FOR AN $AR(p)$ MODEL

Various aspects of estimation in an $ARMA(p, q)$ model simplify when $q = 0$. This simplification can be quite useful, so we consider the special case first. In particular, three estimation methods are examined: empirical estimation, least squares, and maximum likelihood. The $AR(p)$ model is

$$y_t = \alpha + \sum_{s=1}^{p} \phi_s y_{t-s} + e_t$$

where $E(e_t) = 0$, $\text{Var}(e_t) = \sigma^2$. The parameters to be estimated are $\alpha, \phi_1, \phi_2, \ldots, \phi_p$ and σ^2.

As established in Section 2,

$$\hat{E}(y_t | y_{t-1}, \ldots, y_{t-p}) = \mu + \sum_{s=1}^{p} \phi_s(y_{t-s} - \mu) = \alpha + \sum_{s=1}^{p} \phi_s y_{t-s}.$$

Thus, modifying (5.2.7) we get estimates $\hat{\phi}_1, \ldots, \hat{\phi}_p$ that satisfy

$$\hat{\sigma}_y(k) = \sum_{s=1}^{p} \hat{\phi}_s \hat{\sigma}(s - k)$$

for $k = 1, \ldots, p$. In particular,

$$\begin{bmatrix} \hat{\phi}_1 \\ \vdots \\ \hat{\phi}_p \end{bmatrix} = \begin{bmatrix} \hat{\sigma}(0) & \hat{\sigma}(1) & \cdots & \hat{\sigma}(p-1) \\ \hat{\sigma}(1) & \hat{\sigma}(0) & \cdots & \hat{\sigma}(p-2) \\ \vdots & \vdots & & \vdots \\ \hat{\sigma}(p-1) & \hat{\sigma}(p-2) & \cdots & \hat{\sigma}(0) \end{bmatrix}^{-1} \begin{bmatrix} \hat{\sigma}(1) \\ \hat{\sigma}(2) \\ \vdots \\ \hat{\sigma}(p) \end{bmatrix}.$$

Note that from our discussion above

$$\hat{\phi}(p) = \hat{\phi}_p.$$

Moreover, taking $\hat{\mu} = \bar{y}.$, gives

$$\hat{\alpha} = \left(1 - \sum_{s=1}^{p} \hat{\phi}_s \right) \hat{\mu}$$

The second method of estimating $\alpha, \phi_1, \ldots, \phi_p$ is least squares. The least squares estimates are the values that minimize $\sum_{t=p+1}^{n} e_t^2$ or equivalently

$$\sum_{t=p+1}^{n} \left[y_t - \alpha - \sum_{s=1}^{p} \phi_s y_{t-s} \right]^2.$$

The sum is from $t = p + 1$ to n because not all of the y_{t-s}'s are observed for $t \leq p$. Conditional methods use precisely the cases for which we have complete data.

In applying the Gauss-Newton method to obtain least squares estimates, let $v = (y_n, \ldots, y_{p+1})'$, $\xi = (\alpha, \phi_1, \ldots, \phi_p)'$, $z'_t = (1, y_{t-1}, \ldots, y_{t-p})$, and for $t = n, \ldots, p + 1$

$$\alpha + \sum_{s=1}^{p} \phi_s y_{t-s} = f_t(\xi) = z'_t \xi.$$

Note that each $f_t(\xi)$ is a linear function of ξ so $F(\xi)$ is a linear function of ξ. In particular,

$$Z = \begin{bmatrix} z'_n \\ \vdots \\ z'_{p+1} \end{bmatrix}$$

and $F(\xi) = Z\xi$. Example 5.4.2 applies giving

$$\begin{bmatrix} \hat{\alpha} \\ \hat{\phi}_1 \\ \vdots \\ \hat{\phi}_p \end{bmatrix} = \hat{\xi} = (Z'Z)^{-1} Z'v$$

which is obtained from fitting the linear model

$$v = Z\xi + e, \; \mathrm{E}(e) = 0, \; \mathrm{Cov}(e) = \sigma^2 I.$$

Thus, conditional least squares estimates for an $AR(p)$ model are particularly easy to find.

The conditional least squares estimates are also conditional maximum likelihood estimates when the process is Gaussian. The $AR(p)$ model is a special case of the $ARMA(p, q)$ model and conditional least squares estimates are conditional MLE's in the more general model. The equivalence will be established later.

Finally, to estimate σ^2 one can use either

$$\hat{\sigma}^2 = \frac{1}{n - p} \sum_{t=p+1}^{n} \left[y_t - \hat{\alpha} - \sum_{s=1}^{p} \hat{\phi}_s y_{t-s} \right]^2$$

or

$$\mathrm{MSE} = \frac{1}{n - 2p - 1} \sum_{t=p+1}^{n} \left[y_t - \hat{\alpha} - \sum_{s=1}^{p} \hat{\phi}_s y_{t-s} \right]^2.$$

CONDITIONAL LEAST SQUARES ESTIMATION FOR AN $ARMA(p,q)$ MODEL

We now consider conditional least squares estimates for the model

$$y_t = \alpha + \sum_{s=1}^{p} \phi_s y_{t-s} + e_t - \sum_{s=1}^{q} \theta_s e_{t-s} \,.$$

Let $\xi = (\alpha, \phi_1, \ldots, \phi_p, \theta_1, \ldots, \theta_q)'$. Conditional least squares estimates $\hat{\xi} = (\hat{\alpha}, \hat{\phi}_1, \ldots, \hat{\phi}_p, \hat{\theta}_1, \ldots, \hat{\theta}_q)'$ minimize $\sum_{t=p+1}^{n} e_t^2$ or equivalently

$$\mathrm{SSE}_C(\xi) = \sum_{t=p+1}^{n} \left(y_t - \left[\alpha + \sum_{s=1}^{p} \phi_s y_{t-s} - \sum_{s=1}^{q} \theta_s e_{t-s} \right] \right)^2 \,.$$

The difficulty in minimizing this is that the e_t's depend on the other parameters.

To apply the Gauss-Newton method, let $v = (y_n, \ldots, y_{p+1})'$ and

$$f_t(\xi) = \alpha + \sum_{s=1}^{p} \phi_s y_{t-s} - \sum_{s=1}^{q} \theta_s e_{t-s}(\xi) \,.$$

Differentiating

$$df_t(\xi) = \left[\frac{\partial f_t(\xi)}{\partial \alpha}, \frac{\partial f_t(\xi)}{\partial \phi_1}, \ldots, \frac{\partial f_t(\xi)}{\partial \phi_p}, \frac{\partial f_t(\xi)}{\partial \theta_1}, \ldots, \frac{\partial f_t(\xi)}{\partial \theta_q} \right]$$

where

$$\frac{\partial f_t(\xi)}{\partial \alpha} = 1$$

$$\frac{\partial f_t(\xi)}{\partial \psi_s} = y_{t-s} - \sum_{s=1}^{q} \theta_s \frac{\partial e_{t-s}(\xi)}{\partial \phi_s}$$

and

$$\frac{\partial f_t(\xi)}{\partial \theta_s} = -\sum_{s=1}^{q} \left[e_{t-s}(\xi) + \theta_s \frac{\partial e_{t-s}(\xi)}{\partial \theta_s} \right] \,.$$

Given a current estimate $\xi_r = (\alpha^{(r)}, \phi_1^{(r)}, \ldots, \phi_p^{(r)}, \theta_1^{(r)}, \ldots, \theta_q^{(r)})'$, we need to be able to evaluate $f_t(\xi_r)$ and $df_t(\xi_r)$ for $t = n, \ldots, p+1$. In particular, we need to be able to evaluate $e_t(\xi_r)$, $\partial e_t(\xi_r)/\partial \phi_s$, and $\partial e_t(\xi_r)/\partial \theta_s$. To do this, we repeat the assumption made in Section 3 that

$$e_t = 0, \qquad t = 0, -1, -2, \ldots \tag{1}$$

where now we think of these statements as implying that $e_t(\xi) = 0$ for all ξ. Thus we have conditioned on the values of e_t, $t = 0, -1, -2, \ldots$. Because the functions are constant the derivatives are zero, thus

$$\frac{\partial e_t(\xi)}{\partial \alpha} = \frac{\partial e_t(\xi)}{\partial \phi_s} = \frac{\partial e_t(\xi)}{\partial \theta_s} = 0 \qquad t = 0, -1, -2, \ldots \,. \tag{2}$$

Recalling that

$$e_t(\xi) = \sum_{j=1}^{q} \theta_j e_{t-j}(\xi) + y_t - \sum_{j=1}^{p} \phi_j y_{t-j} \tag{3}$$

we have

$$\frac{\partial e_t(\xi)}{\partial \phi_s} = -y_{t-s} + \sum_{j=1}^{q} \theta_j \frac{\partial e_{t-j}(\xi)}{\partial \phi_s} \tag{4}$$

and

$$\frac{\partial e_t(\xi)}{\partial \theta_s} = e_{t-s}(\xi) + \sum_{j=1}^{q} \theta_j \frac{\partial e_{t-j}(\xi)}{\partial \theta_s}. \tag{5}$$

Assumptions (1) and (2) along with (3), (4), and (5) allow us to compute $e_t(\xi_r)$ and the necessary partial derivatives. Computation of the $e_t(\xi_r)$'s assuming (1) was discussed in Section 3. Our discussion there assumed that $\xi = (\alpha, \phi_1, \ldots, \phi_p, \theta_1, \ldots, \theta_q)'$ was known. We simply use those results assuming that $\xi = \xi_r$.

Computation of the partial derivatives is a bit more involved. We illustrate the method for $\partial e_t(\xi)/\partial \theta_s$ with $q = 2$. Assume that for $t = 1, \ldots, n$, $e_t = e_t(\xi_r)$ has already been computed. We simplify notation by writing $\theta_s^{(r)} = \theta_s$ and $\partial e_t(\xi)/\partial \theta_s$ as $\partial e_t/\partial \theta_s$. The method makes repeated use of (1), (2), and (5).

For $t = 1$

$$\begin{aligned}
\frac{\partial e_1}{\partial \theta_1} &= e_0 + \theta_1 \frac{\partial e_0}{\partial \theta_1} + \theta_2 \frac{\partial e_{-1}}{\partial \theta_1} \\
&= 0 + \theta_1 0 + \theta_2 0 \\
&= 0
\end{aligned}$$

$$\begin{aligned}
\frac{\partial e_1}{\partial \theta_2} &= e_{-1} + \theta_1 \frac{\partial e_0}{\partial \theta_2} + \theta_2 \frac{\partial e_{-1}}{\partial \theta_2} \\
&= 0 + \theta_1 0 + \theta_2 0 \\
&= 0.
\end{aligned}$$

For $t = 2$

$$\begin{aligned}
\frac{\partial e_2}{\partial \theta_1} &= e_1 + \theta_1 \frac{\partial e_1}{\partial \theta_1} + \theta_2 \frac{\partial e_0}{\partial \theta_1} \\
&= e_1 + \theta_1 0 + \theta_2 0 \\
&= e_1
\end{aligned}$$

$$\begin{aligned}
\frac{\partial e_2}{\partial \theta_2} &= e_0 + \theta_1 \frac{\partial e_1}{\partial \theta_2} + \theta_2 \frac{\partial e_0}{\partial \theta_2} \\
&= 0 + \theta_1 0 + \theta_2 0 \\
&= 0.
\end{aligned}$$

For $t = 3$

$$\frac{\partial e_3}{\partial \theta_1} = e_2 + \theta_1 \frac{\partial e_2}{\partial \theta_1} + \theta_2 \frac{\partial e_1}{\partial \theta_1}$$
$$= e_2 + \theta_1 e_1$$

$$\frac{\partial e_3}{\partial \theta_2} = e_1 + \theta_1 \frac{\partial e_2}{\partial \theta_2} + \theta_2 \frac{\partial e_1}{\partial \theta_2}$$
$$= e_1 + \theta_1 0 + \theta_2 0$$
$$= e_1 .$$

For $t = 4$

$$\frac{\partial e_4}{\partial \theta_1} = e_3 + \theta_1 \frac{\partial e_3}{\partial \theta_1} + \theta_2 \frac{\partial e_2}{\partial \theta_1}$$
$$= e_3 + \theta_1 [e_2 + \theta_1 e_1] + e_1$$
$$= e_3 + \theta_1 e_2 + (1 + \theta_1^2) e_1$$

$$\frac{\partial e_4}{\partial \theta_2} = e_2 + \theta_1 \frac{\partial e_3}{\partial \theta_2} + \theta_2 \frac{\partial e_2}{\partial \theta_2}$$
$$= e_2 + \theta_1 e_1 + \theta_2 0$$
$$= e_2 + \theta_1 e_1 .$$

This procedure goes on recursively, thus allowing computation of $\partial f_t(\xi_r)/\partial \theta_s$ which is necessary to execute the Gauss-Newton algorithm.

CONDITIONAL MAXIMUM LIKELIHOOD ESTIMATION FOR THE $ARMA(p, q)$ MODEL

It remains to establish the equivalence of conditional least squares and conditional maximum likelihood for Gaussian processes. To do this, we need to find the joint distribution of $y_n, \ldots, y_{p+1}$. The joint distribution is conditional on the unknown parameters ξ. We also condition on $e_p, e_{p-1}, \ldots$ or equivalently $y_p, y_{p-1}, \ldots$. Writing the stationary invertible $ARMA(p, q)$ process as

$$\Phi(B)(y_t - \mu) = \Theta(B)e_t$$

and letting

$$\Psi(B) = \Theta(B)/\Phi(B)$$

we have

$$(y_t - \mu) = \Psi(B)e_t . \tag{6}$$

From (6), with $e_p, e_{p-1}, \ldots$ fixed, each $y_t - \mu$, $t > p$ is a linear function of the random variables $e_t, \ldots, e_{p+1}$. In particular, write

$$\begin{pmatrix} y_n \\ \vdots \\ y_{p+1} \end{pmatrix} = A \begin{pmatrix} e_n \\ \vdots \\ e_{p+1} \end{pmatrix} + \eta$$

The fixed vector η is $\eta = (\eta_n, \ldots, \eta_{p+1})'$ with $\eta_t = \mu + \sum_{s=0}^{\infty} \psi_{t-p+s} e_{p-s}$.

Moreover,

$$A = \begin{bmatrix} a_n' \\ \vdots \\ a_{p+1}' \end{bmatrix}$$

and $a_t' = (0, \ldots, 0, \psi_0, \psi_1, \ldots, \psi_{t-p-1})$ with $\psi_0 = 1$. ($\psi_0 = 1$ because the first coefficients in both $\Phi(B)$ and $\Theta(B)$ are 1.) Observe that A is a nonsingular, upper triangular matrix with 1's down the diagonal. The distribution of $y = (y_n, \ldots, y_{p+1})'$ is

$$y \sim N(\eta, \sigma^2 A A')$$

and the density is

$$
\begin{aligned}
f(y) &= (2\pi)^{-\left(\frac{n-p}{2}\right)} |\sigma^2 A A'|^{-\frac{1}{2}} \exp[-(y-\eta)'(AA')^{-1}(y-\eta)/2\sigma^2] \quad (7) \\
&= (2\pi)^{-\left(\frac{n-p}{2}\right)} (\sigma^2)^{-\frac{n-p}{2}} |A|^{-1} \\
&\quad \times \exp[-\{A^{-1}(y-\eta)\}'\{A^{-1}(y-\eta)\}/2\sigma^2] \\
&= (2\pi)^{-\left(\frac{n-p}{2}\right)} (\sigma^2)^{-\frac{n-p}{2}} \exp[-\{A^{-1}(y-\eta)\}'\{A^{-1}(y-\eta)\}/2\sigma^2]
\end{aligned}
$$

where the last equality follows from the fact that A is upper triangular with 1's on the diagonal so $|A| = 1$.

The problem with (7) is that it is a complicated function of our original parameters: ξ and σ^2. Both A and η depend on the coefficients of $\Psi(B)$. Note however that

$$A^{-1}(y-\eta) = (e_n, \ldots, e_{p+1})'$$

where each e_t is a function of both y and the parameters ξ. Writing the $ARMA(p,q)$ model as

$$e_t(y, \xi) = \sum_{s=1}^{q} \theta_s e_{t-s}(y, \xi) + y_t - \sum_{s=1}^{p} \phi_s y_{t-s} - \alpha$$

we see that

$$
\begin{aligned}
&\{A^{-1}(y-\eta)\}'\{A^{-1}(y-\eta)\} \\
&= \sum_{t=p+1}^{n} \left[y_t - \alpha - \sum_{s=1}^{p} \phi_s y_{t-s} + \sum_{s=1}^{q} \theta_s e_{t-s}(y, \xi) \right]^2 \quad (8) \\
&= \mathrm{SSE}_C(\xi)
\end{aligned}
$$

and the density can be rewritten as

$$f(y) = (2\pi)^{-\frac{(n-p)}{2}} (\sigma^2)^{-\frac{(n-p)}{2}} \exp\left\{ -\frac{1}{2\sigma^2} \mathrm{SSE}_C(\xi) \right\}.$$

Thinking of this as a function of σ^2 and ξ with y fixed, we have the likelihood

$$L(\xi, \sigma^2) = (2\pi)^{-\frac{(n-p)}{2}} (\sigma^2)^{-\frac{(n-p)}{2}} \exp\left\{-\frac{1}{2\sigma^2}\mathrm{SSE}_C(\xi)\right\}. \qquad (9)$$

Clearly, for any σ^2, the value of ξ that minimizes (8) will maximize the likelihood. However, by definition, the minimizing value of (8) is given by the least squares estimates. Given the least squares estimate $\hat{\xi}$, the MLE of σ^2 can then be found as in Christensen (1987, Section II.4) as

$$\hat{\sigma}^2 = \frac{1}{n-p} \sum_{t=p+1}^{n} \left[y_t - \hat{\alpha} - \sum_{s=1}^{p} \hat{\phi}_s y_{t-s} + \sum_{s=1}^{q} \hat{\theta}_s e_{t-s}(y, \hat{\xi}) \right]^2. \qquad (10)$$

Unconditional Estimation for the $ARMA(p, q)$ Model

We begin with maximum likelihood estimation. This discussion will lead naturally into least squares estimation. For unconditional MLE's, we need the density of $Y = (y_n, \cdots, y_1)'$. Again consider the equality (6) however we now redefine A as an $n \times \infty$ matrix

$$A = \begin{bmatrix} a_n' \\ \vdots \\ a_1' \end{bmatrix}$$

where

$$a_t' = (0, \cdots, 0, \psi_0, \psi_1, \psi_2, \cdots)$$

with the first $n - t$ columns equal to 0 and $\psi_0 = 1$. Writing the infinite vector

$$e = (e_n, e_{n-1}, e_{n-2}, \cdots)'$$

equation (6) becomes

$$Y = Ae + \mu J \qquad (11)$$

where J is an n vector of 1's. Recalling that the e_t's are i.i.d. $N(0, \sigma^2)$, we see that

$$Y \sim N(\mu J, \sigma^2 A A')$$

so the unconditional likelihood function is

$$\begin{aligned} f(Y|\xi, \sigma^2) &= (2\pi)^{-\frac{n}{2}} (\sigma^2)^{-\frac{n}{2}} |AA'|^{-\frac{1}{2}} \\ &\quad \times \exp\left[-(Y - \mu J)'(AA')^{-1}(Y - \mu J) \Big/ 2\sigma^2 \right]. \qquad (12) \end{aligned}$$

It is important to note that both μ and A depend on the parameters ξ. When convenient we write $A(\xi)$ for A. Equation (12) can be written in a somewhat simpler form. Note that

$$\hat{E}(e|Y, \xi) \equiv \hat{E}(e|Y) = A'(AA')^{-1}(Y - \mu J).$$

Clearly,

$$
\begin{aligned}
(Y - \mu J)'(AA')^{-1}(Y - \mu J) &= \{\hat{E}(e|Y,\xi)\}'\{\hat{E}(e|Y,\xi)\} \\
&= \sum_{t=-\infty}^{n} \{\hat{E}(e_t|Y,\xi)\}^2
\end{aligned}
$$

so substituting into (12) gives the likelihood

$$
\begin{aligned}
L(\xi,\sigma^2) &= (2\pi)^{-\frac{n}{2}}(\sigma^2)^{-\frac{n}{2}}|A(\xi)A'(\xi)|^{-\frac{1}{2}} \\
&\quad \times \exp\left\{-\hat{E}(e|Y,\xi)'\hat{E}(e|Y,\xi)\Big/2\sigma^2\right\}.
\end{aligned}
\tag{13}
$$

Maximum likelihood estimates maximize the function (13). Least squares estimates minimize

$$
\begin{aligned}
\mathrm{SSE}(\xi) &= (Y - \mu J)'(AA')^{-1}(Y - \mu J) \\
&= \hat{E}(e|Y,\xi)'\hat{E}(e|Y,\xi) \\
&= \sum_{t=-\infty}^{n} \{\hat{E}(e_t|Y,\xi)\}^2.
\end{aligned}
$$

In unconditional estimation, the least squares estimates need not equal the maximum likelihood estimates. However, for moderate to large sample sizes n and parameter values that are not near their boundaries, the least squares estimate of ξ gives a very good approximation to the MLE, cf. Box and Jenkins (1970, Section 7.1.4) and McLeod (1970). This phenomenon occurs because the determinant of the covariance matrix usually plays a very minor role in determining the maximum of the likelihood.

To actually find estimates, an iterative procedure is required. The Gauss-Newton method can be used to find least squares estimates. Because the determinant $|A(\xi)A'(\xi)|$ depends on ξ, some other method, e.g., Newton-Raphson, must be used to obtain MLE's. (Newton-Raphson was discussed in Christensen (1987, Section XV.4.)) In addition, the values $\hat{E}(e_t|Y,\xi)$ need to be evaluated. As in Section 3, we could use

$$
\hat{E}(e_t|Y,\xi) \doteq \hat{E}(e_t|Y_\infty,\xi)
$$

where

$$
\hat{E}(e_t|Y_\infty,\xi) = \begin{cases} 0 & t > n \\ e_t & t \leq n \end{cases}
\tag{14}
$$

along with the assumption that $e_t = 0$, $t \leq 0$. This assumption implies that $y_t = \mu$, $t \leq 0$ and thus we can compute e_t for $t = 1,\ldots,n$ using

$$
e_t = \sum_{s=1}^{q} \theta_s e_{t-s} + y_t - \sum_{s=1}^{p} \phi_s y_{t-s} - \alpha.
$$

These very strong assumptions can be improved upon in practice by incorporating Box and Jenkins' method of *back forecasting (backcasting)*. Backcasting is used to obtain values for $\hat{E}(e_t|Y,\xi)$. It is based on the observation that if w_t and e_t are both white noise processes, the mean and covariance function of a stationary process defined by

$$\Phi(B)(y_t - \mu) = \Theta(B)e_t$$

must be the same as the mean and covariance function of

$$\Phi(F)(y_t - \mu) = \Theta(F)w_t \tag{15}$$

where F is the *forward* shift operator, i.e., $F(w_t) = w_{t+1}$. Thus model (15) is every bit as good a model for the data as the standard $ARMA(p,q)$ model. Now, rather than assuming $e_t = 0$, $t = 0, -1, \ldots$ we assume $w_t = 0$ for $t = n+1, n+2, \ldots$. Rather than using (14), we use

$$\hat{E}(w_t|Y^\infty,\xi) = \begin{cases} w_t & t \geq 1 \\ 0 & t < n \end{cases} \tag{16}$$

where $Y^\infty = (y_1, \cdots, y_n, 0, 0, 0, \cdots)'$. Note that the assumption $w_t = 0$ for $t \geq n+1$ implies that $y_t = \mu$ for $t \geq n+1$. Rewriting (15) as

$$w_t = \sum_{s=1}^{q} \theta_s w_{t+s} + y_t - \sum_{s=1}^{p} \phi_s y_{t+s} - \alpha \tag{17}$$

we can compute $w_n, w_{n-1}, \cdots, w_1, \cdots$ recursively. This in turn allows us to backcast values $\hat{E}(y_t|Y^\infty,\xi)$ for $t < 1$, i.e.,

$$\hat{E}(y_t|Y^\infty,\xi) = \alpha + \sum_{s=1}^{p} \phi_s \hat{E}(y_{t+s}|Y^\infty,\xi)$$

$$+ \hat{E}(w_t|Y^\infty,\xi) - \sum_{s=1}^{q} \theta_s \hat{E}(w_{t+s}|Y^\infty,\xi)$$

where the terms $\hat{E}(w_t|Y^\infty,\xi)$ are computed using (16) and (17) and the values $\hat{E}(y_t|Y^\infty,\xi)$ are computed recursively for $t < 1$ and for $t \geq 1$, $\hat{E}(y_t|Y^\infty,\xi) = y_t$.

The backcasting values $\hat{E}(y_t|Y^\infty,\xi)$, $t < 1$ can be used in the standard $ARMA$ model to improve on the assumption $e_t = 0$, $t < 1$ and its consequence $y_t = \mu$, $t < 1$. Instead of these assumptions we choose a large value Q and assume that $\hat{E}(y_t|Y,\xi) = \hat{E}(y_t|Y^\infty,\xi)$ for $t = 0, 1, \ldots, 1 - Q$ and that $e_t = 0$ for $t < 1 - Q$. As before, the assumption on e_t implies that $y_t = \mu$ for $t < 1 - Q$.

The functions $L(\xi,\sigma^2)$ and $SSE(\xi)$ are now evaluated using

$$\hat{E}(e_t|Y,\xi) = \sum_{s=1}^{q} \theta_s \hat{E}(e_{t-s}|Y,\xi) + \hat{E}(y_t|Y,\xi) - \sum_{s=1}^{p} \phi_s \hat{E}(y_{t-s}|Y,\xi) - \alpha.$$

where

$$\hat{E}(y_t|Y,\xi) \doteq \begin{cases} y_t & t = 1,\ldots,n \\ \hat{E}(y_t|Y^\infty,\xi) & t = 1-Q,\ldots,0 \\ \mu & t \le -Q \end{cases}$$

$$\hat{E}(e_t|Y,\xi) \doteq 0 \qquad t \le -Q$$

and $\hat{E}(e_t|Y,\xi)$ is computed recursively for $t > -Q$.

One problem is that a specific choice of Q must be made. Without getting into details, we mention a basic consideration in the selection of Q. For a stationary $ARMA$ process, the covariance function is given in (5.2.12) as $\sigma(k) = \sigma^2 \sum_{s=0}^{\infty} \psi_s \psi_{s+k}$. Because $\sum_{s=0}^{\infty} \psi_s$ is finite, $\lim_{s\to\infty} \psi_s = 0$. It is intuitively obvious (and not hard to show) that if the correlation between y_{n+k} and the observed data is approaching zero then $\hat{E}(y_{n+k}|Y)$ must approach μ, the mean of the stationary process. If the correlation between the future and the present is zero, there is no basis for predicting the future as anything other than the mean of the process

The same phenomenon occurs in back forecasting. For a stationary process, the back forecasts should settle down around $\hat{\mu}$ for times t that are large and negative. Moreover, when the backcasts have approached μ, the predictions of e_t must be near zero. For a given Q, by checking whether these phenomena occur for t near $-Q$ one can decide whether Q is sufficiently large.

Brockwell and Davis (1987, Chapter 8) present methods for unconditional maximum likelihood and least squares estimation based on the innovations algorithm, cf. Exercises 5.5 and 5.6. These use the exact value of $\hat{E}(e_t|Y,\xi)$ rather than the approximation $\hat{E}(e_t|Y_\infty,\xi)$. Brockwell and Davis also present asymptotic distributional and inferential results.

EXERCISE 5.6.   Let $e(y_t|y_{t-1},\cdots,y_1;\xi) = y_t - \hat{E}(y_t|y_{t-1},\cdots,y_1,\xi)$ from an $ARMA(p,q)$ model with parameters (ξ,σ^2) and let $p_t(\xi,\sigma^2)$ be the corresponding prediction variance. Use the results of Exercise 5.5 to show that

$$L(\xi,\sigma^2) = (2\pi)^{-\frac{n}{2}} \left[\prod_{t=1}^{n} p_t(\xi,\sigma^2)\right]^{-\frac{1}{2}}$$

$$\times \exp\left\{ -y_1^2/2\sigma(0)^2 - \sum_{t=1}^{n} [e(y_t|y_{t-1},\cdots,y_1;\xi)]^2 \Big/ 2p_t(\xi,\sigma^2) \right\}.$$

Hint: Show that there is a nonsingular linear transformation between the data vector Y and the vector of sequential prediction errors.

ESTIMATION FOR $ARIMA(p,d,q)$ MODELS

As in Section 3, $ARIMA(p,d,q)$ models are analyzed by considering the corresponding $ARMA(p,q)$ model for

$$z_t = \nabla^d y_t .$$

V.6   Model Selection

Box and Jenkins (1970) suggest performing model selection by examination of the empirical correlations and partial correlations. In addition, general model selection criteria such as the Akaike information criterion (AIC) and Schwarz's asymptotic Bayesian modification of the AIC can be applied to time domain models (cf. Akaike, 1973 and Schwarz, 1978).

BOX-JENKINS

The Box-Jenkins approach is based on two facts. First, for an $AR(p)$ process, $\phi(k) = 0$ for $k > p$. Second, for a $MA(q)$ process, $\rho(k) = 0$ for $k > q$. Empirical estimates $\hat{\phi}(k)$ and $\hat{\rho}(k)$ are obtained as in the previous section.

Recall that the stationary invertible $ARMA(p, q)$ model

$$\Phi(B)(y_t - \mu) = \Theta(B)e_t$$

can be written as an infinite moving average

$$y_t - \mu = \frac{\Theta(B)}{\Phi(B)}e_t$$

and also as an infinite autoregressive process

$$\frac{\Phi(B)}{\Theta(B)}(y_t - \mu) = e_t \,.$$

Moreover, the infinite sum $\Theta(1)/\Phi(1) = 1/\left[\Phi(1)/\Theta(1)\right]$ is finite so the terms in each sequence converge to zero fairly quickly.

If $q - 0$, $\hat{\rho}(k)$ should gradually approach zero while $\hat{\phi}(k)$ drops off precipitously after $k = p$. Similarly if $p = 0$, $\hat{\phi}(k)$ should gradually approach zero while $\hat{\rho}(k)$ drops off after $k = p$. If both p and q are positive, then the first p terms of $\Phi(B)/\Theta(B)$ will tend to dominate so $\hat{\phi}(k)$ should drop off substantially after $k = p$. Similarly, the first q terms of $[\Phi(B)]^{-1}\Theta(B)$ should dominate leading $\hat{\rho}(k)$ to drop off substantially after $k = q$.

For example, if $|\hat{\phi}(k)|$ drops precipitously to near zero after $k = 3$, and if $|\hat{\rho}(k)|$ decreases gradually to zero, an $AR(3)$ model is suggested. If $|\hat{\rho}(k)|$ drops quickly after $k = 2$ and $|\hat{\phi}(k)|$ decreases gradually, a $MA(2)$ model is suggested. If $|\hat{\phi}(k)|$ drops after $k = 3$ and $|\hat{\rho}(k)|$ drops after $k = 2$, an $ARMA(3, 2)$ model is suggested.

The standard errors of $\hat{\rho}(k)$ and $\hat{\phi}(k)$ can be used to help decide which correlations are important. For white noise, i.e., $\rho(k) = 0$ for all $k \geq 1$, the large sample standard deviation of $\hat{\rho}(k)$ is estimated by

$$\text{SE}(\hat{\rho}(k)) = \frac{1}{\sqrt{n}}\left[1 + 2\sum_{j=1}^{k-1}\hat{\rho}(j)^2\right]^{\frac{1}{2}},$$

cf. Bartlett (1946). Similarly, Quenille (1949) has shown that for large samples from an $AR(p)$ the standard error of $\hat{\phi}(k)$ for $k > p$ is

$$\text{SE}(\hat{\phi}(k)) \doteq 1/\sqrt{n}.$$

To identify important correlations, it is useful to plot $(k, \hat{\rho}(k))$ and $(k, \hat{\phi}(k))$. Values with absolute values greater than, say, twice the standard error are likely to be nonzero.

Nonstationarity can sometimes be identified by graphing the time series. A stationary process should display constant variability about its mean value μ. If the variability in the plot seems to increase with time, a log transformation may alleviate the problem. If a trend appears in the plot, differencing (i.e., considering $\nabla^d y_t$) would be in order. Once d has been determined so that the plot looks stationary, p and q can be identified as above yielding an $ARIMA(p, d, q)$ model.

As a supplement to visual inspection of the graph, there are several statistics that give suggestions of nonstationarity. First, a value of $|\hat{\phi}(1)|$ near 1 suggests possible nonstationarity. Also, $\hat{\rho}(k)$ approaching zero very slowly suggests nonstationarity. Finally, nonstationarity is suggested in the frequency domain by the existence of one or more small frequencies that are very important. In practice, the differencing parameter d is rarely taken to be greater than 2.

Having identified a time domain model, we can attempt to check its appropriateness. In any time domain model, e_t is a white noise process. Assume the most general model, an $ARIMA(p, d, q)$. The e_t process can be estimated by

$$\hat{e}_t = \sum_{s=1}^{q} \hat{\theta}_s \hat{e}_{t-s} + \nabla^d(y_t - \hat{\mu}) - \sum_{s=1}^{p} \hat{\phi}_s \nabla^d(y_{t-s} - \hat{\mu})$$

where it is assumed that $\hat{e}_t = 0$, $t = 0, -1, -2, \ldots$. Brockwell and Davis (1987, Section 9.4) suggest the use of standardized residuals based only on previous data,

$$r_t = \left[y_t - \hat{E}(y_t | y_{t-1}, \cdots, y_1; \hat{\xi}) \right] \Big/ \sqrt{p_t(\hat{\xi}, \hat{\sigma}^2)}$$

where $p_t(\xi, \sigma^2)$ is the prediction variance as found in Section 3. These residuals arise naturally when using the innovations algorithm to obtain unconditional maximum likelihood estimates (cf. Exercise 5.6) and appear to be an improvement over the unstandardized residuals $\hat{e}_t$.

The residuals can be plotted against time to detect evidence of non-stationarity. For example, the residuals may display a trend or increasing variability.

The correlation and partial correlation functions for white noise are

$$\rho(0) \equiv \phi(0) = 1$$
$$\rho(k) = \phi(k) = 0 \qquad k = 1, 2, \ldots .$$

The estimated correlations and partial correlations for the residual process should be similar if the model is correct.

The spectral density of white noise is $f(v) = \sigma^2$. If the model is correct, the residual process should not have any important frequencies.

Finally, to check whether the white noise is Gaussian, a normal plot of the residuals can be used, cf. Christensen (1987, Section XIII.2).

MODEL SELECTION CRITERIA

If one has identified several candidate models, a formal model selection criterion can be used to identify the best of these. The criteria to be considered require a maximum likelihood fit of the model, so computing the criteria for large numbers of models is expensive. (Using approximate fits for large numbers of models may be a reasonable practical procedure.)

The AIC and Schwarz's large sample Bayesian modification, the Bayesian information criterion (BIC), are discussed in Clayton, Geisser, and Jennings (1986). As applied to Gaussian time domain models that include a mean, these criteria can be estimated by

$$\mathrm{AIC} = \log \hat\sigma^2 + 2(p + q + 1)/n$$

and

$$\mathrm{BIC} = \log \hat\sigma^2 + (p + q + 1)\log(n)/n\,.$$

The model that minimizes the criterion is the best fitting model. In practice, one should use these criteria to identify a small group of best models. For each of these models, the residual process should be evaluated to determine whether a Gaussian white noise assumption appears reasonable. For multivariate autoregressions, a simulation by Lütkepohl (1985) suggests that the BIC outperforms the AIC and many other criteria in terms of correct model identification and minimization of prediction errors. Clayton, Geisser, and Jennings (1986) examine simulations for data that are not time series. Hurvich and Tsai (1989) discuss correcting the AIC for bias.

The AIC is not asymptotically consistent. The BIC criterion is a modification of the AIC specifically developed to achieve consistency. It is not surprising that the BIC outperforms criteria that are not consistent for large samples where the true model is one of the models being considered. In practice however, our models are only approximations to reality. The appropriate question is not "Which criterion selects the correct model most often?" but "Which criterion selects the best approximation most often?"

AN EXAMPLE

EXAMPLE 5.6.1.   Consider again the coal production data from Example 4.2.1. The data are given in Table IV.1 and are displayed in Figure IV.1. Figures V.1 and V.2 give the estimated correlation and partial correlation

```
            -1.0 -0.8 -0.6 -0.4 -0.2  0.0  0.2  0.4  0.6  0.8  1.0
  k    RHO    +----+----+----+----+----+----+----+----+----+----+
  1   0.769                           |     XXXX|XXXXXXXXXXXXXXX
  2   0.592                        |        XXXXXXXXX|XXXXXX
  3   0.552                     |           XXXXXXXXXXX|XXX
  4   0.411                  |              XXXXXXXXXXX |
  5   0.296                  |              XXXXXXXX    |
  6   0.204               |                 XXXXXX      |
  7   0.115               |                 XXXX        |
  8   0.030               |                 XX          |
  9  -0.046               |               XX            |
 10  -0.071               |               XXX           |
 11  -0.131               |              XXXX            |
 12  -0.198               |             XXXXXX           |
 13  -0.201               |             XXXXXX           |
 14  -0.202               |             XXXXXX           |
 15  -0.231               |            XXXXXXX           |
 16  -0.216               |             XXXXXX           |
 17  -0.177               |             XXXX             |
```

FIGURE V.1. Estimated Autocorrelations for the Coal Production Data

functions. Standard errors for the correlations can be based on Bartlett and Quenille's asymptotic formulae; the resulting 95% rejection regions for testing that the correlations are zero are displayed as vertical bars in the figures. Note that multiple tests of the correlations are being performed without controlling the overall error rate; thus any marginally significant correlations are still questionable. The autocorrelations are dying out quite slowly and the dominant partial autocorrelation is $\hat{\phi}(1) = .769$. These traits are consistent with a nonstationary process.

In an attempt to eliminate nonstationarity, we consider the first difference process $y_{i+1} - y_i$, $i = 1, \ldots, 60$. The data are plotted in Figure V.3. The estimated autocorrelation and partial autocorrelation functions are given in Figures V.4 and V.5. Neither plot gives any strong indication of nonstationarity so we proceed to model the first differences.

The small correlation and partial correlation values at $k = 1$ and the relatively large values at $k = 2, 3$ suggest that the $ARIMA$ models $(0, 1, 0)$, $(2, 1, 2)$, and $(3, 1, 3)$ should be examined. Summary statistics for the three models along with $(1,1,1)$ and several others are contained in Table V.1. All of the models in Table V.1 were fitted by unconditional least squares with backcasting. Recall that this is often a good approximation to the maximum likelihood fit. The estimated AIC and BIC values require the maximum likelihood estimate of σ^2; this was approximated by the sum of squares error divided by 60. Note that relatively small changes in the AIC and BIC criteria correspond to substantial changes in the MSE. The importance of small changes in the information criteria should not be discounted. Of the

```
                -1.0 -0.8 -0.6 -0.4 -0.2  0.0  0.2  0.4  0.6  0.8  1.0
 k     PHI      +----+----+----+----+----+----+----+----+----+----+
 1    0.769                             |      XXXXXX|XXXXXXXXXXXXXX
 2    0.002                             |      X     |
 3    0.236                             |      XXXXXXX
 4   -0.227                       XXXXXXX            |
 5    0.031                             |     XX     |
 6   -0.136                             |  XXXX      |
 7    0.018                             |     X      |
 8   -0.106                             |  XXXX      |
 9   -0.020                             |    XX      |
10    0.026                             |     XX     |
11   -0.112                             |  XXXX      |
12   -0.048                             |    XX      |
13    0.001                             |     X      |
14    0.009                             |     X      |
15   -0.070                             |   XXX      |
16    0.043                             |     XX     |
17    0.008                             |     X      |
```

FIGURE V.2. Estimated Partial Autocorrelations for the Coal Production Data

four models that were considered with $p = q$, (2,1,2) fits the best as judged by the MSE and the AIC and fits second best as judged by the BIC. The best BIC model is the rather uninteresting (0,1,0). The unconditional least squares estimates of the parameters in the (2,1,2) model are given below.

Parameter	Estimate	S.E.	Ratio
ϕ_1	−0.8029	0.3075	−2.61
ϕ_2	0.4551	0.2576	−1.77
θ_1	−0.8029	0.3400	−2.36
θ_2	−0.0414	0.3301	−0.13
α	10.61	12.71	0.83
σ^2	2848		

It is particularly odd that, as measured by the ratio of the estimate to the standard error, the least significant of the parameters is θ_2. This seemingly contradicts the fact, apparent from Table V.1, that the $ARIMA(2,1,1)$ model gives a substantially worse fit than the (1,1,2) model. This phenomenon is probably associated with the fact that the parameter estimates are highly correlated. The asymptotic correlation matrix is given below.

	$\hat{\phi}_1$	$\hat{\phi}_2$	$\hat{\theta}_1$	$\hat{\theta}_2$	$\hat{\alpha}$
$\hat{\phi}_1$	1.000	0.801	0.918	0.867	−0.028
$\hat{\phi}_2$	0.801	1.000	0.731	0.863	−0.018
$\hat{\theta}_1$	0.918	0.731	1.000	0.920	−0.030
$\hat{\theta}_2$	0.867	0.863	0.920	1.000	−0.027
$\hat{\alpha}$	−0.028	−0.018	−0.030	−0.027	1.000

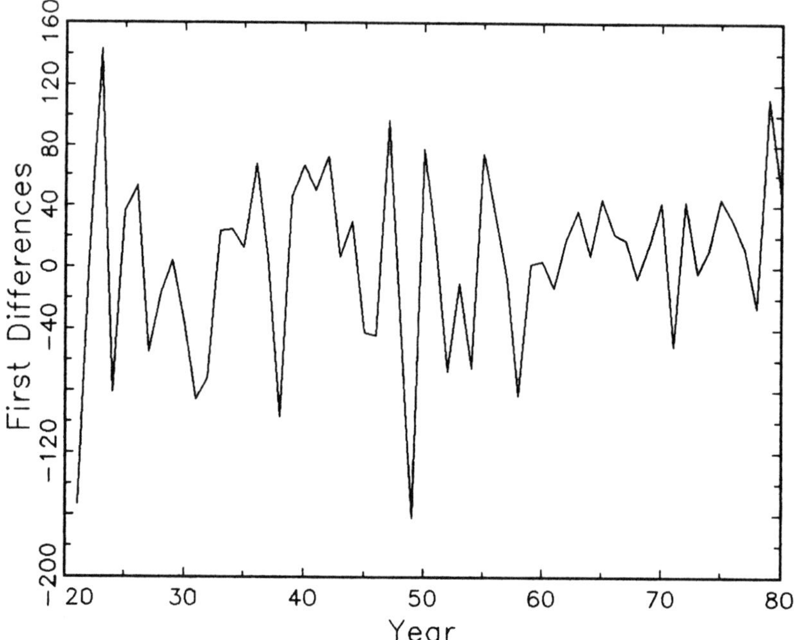

FIGURE V.3. First Difference of the Coal Production Data

TABLE V.1. Coal Production Statistics

MODEL	DFE	SSE	MSE	AIC	BIC
(0,1,0)	59	200,191	3393	8.15	8.18
(1,1,1)	57	194,752	3417	8.19	8.29
(2,1,2)	55	156,634	2848	8.03	8.21
(3,1,3)	53	155,360	2931	8.09	8.34
(2,1,1)	56	189,977	3392	8.19	8.33
(1,1,2)	56	163,739	2924	8.05	8.18
(1,1,2)*	57	165,906	2911	8.02	8.13
(0,1,2)*	58	174,920	3016	8.04	8.11

* indicates a model fitted without a constant

From the summary statistics in Table V.1, all of the models (0,1,2)*, (1,1,2)*, (1,1,2), and (2,1,2) appear reasonable. The models (0,1,2)* and (1,1,2)* are *ARIMA* models in which the mean is assumed to be zero. Having had the benefit of looking at the seven years of additional data displayed in Figure IV.10, I am inclined to believe that the series has an increasing trend so my choice for further consideration is (1,1,2). Recall that a linear trend corresponds to a nonzero mean in the first difference process. The unconditional least squares estimates for (1,1,2) are given below.

```
              -1.0 -0.8 -0.6 -0.4 -0.2  0.0  0.2  0.4  0.6  0.8  1.0
  k    RHO    +----+----+----+----+----+----+----+----+----+----+
  1  -0.067                        |  XXX        |
  2  -0.282                      X | XXXXXX       |
  3   0.245                        |      XXXXXXX|
  4  -0.043                        |   XX        |
  5  -0.044                        |   XX        |
  6   0.042                        |    XX       |
  7   0.019                        |    X        |
  8  -0.032                        |   XX        |
  9  -0.099                        |  XXX        |
 10   0.038                        |    XX       |
 11   0.108                        |    XXXX     |
 12  -0.168                        | XXXXX       |
 13  -0.029                        |   XX        |
 14   0.105                        |    XXXX     |
 15  -0.118                       |  XXXX         |
 16  -0.026                       |   XX          |
 17   0.113                       |    XXXX       |
```

FIGURE V.4. Estimated Autocorrelations for the First Difference of the Coal Production Data

```
              -1.0 -0.8 -0.6 -0.4 -0.2  0.0  0.2  0.4  0.6  0.8  1.0
  k    PHI    +----+----+----+----+----+----+----+----+----+----+
  1  -0.067                        |  XXX        |
  2  -0.287                      X | XXXXXX       |
  3   0.221                        |      XXXXXXX
  4  -0.114                        | XXXX        |
  5   0.097                        |    XXX      |
  6  -0.067                        |  XXX        |
  7   0.085                        |    XXX      |
  8  -0.069                        |  XXX        |
  9  -0.074                        |  XXX        |
 10  -0.005                        |    X        |
 11   0.091                        |    XXX      |
 12  -0.147                        | XXXXX       |
 13   0.016                        |    X        |
 14  -0.031                        |   XX        |
 15  -0.040                        |   XX        |
 16  -0.031                        |   XX        |
 17   0.060                        |    XXX      |
```

FIGURE V.5. Estimated Partial Autocorrelations for the First Difference of the Coal Production Data

Parameter	Estimate	S.E.	Ratio
ϕ_1	-0.4249	0.2289	-1.86
θ_1	-0.4658	0.2241	-2.08
θ_2	0.3773	0.1562	2.42
α	6.305	7.607	0.83

$$
\begin{aligned}
\text{dfE} &= 56 \\
\text{SSE} &= 163,739 \text{ (back forecasts excluded)} \\
\text{MSE} &= 2924.
\end{aligned}
$$

The correlation matrix for the estimated parameters is

	$\hat{\phi}_1$	$\hat{\theta}_1$	$\hat{\theta}_2$	$\hat{\alpha}$
$\hat{\phi}_1$	1.000	0.830	0.579	-0.038
$\hat{\theta}_1$	0.830	1.000	0.823	-0.053
$\hat{\theta}_2$	0.579	0.823	1.000	-0.059
$\hat{\alpha}$	-0.038	-0.053	-0.059	1.000

The residuals from this fitted model should be evaluated to see if they are consistent with a white noise error process. The correlation and partial correlation functions are given in Figures V.6 and V.7. Neither seems unreasonable. A rankit (normal scores) plot looks reasonably linear and a frequency analysis of the residuals gives spectral estimates that are consistent with the error process being white noise. The primary problem with the residuals is that the variability seems to decrease after 1960. The plot of residuals versus year is given in Figure V.8.

Assuming that the possible heteroscedasticity is not a serious problem, what have we accomplished? Not a great deal. The model (1,1,2) does not really explain the data very well. Fuller (1976, p. 364) suggests using F statistics for approximate tests of model adequacy. These are certainly reasonable statistics for model comparisons. The problem with their use is in determining an appropriate reference distribution. Testing (1,1,2) against (0,1,0) gives

$$
F = \frac{(200,191 - 163,739)/3}{2924} = 4.155.
$$

If we compare the test statistic to an $F(3, 56)$ distribution we obtain a rough P value of .01. For comparing a model to the model of no $ARMA$ structure, this is not particularly impressive. The R^2 type measure

$$
\frac{200,191 - 163,739}{200,191} = .182
$$

indicates that the (1,1,2) model accounts for only 18% of the variability in the data. This, by itself, does not mean that the fitted model is a poor one; we could have the perfect model but a process that is subject to a great

```
            -1.0 -0.8 -0.6 -0.4 -0.2  0.0  0.2  0.4  0.6  0.8  1.0
  k    RHO    +----+----+----+----+----+----+----+----+----+----+
  1    0.002                          |     X     |
  2    0.012                          |     X     |
  3    0.113                          |    XXXX   |
  4    0.030                          |     XX    |
  5   -0.044                          |    XX     |
  6    0.033                          |     XX    |
  7   -0.022                          |    XX     |
  8   -0.033                          |    XX     |
  9   -0.101                          |   XXXX    |
 10    0.002                         |      X      |
 11    0.045                         |      XX     |
 12   -0.144                         |  XXXXX      |
 13   -0.039                         |     XX      |
 14    0.029                         |      XX     |
 15   -0.085                         |    XXX      |
 16   -0.049                         |     XX      |
 17    0.072                         |       XXX   |
```

FIGURE V.6. Estimated Autocorrelations for the Residuals of the (1,1,2) model


```
            -1.0 -0.8 -0.6 -0.4 -0.2  0.0  0.2  0.4  0.6  0.8  1.0
  k    PHI    +----+----+----+----+----+----+----+----+----+----+
  1    0.002                         |     X     |
  2    0.012                         |     X     |
  3    0.113                         |    XXXX    |
  4    0.029                         |     XX     |
  5   -0.047                         |    XX      |
  6    0.020                         |     X      |
  7   -0.028                         |    XX      |
  8   -0.025                         |    XX      |
  9   -0.105                         |   XXXX     |
 10    0.005                         |     X      |
 11    0.060                         |     XXX    |
 12   -0.126                         |  XXXX      |
 13   -0.038                         |    XX      |
 14    0.015                         |     X      |
 15   -0.054                         |    XX      |
 16   -0.039                         |    XX      |
 17    0.053                         |     XX     |
```

FIGURE V.7. Estimated Partial Autocorrelations for the Residuals of the (1,1,2) model

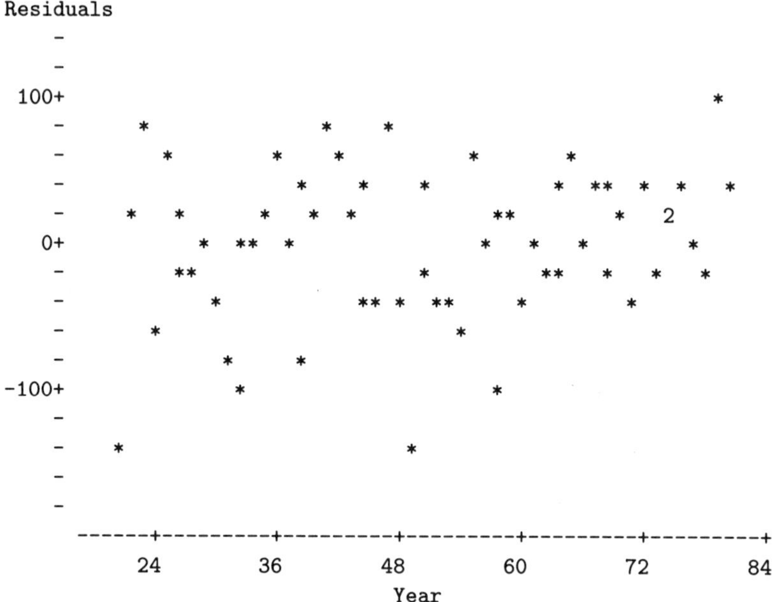

FIGURE V.8. Time Plot for the Residuals of the (1,1,2) Model

deal of variability. However, whether the model is appropriate or not, we cannot expect to obtain accurate forecasts from it.

One of the most important uses of time domain models is to predict the future. Table V.2 contains the actual coal production figures for 1981 through 1987 along with the forecasts based on the (1,1,2) model fitted both with and without a nonzero mean. Ninety-five percent prediction intervals are given for both models. While neither model gives very accurate predictions, the model that includes a trend does a bit better. The model without a trend gives predictions that are settling down around 785. For both models, the prediction intervals are so large that, in spite of the poor point predictions, the actual coal production values fall inside them. This results from the large estimates of σ^2 associated with the models.

The data from 1921 to 1987 are plotted in Figure V.9 along with the predicted values from (1,1,2). For the data from 1921 to 1980, which was used in estimation, the model consistently predicts the next observation to be similar to the last one. In other words, the predicted values are very similar to the data except that they are shifted to the right by one year. This phenomenon makes Figure V.9 appear misleadingly good. Visually, people tend to evaluate the proximity of two curves by their orthogonal distance. In Figure V.9 this is quite small because of the shifting phenomenon described above. In fact, for the purpose of prediction, we need to evaluate

TABLE V.2. *ARIMA* Model Forecasts

Year	Actual	ARIMA (1,1,2)	95% Limits Lower	95% Limits Upper	ARIMA (1,1,2)*	95% Limits Lower	95% Limits Upper
81	818.4	786.9	680.9	892.9	782.6	676.8	888.4
82	833.5	795.8	642.8	948.8	786.1	632.5	939.8
83	777.9	798.3	630.7	966.0	784.6	615.4	953.8
84	891.8	803.6	615.0	992.1	785.3	594.5	976.1
85	879.0	807.6	603.3	1012.0	785.0	577.8	992.2
86	886.1	812.2	592.1	1032.4	785.1	561.6	1008.6
87	912.7	816.6	582.2	1051.0	785.1	546.8	1023.4

The model (1,1,2)* is fitted with a mean of zero.

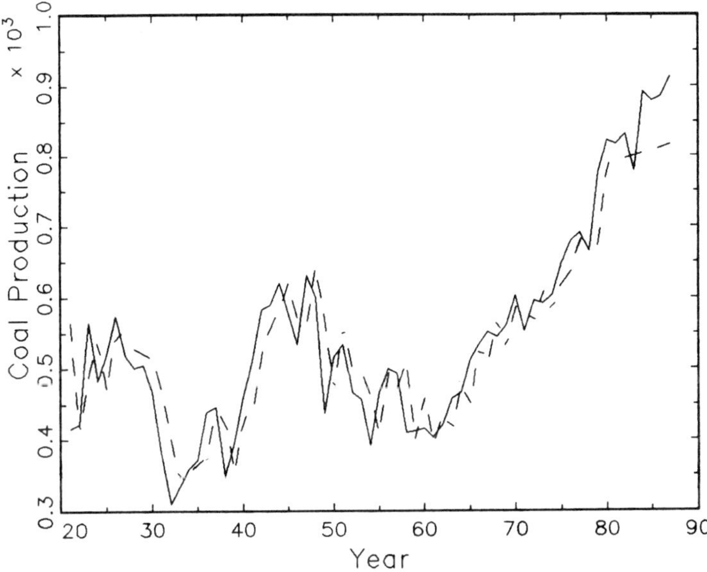

FIGURE V.9. Coal Production (solid line) and Predicted Values (dashed line) for 1921 to 1987

the vertical distance between the two curves at each time point. Thus, although it is less directly relevant to the prediction problem, the plot of the residuals, Figure V.8, gives a more accurate picture of the quality of the forecasts. The one-year shift phenomenon in Figure V.9 is a direct result of the poor fitting (1,1,2) model. We are actually fitting an $ARMA(1,2)$ model to $z_t = y_t - y_{t-1}$, $t = 2, \ldots, 61$. The predicted differences are say, $\hat{z}_t$ and the predicted values are $\hat{y}_t = \hat{z}_t + y_{t-1}$. As compared to the y_{t-1} values, the predicted differences are relatively close to zero. Thus, the prediction plot is essentially the data plot lagged by one.

The quality of the predictions goes down markedly after 1980. It is always harder to predict the future than the past and our model is not particularly effective. Actually, both Figures V.8 and V.9 suggest that the behavior of the series since 1960 may be inconsistent with the previous data. Thus, it is not surprising that the predictions for 1981 to 1987 are not terribly good. Perhaps a better approach would be to analyze only the data from 1960 forward. Unfortunately, 21 or even 28 observations make a rather inadequate data base for time series analysis.

In general, our only hope for prediction is that the future will behave like the relevant past. If there is no relevant past or if we include the irrelevant past, as we may have in this example, we are left to swim in the oceans of uncertainty, or worse, to follow random paths in the deserts of decision making that lead nowhere but give the illusion of progress. (Okay, okay, I'll take my tongue out of my cheek now.)

V.7   Seasonal Adjustment

Consider a time series consisting of monthly flypaper sales. It is reasonable that sales may be related to the previous couple of months, but sales may very well be related to the corresponding values in the previous year. An appropriate autoregressive model might be

$$y_t = \alpha + \phi_1 y_{t-1} + \phi_2 y_{t-2} + \phi_{1,1} y_{t-12} - \phi_1 \phi_{1,1} y_{t-13} - \phi_2 \phi_{1,1} y_{t-14} + e_t$$

or equivalently

$$y_t = \alpha + \phi_1 y_{t-1} + \phi_2 y_{t-2} + \phi_{1,1}[y_{t-12} - \phi_1 y_{t-13} - \phi_2 y_{t-14}] + e_t . \tag{1}$$

Write

$$\Phi(B) = 1 - \phi_1 B - \phi_2 B^2$$

and

$$\Phi_1(B^{12}) = 1 - \phi_{1,1} B^{12} ,$$

then model (1) is

$$\Phi_1(B^{12})\Phi(B)(y_t - \mu) = e_t .$$

This is referred to as a *multiplicative seasonal autoregressive* model and is written $AR(2) \times (1)_{12}$.

In general, if the seasonal effects occur every T time units, we can define a *multiplicative seasonal autoregressive integrated moving average* model $ARIMA(p, d, q) \times (P, D, Q)_T$

$$\Phi_P(B^T)\Phi(B)\nabla_T^D \nabla^d (y_t - \mu) = \Theta_Q(B^T)\Theta(B)e_t$$

where

$$\Phi_P(B^T) = 1 - \sum_{s=1}^{P} \phi_{s,P} B^{Ts}$$

$$\nabla_T^D = (1 - B^T)^D$$

and

$$\Theta_Q(B^T) = 1 - \sum_{s=1}^{Q} \theta_{s,Q} B^{Ts}.$$

These models are discussed in detail in Box and Jenkins (1970).

EXAMPLE 5.7.1.   Table V.3 consists of data from a daily census made of the inpatients at the University of Wisconsin Hospital between April 21 and July 29, 1974. These data were previously presented by Pandit and Wu (1983, Appendix A). Figure V.10 presents the data graphically. There is a clear periodicity in the data. While this suggests that a frequency analysis of the data may be particularly enlightening, our object is to illustrate techniques used for identifying multiplicative seasonal ARIMA models. Figures V.11 and V.12 present the estimated correlation function and the estimated partial correlation function for the series. Note that the correlation function displays periodic behavior with a period of 7. This makes sense in terms of a weekly cycle. Moreover, the correlations die out very slowly. Together these suggest the appropriateness of a seasonal difference of order 7.

It should be noted that the need for seasonal differencing is often accompanied by the need for regular differencing. Moreover, the need for seasonal differencing is often hidden until the regular differencing is completed.

TABLE V.3. University of Wisconsin Hospital Data

				Day			
Week	Su	M	Tu	W	Th	F	Sa
I	397	462	486	483	477	438	407
II	421	480	484	486	479	415	400
III	419	477	510	503	500	435	408
IV	417	478	497	500	512	450	421
V	423	471	496	478	463	413	396
VI	366	375	444	469	480	439	402
VII	442	492	507	518	493	439	399
VIII	428	476	499	488	460	419	380
IX	406	472	502	495	490	443	398
X	417	490	505	499	484	430	384
XI	392	452	455	426	414	405	379
XII	410	485	514	525	511	461	436
XIII	444	488	494	510	493	429	392
XIV	420	466	476	494	484	423	388
XV	411	472					

Figures V.13 and V.14 give the correlation function and partial correlation function for the series generated by taking differences of order 7. The correlations display the classic pattern of an autoregressive process with $p > 1$. The only problem is that they are dying out very slowly. The partial

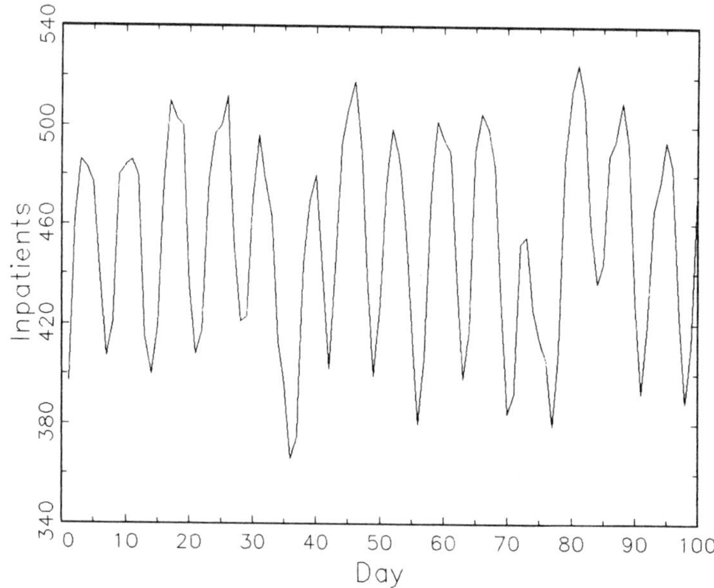

FIGURE V.10. Plot of Hospital Data

```
        -1.0 -0.8 -0.6 -0.4 -0.2  0.0  0.2  0.4  0.6  0.8  1.0
        +----+----+----+----+----+----+----+----+----+----+
 1   0.615                              XXXXXXXXXXXXXXX
 2  -0.048                              XX
 3  -0.503                   XXXXXXXXXXXXX
 4  -0.541                   XXXXXXXXXXXXX
 5  -0.168                          XXXX
 6   0.359                              XXXXXXXXX
 7   0.615                              XXXXXXXXXXXXXXX
 8   0.291                              XXXXXXX
 9  -0.281                         XXXXXXXX
10  -0.676               XXXXXXXXXXXXXXXXX
11  -0.661               XXXXXXXXXXXXXXXX
12  -0.238                          XXXXXX
13   0.301                              XXXXXXXXX
14   0.581                              XXXXXXXXXXXXXXX
15   0.335                              XXXXXXXXX
16  -0.148                          XXXXX
17  -0.482                    XXXXXXXXXXXX
18  -0.484                    XXXXXXXXXXXX
19  -0.109                          XXXX
20   0.383                              XXXXXXXXXXX
```

FIGURE V.11. Correlation Function for the Hospital Data

```
          -1.0 -0.8 -0.6 -0.4 -0.2  0.0  0.2  0.4  0.6  0.8  1.0
              +----+----+----+----+----+----+----+----+----+----+
  1    0.615                              XXXXXXXXXXXXXXXX
  2   -0.687                XXXXXXXXXXXXXXXXXX
  3   -0.092                              XXX
  4   -0.072                              XXX
  5    0.250                              XXXXXX
  6    0.277                              XXXXXXX
  7    0.024                              XX
  8   -0.510                XXXXXXXXXXXXXX
  9   -0.085                              XXX
 10   -0.232                         XXXXXX
 11   -0.070                              XXX
 12   -0.022                              XX
 13   -0.016                              X
 14    0.080                              XXX
 15   -0.158                          XXXX
 16    0.082                              XXX
 17   -0.048                              XX
 18   -0.165                          XXXX
 19    0.009                              X
 20   -0.005                              X
```

FIGURE V.12. Partial Correlation Function for the Hospital Data

correlations show an extremely large value for ϕ_1. These conclusions are consistent with the need for a regular difference.

Figures V.15 and V.16 give the correlation function and partial correlation function for the series generated by taking a seasonal difference of order 7 and a regular difference of order 1. The correlations suggest the need for a moving average term of order 7. Thus the suggested moving average portion of the model is $\Theta(B) = 1$ and $\Theta_1(B^7) = 1 - \theta_{1,1}B^7$. The partial correlations are large for $k = 6, 7, 13$ and not small for $k = 1$. One possible way to model this would be to take $\Phi(B) = 1 - \phi_1 B - \phi_6 B^6$ and $\Phi_1(B^7) = 1 - \phi_{1,1}B^7$. All together, the model is an $ARIMA([1,6],1,0) \times (1,1,1)_7$ where the notation $[1,6]$ is used to indicate the peculiar nature of $\Phi(B)$. Although this is a model that involves only four parameters in addition to σ^2, it is really quite complex. The model for y_t involves y_{t-k} for $k = 1, 2, 6, 7, 8, 9, 13,$ 14, 15, 16, 20, and 21.

While we halt this example now, it should be recognized that this is only the beginning of an analysis of this data. The model needs to be fitted, the residuals checked, other models need to be investigated. A plot of the regularly and seasonally differenced series is also quite interesting. There is a strange increase in variability that occurs around 35 days and again around 70 days. This may just be an oddity of the data or it may be an indication of some structure in the hospital that generates a five-week seasonal pattern. More investigation of hospital practices or more data

```
             -1.0 -0.8 -0.6 -0.4 -0.2  0.0  0.2  0.4  0.6  0.8  1.0
             +----+----+----+----+----+----+----+----+----+----+
    1   0.830                              XXXXXXXXXXXXXXXXXXXXX
    2   0.590                              XXXXXXXXXXXXXXX
    3   0.393                              XXXXXXXXXX
    4   0.158                              XXXX
    5  -0.094                           XXX
    6  -0.348                    XXXXXXXXX
    7  -0.531               XXXXXXXXXXXXX
    8  -0.559               XXXXXXXXXXXXX
    9  -0.558               XXXXXXXXXXXXX
   10  -0.543               XXXXXXXXXXXXX
   11  -0.443                 XXXXXXXXXXX
   12  -0.310                   XXXXXXXX
   13  -0.202                     XXXXX
   14  -0.076                        XXX
   15   0.062                           XXX
   16   0.173                           XXXXX
   17   0.267                           XXXXXXX
   18   0.302                           XXXXXXXX
   19   0.293                           XXXXXXXX
```

FIGURE V.13. Correlation Function for the Seasonally Differenced Hospital Data

```
             -1.0 -0.8 -0.6 -0.4 -0.2  0.0  0.2  0.4  0.6  0.8  1.0
             +----+----+----+----+----+----+----+----+----+----+
    1   0.830                              XXXXXXXXXXXXXXXXXXXXX
    2  -0.317                  XXXXXXXXX
    3   0.039                           XX
    4  -0.348                  XXXXXXXXX
    5  -0.167                      XXXX
    6  -0.354                  XXXXXXXXX
    7  -0.035                          XX
    8   0.168                           XXXXX
    9  -0.195                     XXXXX
   10  -0.042                         XX
   11  -0.043                         XX
   12  -0.133                      XXXX
   13  -0.236                    XXXXXX
   14   0.072                           XXX
   15   0.071                           XXX
   16  -0.103                      XXXX
   17   0.079                           XXX
   18  -0.118                      XXXX
   19  -0.121                      XXXX
```

FIGURE V.14. Partial Correlation Function for the Seasonally Differenced Hospital Data

```
          -1.0 -0.8 -0.6 -0.4 -0.2  0.0  0.2  0.4  0.6  0.8  1.0
             +----+----+----+----+----+----+----+----+----+----+
  1    0.215                                XXXXXX
  2   -0.115                               XXXX
  3    0.112                               XXXX
  4    0.049                               XX
  5    0.026                               XX
  6   -0.226                            XXXXXXX
  7   -0.463                       XXXXXXXXXXXXX
  8   -0.086                             XXX
  9   -0.058                             XX
 10   -0.252                           XXXXXXX
 11   -0.100                            XXXX
 12    0.071                               XXX
 13   -0.044                             XX
 14   -0.030                             XX
 15    0.072                               XXX
 16    0.056                               XX
 17    0.171                               XXXXX
 18    0.115                               XXXX
 19    0.004                               X
```

FIGURE V.15. Correlation Function for the Regularly and Seasonally Differenced Hospital Data

would be needed to verify a five-week cycle. Of course, the odd behavior of the series could just mean that differencing has not achieved stationarity.

V.8   The Multivariate State-Space Model and the Kalman Filter

The *state-space model* provides a quite general paradigm for modeling multivariate time series. At each time t, a $q_t \times 1$ vector of observations Y_t is observed. We assume that Y_t satisfies a linear model

$$Y_t = X_t \beta_t + e_t \tag{1}$$

in which all of the components are allowed to vary with time but X_t always has p columns. The error vector e_t is assumed to satisfy

$$E(e_t) = 0$$

and

$$\mathrm{Cov}(e_t) = V_t .$$

The linear model (1) is often called the *observation equation*.

```
          -1.0 -0.8 -0.6 -0.4 -0.2  0.0  0.2  0.4  0.6  0.8  1.0
            +----+----+----+----+----+---+----+----+----+----+----+
  1    0.215                               XXXXXX
  2   -0.169                             XXXXX
  3    0.191                               XXXXXX
  4   -0.055                             XX
  5    0.081                              XXX
  6   -0.310                         XXXXXXXXX
  7   -0.368                         XXXXXXXXXX
  8    0.014                               X
  9   -0.155                             XXXX
 10   -0.123                             XXXX
 11   -0.042                             XX
 12    0.066                              XXX
 13   -0.292                         XXXXXXXX
 14   -0.197                            XXXXX
 15   -0.029                             XX
 16   -0.193                            XXXXXX
 17   -0.005                              X
 18   -0.001                              X
 19   -0.026                             XX
```

FIGURE V.16. Partial Correlation Function for the Regularly and Seasonally Differenced Hospital Data

Dependencies between times are modeled through the $p \times 1$ vector β_t. Rather than assuming that β_t is fixed, we assume that β_t satisfies a multivariate autoregressive model. Let Φ be a $p \times p$ matrix. We assume that

$$\beta_t = \Phi\beta_{t-1} + \varepsilon_t \tag{2}$$

where ε_t is a $p \times 1$ error vector with

$$E(\varepsilon_t) = 0$$

and

$$\mathrm{Cov}(\varepsilon_t) = \Sigma_t \,.$$

The autoregressive model (2) is referred to as the *state* equation.

Finally, for $i = 1, \ldots, t$ and $j = 1, \ldots, t$, assume that

$$\mathrm{Cov}(e_i, e_j) = 0 \qquad i \neq j\,,$$
$$\mathrm{Cov}(\varepsilon_i, \varepsilon_j) = 0 \qquad i \neq j\,,$$

and

$$\mathrm{Cov}(e_i, \varepsilon_j) = 0 \qquad \text{all } i, j\,.$$

To start the sequence off, we assume that β_0 has

$$E(\beta_0) \;=\; \tilde{\beta}_0$$

$$\begin{aligned}
\mathrm{Cov}(\beta_0) &= P_0 \\
\mathrm{Cov}(\beta_0, e_i) &= 0 \qquad i = 1,\ldots,t \\
\mathrm{Cov}(\beta_0, \varepsilon_i) &= 0 \qquad i = 1,\ldots,t .
\end{aligned}$$

Traditionally, prediction of the unobservable vector β_t has been viewed as the primary goal of state-space model analysis. The *Kalman filter* is a recursive procedure for predicting β_t on the basis of $Y_t, Y_{t-1}, \ldots, Y_1$.

At first glance, equation (2) may seem rather restrictive. It appears to be a matrix version of a first-order autoregressive process. The fact that it appears to be first-order may seem restrictive. In fact, because (2) involves matrices, there is nothing intrinsically first order about it.

EXAMPLE 5.8.1.   A very basic model in many applications is that observations are the sum of a signal μ_t plus some noise e_t, i.e.,

$$y_t = \mu_t + e_t .$$

If the signal is the result of an autoregressive process, a state-space model is appropriate. Suppose μ_t is an $AR(3)$ process, i.e.,

$$\mu_t = \phi_1 \mu_{t-1} + \phi_2 \mu_{t-2} + \phi_3 \mu_{t-3} + \varepsilon_t .$$

The state equation is

$$\begin{pmatrix} \mu_t \\ \mu_{t-1} \\ \mu_{t-2} \end{pmatrix} = \begin{bmatrix} \phi_1 & \phi_2 & \phi_3 \\ 1 & 0 & 0 \\ 0 & 1 & 0 \end{bmatrix} \begin{pmatrix} \mu_{t-1} \\ \mu_{t-2} \\ \mu_{t-3} \end{pmatrix} + \begin{pmatrix} \varepsilon_t \\ 0 \\ 0 \end{pmatrix} .$$

The observation equation is

$$y_t = (1,0,0) \begin{pmatrix} \mu_t \\ \mu_{t-1} \\ \mu_{t-2} \end{pmatrix} + e_t .$$

Thus $X_t = (1,0,0)$ and $\beta_t = (\mu_t, \mu_{t-1}, \mu_{t-2})'$. Note that, in practice, it would be unusual for the values ϕ_1, ϕ_2, and ϕ_3 to be known. In the development of the Kalman filter, these are assumed to be known. Typically, they will be estimated from the data and the estimates will be substituted for the true parameters to make predictions.

The state-space model was originally introduced to track missiles using satellite observations of their positions. The satellite observations Y_t are subject to error. Based on a first-order differential equation, the actual position of the missile is modeled by a first-order autoregressive process.

The following example illustrates a model in which the matrix Φ is completely known.

EXAMPLE 5.8.2.   Phadke (1981) and Meinhold and Singpurwalla (1983) present a model useful in quality control. The number of defective items in a process is transformed into a value y_t. (The transformation is used to make the data distribution approximate a normal distribution.) The transformed number of defectives is modeled as a signal plus noise

$$y_t = \mu_t + e_t.$$

However, the signal is generated in an unusual fashion. The signal is subject to a drift. The underlying signal for defectives is a parameter θ_t determined by

$$\theta_t = \theta_{t-1} + w_{t,1}$$

where $w_{t,1}$ is an error term that is uncorrelated with other error terms. However, the signal for any particular observation is subject to additional error, i.e.,

$$\mu_t = \theta_t + w_{t,2}$$

where $w_{t,2}$ is an uncorrelated error term. Upon observing that

$$\mu_t = \theta_{t-1} + w_{t,1} + w_{t,2},$$

we see that the state equation is

$$\begin{pmatrix} \theta_t \\ \mu_t \end{pmatrix} = \begin{pmatrix} 1 & 0 \\ 1 & 0 \end{pmatrix} \begin{pmatrix} \theta_{t-1} \\ \mu_{t-1} \end{pmatrix} + \begin{pmatrix} w_{t,1} \\ w_{t,1} + w_{t,2} \end{pmatrix}$$

and the observation equation is

$$y_t = (0, 1) \begin{pmatrix} \theta_t \\ \mu_t \end{pmatrix} + e_t.$$

The matrix Φ in the state equation is completely known.

The state-space model and the Kalman filter were originally introduced by Kalman (1960) and Kalman and Bucy (1961). This was done in the engineering literature. Harrison and Stevens (1971, 1976) first presented the method as a Bayesian procedure based on Φ being known and a multivariate normal distribution for $(\beta_0', e_1', \ldots, e_t', \varepsilon_1', \ldots, \varepsilon_t')'$. Meinhold and Singpurwalla (1983) give a nice exposition of this approach. The approach taken here is based on linear expectations. Because linear expectations are the conditional expectations for multivariate normals, the derivation given here is also valid for multivariate normals. Deely and Lindley (1981) have extended the Bayesian analysis by incorporating the fact that the parameters of the state-space model are unknown. Diderrich (1985) has examined the relationship between the Kalman filter and Goldberger-Theil estimators. Wegman (1982) relates the Kalman filter to stochastic differential equations. Shumway (1988, p. 174) gives a number of references to various applications of the Kalman filter.

THE KALMAN FILTER

The Kalman filter is a procedure for predicting β_t based on the data $Y_1, \ldots, Y_t$. It is a recursive procedure in that the prediction of β_t is based on modifying the predictor of β_{t-1}. Let $x' = (Y'_{t-1}, Y'_{t-2}, \ldots, Y'_1)$. We wish to find $\hat{E}(\beta_t | Y_t, x)$. By Proposition 3.1.8,

$$\hat{E}(\beta_t | Y_t, x) = \hat{E}(\beta_t | x)$$
$$+ \operatorname{Cov}\left(\beta_t, Y_t - \hat{E}(Y_t | x)\right) [\operatorname{Cov}(Y_t - \hat{E}(Y_t | x))]^{-1} [Y_t - \hat{E}(Y_t | x)]. \quad (3)$$

We now proceed to identify the various parts of this equation.

First,

$$\hat{E}(\beta_t | x) = \hat{E}(\Phi \beta_{t-1} + \varepsilon_t | x)$$
$$= \Phi \hat{E}(\beta_{t-1} | x) + \hat{E}(\varepsilon_t | x).$$

However, ε_t is uncorrelated with the earlier errors and β_0 while x is a linear function of the earlier errors and β_0; thus $\operatorname{Cov}(\varepsilon_t, x) = 0$. By Proposition 3.1.5, $\hat{E}(\varepsilon_t | x) = 0$ and

$$\hat{E}(\beta_t | x) = \Phi \hat{E}(\beta_{t-1} | x). \quad (4)$$

Next, examine $Y_t - \hat{E}(Y_t | x)$.

$$\hat{E}(Y_t | x) = \hat{E}(X_t \beta_t + e_t | x)$$
$$= X_t \hat{E}(\beta_t | x) + \hat{E}(e_t | x)$$
$$= X_t \Phi \hat{E}(\beta_{t-1} | x)$$

where we have used (4) and the fact that $\operatorname{Cov}(e_t, x) = 0$. Thus,

$$Y_t - \hat{E}(Y_t | x) = Y_t - X_t \Phi \hat{E}(\beta_{t-1} | x). \quad (5)$$

The covariance matrix of this prediction error can be computed as follows. Using (1) and (2) and the fact that errors are uncorrelated with previous events

$$\operatorname{Cov}[Y_t - \hat{E}(Y_t | x)] = \operatorname{Cov}[Y_t - X_t \Phi \hat{E}(\beta_{t-1} | x)]$$
$$= \operatorname{Cov}[(X_t \beta_t + e_t) - X_t \Phi \hat{E}(\beta_{t-1} | x)]$$
$$= \operatorname{Cov}[(X_t [\Phi \beta_{t-1} + \varepsilon_t] + e_t) - X_t \Phi \hat{E}(\beta_{t-1} | x)] \quad (6)$$
$$= \operatorname{Cov}[X_t \Phi (\beta_{t-1} - \hat{E}(\beta_{t-1} | x)) + X_t \varepsilon_t + e_t]$$
$$= X_t \Phi \operatorname{Cov}[\beta_{t-1} - \hat{E}(\beta_{t-1} | x)] \Phi' X'_t + X_t \Sigma_t X'_t + V_t$$
$$= X_t [\Phi P_{t-1} \Phi' + \Sigma_t] X'_t + V_t$$

where

$$P_{t-1} = \operatorname{Cov}[\beta_{t-1} - \hat{E}(\beta_{t-1} | x)]. \quad (7)$$

Finally, we need to compute $\mathrm{Cov}[\beta_t, Y_t - \hat{E}(Y_t|x)]$. We begin the computation by mentioning four facts. First, β_t is a linear function of β_0, ε_t and the errors previous to time t, thus

$$\mathrm{Cov}(\beta_t, e_t) = 0.$$

Second, β_{t-1} is a linear function of β_0 and errors previous to time t, so

$$\mathrm{Cov}(\beta_{t-1}, \varepsilon_t) = 0.$$

Third, by Proposition 3.1.10

$$\mathrm{Cov}(\beta_{t-1}, \beta_{t-1} - \hat{E}(\beta_{t-1}|x)) = P_{t-1}.$$

Fourth, because $\mathrm{Cov}(\varepsilon_t, x) = 0$ and $\hat{E}(\beta_{t-1}|x)$ is a linear function of x,

$$\mathrm{Cov}(\varepsilon_t, \beta_{t-1} - \hat{E}(\beta_{t-1}|x)) = 0.$$

The computation goes as follows.

$$
\begin{aligned}
\mathrm{Cov}&[\beta_t, Y_t - \hat{E}(Y_t|x)] \\
&= \mathrm{Cov}[\beta_t, X_t\Phi(\beta_{t-1} - \hat{E}(\beta_{t-1}|x)) + X_t\varepsilon_t + e_t] \\
&= \mathrm{Cov}[\beta_t, X_t\Phi(\beta_{t-1} - \hat{E}(\beta_{t-1}|x))] + \mathrm{Cov}[\beta_t, X_t\varepsilon_t] \\
&= \mathrm{Cov}[\Phi\beta_{t-1} + \varepsilon_t, X_t\Phi(\beta_{t-1} - \hat{E}(\beta_{t-1}|x))] \\
&\quad + \mathrm{Cov}[\Phi\beta_{t-1} + \varepsilon_t, X_t\varepsilon_t] \\
&= \mathrm{Cov}[\Phi\beta_{t-1}, X_t\Phi(\beta_{t-1} - \hat{E}(\beta_{t-1}|x))] + \mathrm{Cov}[\varepsilon_t, X_t\varepsilon_t] \\
&= \Phi\mathrm{Cov}[\beta_{t-1}, \beta_{t-1} - \hat{E}(\beta_{t-1}|x)]\Phi'X_t' + \Sigma_t X_t' \\
&= (\Phi P_{t-1}\Phi' + \Sigma_t)X_t'.
\end{aligned}
\tag{8}
$$

Substituting (4), (5), (6), and (8) into (3) gives the standard form for the Kalman filter. Let

$$R_t = \Phi P_{t-1}\Phi' + \Sigma_t \tag{9}$$

then

$$
\begin{aligned}
\hat{E}(\beta_t|Y_t, x) = {}& \\
&\Phi\hat{E}(\beta_{t-1}|x) + R_t X_t'[X_t R_t X_t' + V_t]^{-1}[Y_t - X_t\Phi\hat{E}(\beta_{t-1}|x)].
\end{aligned}
\tag{10}
$$

Equation (10) is a formula for predicting β_t given Y_t, the prediction of β_{t-1}, and the matrix P_{t-1} defined in (7). P_{t-1} enters through equation (9). Note that the only matrix assumed to be nonsingular is $X_t R_t X_t' + V_t$. Neither V_t nor Σ_t are assumed nonsingular. This is important in some applications, e.g., Example 5.8.1.

To use (10) recursively, we need a recursive formula for P_t. From Proposition 3.1.9,

$$
\begin{aligned}
P_t &= \mathrm{Cov}[\beta_t - \hat{E}(\beta_t|Y_t, x)] \\
&= \mathrm{Cov}[\beta_t - \hat{E}(\beta_t|x)] - \mathrm{Cov}[\beta_t, Y_t - \hat{E}(Y_t|x)] \\
&\quad \times [\mathrm{Cov}(Y_t - \hat{E}(Y_t|x))]^{-1}\mathrm{Cov}[Y_t - \hat{E}(Y_t|x), \beta_t].
\end{aligned}
$$

Parts of this have already been computed in (6) and (8). The only new computation is

$$\text{Cov}[\beta_t - \hat{E}(\beta_t|x)] \;=\; \text{Cov}[\Phi(\beta_{t-1} - \hat{E}(\beta_{t-1}|x)) + \varepsilon_t]$$
$$\;=\; \text{Cov}[\Phi(\beta_{t-1} - \hat{E}(\beta_{t-1}|x))] + \text{Cov}[\varepsilon_t]$$
$$\;=\; \Phi P_{t-1}\Phi' + \Sigma_t \,.$$

Thus, using (9), we get the recursive formula

$$P_t = R_t - R_t X_t'[X_t R_t X_t' + V_t]^{-1} X_t R_t \,. \tag{11}$$

Note that the recursive nature of (11) lies in the fact that R_t is a function of P_{t-1}. The matrix P_t is useful in updating predictions but also provides standard errors of prediction for linear combinations of β_t.

To start the recursive process based on (9), (10), and (11), we need an initial prediction for β_0 and the covariance matrix for the error of that prediction. The initial prediction is based on no data so $\hat{E}(\beta_0) = \text{E}(\beta_0) = \tilde{\beta}_0$ and $\text{Cov}(\beta_0 - \hat{E}(\beta_0)) = \text{Cov}(\beta_0) = P_0$.

Having developed a procedure for obtaining $\hat{E}(\beta_t|Y_t, x)$, prediction of the future given $Y_1, \ldots, Y_t$ is easily performed

$$\hat{E}(\beta_{t+1}|Y_t, x) \;=\; \hat{E}(\Phi\beta_t + \varepsilon_{t+1}|Y_t, x)$$
$$\;=\; \Phi\hat{E}(\beta_t|Y_t, x) \,.$$

The prediction covariance is

$$\text{Cov}[\beta_{t+1} - \hat{E}(\beta_{t+1}|Y_t, x)] \;=\; \text{Cov}[\Phi\{\beta_t - \hat{E}(\beta_t|Y_t, x)\} + \varepsilon_{t+1}]$$
$$\;=\; \Phi P_t\Phi' + \Sigma_{t+1}$$
$$\;=\; R_{t+1} \,.$$

Similarly,

$$\hat{E}(Y_{t+1}|Y_t, x) \;=\; \hat{E}(X_{t+1}\beta_{t+1} + e_{t+1}|Y_t, x)$$
$$\;=\; X_{t+1}\hat{E}(\beta_{t+1}|Y_t, x)$$
$$\;=\; X_{t+1}\Phi\hat{E}(\beta_t|Y_t, x)$$

with prediction covariance matrix

$$\text{Cov}[Y_{t+1} - \hat{E}(Y_{t+1}|Y_t, x)] \;=\; \text{Cov}[X_{t+1}\{\beta_{t+1} - \hat{E}(\beta_{t+1}|Y_t, x)\} + e_{t+1}]$$
$$\;=\; X_{t+1}\text{Cov}[\beta_{t+1} - \hat{E}(\beta_{t+1}|Y_t, x)]X_{t+1}' + V_t$$
$$\;=\; X_{t+1}R_{t+1}X_{t+1}' + V_t \,.$$

To predict events further in the future,

$$\hat{E}(\beta_{t+r}|Y_t, x) = \Phi^r \hat{E}(\beta_t|Y_t, x)$$

with the prediction covariance matrix obtained recursively using

$$\mathrm{Cov}[\beta_{t+r} - \hat{E}(\beta_{t+r}|Y_t, x)] = \Phi\mathrm{Cov}[\beta_{t+r-1} - \hat{E}(\beta_{t+r-1}|Y_t, x)]\Phi' + \Sigma_{t+r}.$$

The covariance matrix depends only on P_t, Φ, and $\Sigma_{t+1}, \ldots, \Sigma_{t+r}$. To predict future observations

$$
\begin{aligned}
\hat{E}(Y_{t+r}|Y_t, x) &= X_{t+r}\hat{E}(\beta_{t+r}|Y_t, x) \\
&= X_{t+r}\Phi^r\hat{E}(\beta_t|Y_t, x).
\end{aligned}
$$

Again, the prediction covariance matrix is obtained recursively

$$\mathrm{Cov}[Y_{t+r} - \hat{E}(Y_{t+r}|Y_t, x)] = X_{t+r}\mathrm{Cov}[\beta_{t+r} - \hat{E}(\beta_{t+r}|Y_t, x)]X'_{t+r} + V_{t+r}.$$

Note that this covariance matrix does not involve any of $X_{t+1}, \ldots, X_{t+r-1}$.

If $\beta_0, e_1, \ldots, e_t$, and $\varepsilon_1, \ldots, \varepsilon_t$ are taken to have a joint normal distribution, then the posterior distribution of β_t given the data (Y_t, x) is multivariate normal with mean $\hat{E}(\beta_t|Y_t, x)$ and covariance matrix P_t. The subjective Bayesian aspect of this analysis lies entirely in the a priori assumption that

$$\beta_0 \sim N(\tilde{\beta}_0, P_0).$$

Parameter Estimation

For the purposes of prediction, the matrices Φ and X_i, V_i, Σ_i, $i = 1, \ldots, t$ were assumed to be known. This is similar to the approach taken for prediction in time domain models. In practice, as with time domain models, many of these parameters must be estimated. In particular, the matrix Φ defining the multivariate autoregressive process and the covariances matrices V_i and Σ_i generally need to be estimated. The design matrices X_i are typically known. The estimation of the covariance matrices present special problems in that they include many more parameters than there are observations. The covariance matrices must be modeled in some way if estimates are to be obtained. Suppose the data $Y_1, \ldots, Y_n$ have been observed. One commonly used and particularly simple model is that for some unknown positive definite matrices V and Σ

$$V_1 = V_2 = \cdots = V_n = V$$

and

$$\Sigma_1 = \Sigma_2 = \cdots = \Sigma_n = \Sigma.$$

The assumption that the observation covariance matrices V_i are equal implies that the number of observations available at each time is the same. Although model (2) implies that the dimensions of β_t and ε_t remain constant, in general, there is no such restriction on Y_t, X_t, and e_t.

The standard estimation method seems to be maximum likelihood based on the assumption that $(\beta_0, e_1, \ldots, e_n, \varepsilon_1, \ldots, \varepsilon_n)'$ has a multivariate normal distribution. It follows that the data $Y_1, Y_2, \ldots, Y_t$ have a multivariate normal distribution. The joint density can be written as the product of conditional densities, i.e.,

$$f(Y_1, \ldots, Y_t) = f(Y_1) f(Y_2|Y_1) f(Y_3|Y_1, Y_2) \cdots f(Y_t|Y_1, Y_2, \ldots, Y_{t-1}) .$$

In fact, assuming that β_0 is fixed, all of the conditional distributions are normals with means and covariances given by the Kalman filter. For example,

$$Y_t|Y_1, \ldots, Y_{t-1} \sim N(\hat{E}(Y_t|x), \mathrm{Cov}[Y_t - \hat{E}(Y_t|x)]) ,$$

where $\hat{E}(Y_t|x)$ is given above (5) and $\mathrm{Cov}[Y_t - \hat{E}(Y_t|x)]$ is given by (6). Write

$$\hat{e}_t = Y_t - \hat{E}(Y_t|Y_1, \ldots, Y_{t-1})$$

and

$$Q_t = \mathrm{Cov}[Y_t - \hat{E}(Y_t|Y_1, \ldots, Y_{t-1})] .$$

The likelihood function is the product of the conditional normal densities, i.e.,

$$L(\Phi, V_1, \ldots, V_n, \Sigma_1, \ldots, \Sigma_n) =$$
$$\left(\prod_{t=1}^{n} (2\pi)^{-q_t/2} \right) \left(\prod_{t=1}^{n} |Q_t|^{-1/2} \right) \exp\left[-\frac{1}{2} \sum_{t=1}^{n} \hat{e}_t' Q_t^{-1} \hat{e}_t \right] .$$

As usual, it is easier to maximize the log-likelihood,

$$\ell(\Phi, V_1, \ldots, V_n, \Sigma_1, \ldots, \Sigma_n) =$$
$$- \log(2\pi) \sum_{t=1}^{n} \frac{q_t}{2} - \frac{1}{2} \sum_{t=1}^{n} \log(|Q_t|) - \frac{1}{2} \sum_{t=1}^{n} \hat{e}_t' Q_t \hat{e}_t .$$

Remembering that parametric models for $V_1, \ldots, V_n$ and $\Sigma_1, \ldots, \Sigma_n$ are mandatory, it is important to note that this log-likelihood function is valid for any such models. If $V_i = V_i(\eta)$ and $\Sigma_i = \Sigma_i(\zeta)$, $i = 1, \ldots, n$, then

$$\ell(\Phi, \eta, \zeta) = \ell(\Phi, V_1(\eta), \ldots, V_n(\eta), \Sigma_1(\zeta), \ldots, \Sigma_n(\zeta)) .$$

Similarly, some or all of the components of Φ can be known without materially changing the likelihood function.

In practice, with almost any parameterization, the log-likelihood will be a very complicated nonquadratic function of the parameters. (Quadratic functions are easy to maximize.) Some sort of iterative maximization technique is generally required. Much of the work on this maximization problem has used the covariance model that assumes equal covariance matrices. Shumway and Stoffer (1982) and Shumway (1988) discuss maximization of $\ell(\Phi, V, \Sigma)$ using the EM algorithm. The EM algorithm is presented in Dempster, Laird, and Rubin (1977). The Newton-Raphson algorithm has also been used to maximize $\ell(\Phi, V, \Sigma)$. This approach is discussed by Gupta and Mehra (1974), Jones (1980), and Ansley and Kohn (1984).

MISSING VALUES

The assumption of equal covariance matrices seems to be the standard way of modeling the V_t's and Σ_t's. As mentioned earlier, equal V_t's implies that q_t, the dimension of Y_t, is the same for all t. This does not seem like a particularly harsh requirement. The vector Y_t will often consist of observations on q_t dependent variables. The assumption that, at each time t, the same number of variables are available to be observed is not very restrictive. The only problem is that some observations may be missing at some times. The generality of the state-space model allows missing values to be handled easily.

Suppose the $q \times 1$ vector Z_t is to be observed at time t. Assume an observation equation

$$Z_t = W_t \beta_t + \xi_t$$

where

$$\mathrm{Cov}(\xi_t) = V.$$

However, the actual observation is Y_t which consists of some of the components of Z_t. If no components are missing $Y_t = Z_t$, $X_t = W_t$, and $e_t = \xi_t$. If some components are lost, write

$$Z_t = \begin{bmatrix} Y_t \\ Y_{tM} \end{bmatrix},$$

$$W_t = \begin{bmatrix} X_t \\ X_{tM} \end{bmatrix},$$

$$\xi_t = \begin{bmatrix} e_t \\ e_{tM} \end{bmatrix},$$

and

$$\mathrm{Cov}(\xi_t) = \begin{bmatrix} V_t & V_{tM} \\ V_{Mt} & V_{MM} \end{bmatrix}.$$

There is no real loss of generality in assuming that the last components of Z_t are the ones that were not observed. The observation equation for Z_t implies an observation equation for the actual observations,

$$Y_t = X_t \beta_t + e_t$$

for which

$$\mathrm{Cov}(e_t) = V_t.$$

In estimating the parameters Φ, V, and Σ, the likelihood allows V_t to be a function of the parameters. Because V_t is a submatrix of V, V_t is clearly a function of the parameter matrix V. In fact, the likelihood function is not much more complicated than if none of the components of Z_t were lost.

Of course, the loss of components has no affect on the Kalman filter. The filter is based on Y_t and treats V_t as known. The lost components are simply ignored. The references given earlier for maximizing the likelihood also discuss the missing value problem.

V.9   Additional Exercises

EXERCISE 5.9.1.   Which of the following processes are stationary? Which are invertible?

a) $(1 - 1.5B + .54B^2)y_t = (1 - .5B)e_t$
b) $(1 - \frac{5}{8}B)y_t = (1 + .1B - .56B^2)e_t$
c) $(1 - 1.6B + .55B^2)y_t = (1 - .6B)e_t$
d) $(1 - B + .2475B^2)y_t = (1 - B - \frac{7}{36}B^2)e_t$
e) $(1 - .7B - .66B^2 + .432B^3)y_t = (1 - .8B + .07B^2)e_t$
f) $(1 - .8B + .07B^2)y_t = (1 - .1B)e_t$

EXERCISE 5.9.2.   Consider the $AR(1)$ model

$$y_t = 10 - .7y_{t-1} + e_t, \quad \sigma^2 = 3$$

a) Find μ, $\sigma(0)$, $\sigma(1)$, $\sigma(2)$, $\sigma(3)$.
b) If $y_4 = 12$, will y_5 tend to be less than or greater than μ?
c) Is the process stationary?
d) Give $\hat{E}(y_{4+k}|Y)$ and the prediction variance for $k = 1, 2, 3$ and $Y = (12, 11, 11, 12)'$.

EXERCISE 5.9.3.   Consider the $MA(1)$ model

$$y_t = 10 + e_t - .4e_{t-1}, \quad \sigma^2 = 3$$

a) Find μ, $\sigma(0)$, $\sigma(1)$, $\sigma(2)$, $\sigma(3)$.
b) If $y_4 = 12$, will y_5 tend to be less than or greater than μ?
c) Is the process stationary?
d) Give $\hat{E}(y_{4+k}|Y_\infty)$ and the prediction variance for $k = 1, 2, 3$ and $Y = (12, 11, 11, 12)'$.

EXERCISE 5.9.4.   Consider the $ARMA(1, 1)$ model

$$y_t = 10 - .7y_{t-1} + e_t - .4e_{t-1}, \quad \sigma^2 = 3$$

a) Find μ, $\sigma(0)$, $\sigma(1)$, $\sigma(2)$, $\sigma(3)$.
b) If $y_4 = 12$, will y_5 tend to be less than or greater than μ?
c) Is the process stationary?
d) Give $\hat{E}(y_{4+k}|Y_\infty)$ and the prediction variance for $k = 1, 2, 3$ and $Y = (12, 11, 11, 12)'$.

EXERCISE 5.9.5.   Do a time domain analysis of the sunspot data in Exercise 4.9.5. This should include estimation, model fitting, and checking of assumptions.

EXERCISE 5.9.6.   Use a multiplicative seasonal model to analyze the international air passenger data of Exercise 4.9.6. This should include estimation, model fitting, and checking of assumptions.

EXERCISE 5.9.7.   For an $MA(1)$ process find an estimate of θ_1 in terms of $\hat{\rho}(1)$. Are any restrictions on $\hat{\rho}(1)$ needed?

EXERCISE 5.9.8.   For an $AR(2)$ process use the Yule-Walker equations with $\hat{\sigma}_y(\cdot)$ replacing $\sigma_y(\cdot)$ to obtain estimates

$$\hat{\phi}_1 = \frac{\hat{\rho}(1)\,(1 - \hat{\rho}(1))}{1 - \hat{\rho}(1)^2}$$

and

$$\hat{\phi}_2 = \frac{\hat{\rho}(2) - \hat{\rho}(1)^2}{1 - \hat{\rho}(1)^2}.$$

EXERCISE 5.9.9.   Show that the variance for predicting k steps ahead in an $AR(1)$ process is

$$\sigma^2 \frac{1 - \phi_1^{2k}}{1 - \phi_1^2}.$$

EXERCISE 5.9.10.   Show, for an $MA(1)$ process with parameter θ_1, that

$$\hat{E}(y_{n+1}|Y_\infty) = -\sum_{s=0}^{\infty} \theta_1^s y_{n-s}.$$

EXERCISE 5.9.11.   Let e_t be a second-order stationary process. Show that

$$y_t = e_t - \theta_1 e_{t-1}$$

and

$$w_t = e_t - \frac{1}{\theta_1} e_{t-1}.$$

have the same correlation function.

EXERCISE 5.9.12.   What are the largest possible values of $|\rho(1)|$ and $|\rho(2)|$ for an $MA(2)$ process?

EXERCISE 5.9.13.   Find the ψ_i's, μ and $\sigma(k)$ for the following processes.
a) the $AR(3)$ process

$$(1 - .9B)(1 + .8B)(1 - .6B)y_t = e_t,$$

b) the $ARMA(2,1)$ process

$$(1 - .9B)(1 - .6B)y_t = (1 - .5)e_t.$$

EXERCISE 5.9.14.   Show that if all the roots x_0 of $\Phi(x)$ have $|x_0| > 1$ then there exists a polynomial

$$\Psi(x) = \sum_{i=0}^{\infty} \psi_i x^i$$

with

$$\sum_{i=0}^{\infty} |\psi_i| < \infty$$

and, for $x \subset [\ 1, 1]$,
$$[\Psi(x)]\,[\Phi(x)] = 1.$$

Hint: Do a Taylor's expansion of $1/\Phi(x)$ about 0.

EXERCISE 5.9.15.   Generate 25 realizations of an $ARMA(1,1)$ process for $\phi_1 = -.8, -.2, 0, .2, .8$ and $\theta_1 = -.8, -.2, 0, .2, .8$.

Chapter VI

Linear Models for Spatial Data: Kriging

Just as data collected sequentially in time may be correlated, data collected at known locations in space may be correlated. For example, deposits of high-quality copper are more likely to occur near other high-quality deposits. The levels of lead contamination in the soil around a smelter are likely to be correlated. The prevalence of AIDS viewed geographically is correlated. There are innumerable situations in which data are collected at various locations in space and thus innumerable potential applications for methods of analysis for spatial data. One branch of statistics concerned with the analysis of such data is known as *geostatistics*. The practical application of geostatistics was developed in relative isolation from the mainstream of statistics. Not surprisingly, it uses some terminology that is unfamiliar to classically trained statisticians. David (1977) and Journel and Hüijbregts (1978) give details of the geostatistical approach using geostatistical terminology. Ripley (1981) takes a point of view that is probably more familiar to most statisticians. He uses ideas of prediction for stochastic processes that are closely related to time series methods. See also Cliff and Ord (1981). In this chapter we present both traditional and geostatistical terminologies.

We begin by discussing the modeling of spatial data in terms of stochastic processes. In Section 2 linear models and best linear unbiased predictors for spatial data are presented. Best linear unbiased prediction is known in the geostatistics literature as *kriging*. The methods of kriging were developed in France by Matheron (1965, 1969). He was originally inspired by the contributions of D. G. Krige. The French work was performed independently of Goldberger (1962) who first derived general best linear unbiased predictors for linear models. The relationship between prediction based on covariances and prediction based on an alternative measure of variability, the semivariogram, is examined in Section 3. The role of measurement error is considered in Section 4. Section 5 looks at the effects of estimating the covariances when performing best linear unbiased prediction. Section 6 gives models for covariance functions and semivariograms. Estimation of covariances and the semivariogram is considered in the final section.

VI.1   Modeling Spatial Data

Spatial data can be considered as a realization of a stochastic process (*random field*)

$$y(u), \qquad u \in D \subset \mathbf{R}^d.$$

Here u is a location in D. Most often d, the dimension of the space, is 1, 2, or 3. For every value of u, $y(u)$ is a random variable. To analyze spatial data, we need to model $y(u)$. We begin by assuming that for any u, $\mathrm{E}\,(y(u))$ and $\mathrm{Var}\,(y(u))$ exist. It follows that $y(u)$ can be decomposed as

$$y(u) = m(u) + e(u)$$

where $m(u)$ is the fixed mean function of $y(u)$, i.e.,

$$\mathrm{E}(y(u)) = m(u)$$

and $e(u)$ is a stochastic error process with

$$\mathrm{E}(e(u)) = 0.$$

In particular, $e(u) = y(u) - m(u)$. Our approach to modeling $y(u)$ involves modeling both $m(u)$ and $e(u)$.

Begin by assuming a linear structure for $m(u)$. This is known in geostatistics as the *universal kriging* model. In particular, assume that there are, say, p known functions of u, $x_1(u), x_2(u), \ldots, x_p(u)$ so that the mean function satisfies

$$m(u) = \sum_{j-1}^{p} \beta_j x_j(u)$$

for some fixed unknown parameters $\beta_1, \ldots, \beta_p$. A special case of the universal kriging model is the *ordinary kriging* model

$$m(u) = \mu$$

for an unknown parameter μ.

A mathematically simpler but typically unrealistic model is simply to assume that $m(u)$ is known. *Simple kriging* is the special case

$$m(u) = \mu_0$$

where μ_0 is a known value. The case with $m(u)$ known will not be treated further. Simple modifications of the procedures outlined below provide data analysis for the case with $m(u)$ known.

In modeling the error process, we will be largely interested in its second-order properties. The covariance function is

$$\sigma(u, w) = \mathrm{Cov}\,(e(u), e(w)).$$

Note also that

$$\sigma(u, w) = \operatorname{Cov}\left(y(u), y(w)\right).$$

Often the covariance function is modeled in terms of an unknown parameter vector θ. In that case, write $\sigma(u, w; \theta)$. Some of the common assumptions made about $e(u)$ are that it is (1) second-order stationary, (2) strictly stationary, (3) intrinsically stationary, (4) increment stationary, or (5) isotropic. These terms are defined below.

STATIONARITY

We restrict attention to processes for which means and variances exist. A process $y(u)$ is said to be *strictly stationary* if for any value k, any locations $u_1, \ldots, u_k$, any (Borel) sets $C_1, \ldots, C_k$, and any vector $h \in \mathbf{R}^d$

$$\Pr[y(u_1) \in C_1, \ldots, y(u_k) \in C_k] =$$
$$\Pr[y(u_1 + h) \in C_1, \ldots, y(u_k + h) \in C_k]. \tag{1}$$

The process is stationary in the sense that the joint distribution of the process evaluated at any set of points is not changed if all of the points are moved in the same way. In particular,

$$E(y(u)) = E(y(u + h))$$

for any vector h so $\mathrm{E}\left(y(u)\right)$ must be a constant, i.e.,

$$E(y(u)) = \mu. \tag{2}$$

Also, for two locations u and w and any vector h

$$\sigma(u, w) = \sigma(u + h, w + h).$$

In particular, let $h = -w$ so

$$\sigma(u, w) = \sigma(u - w, 0)$$

and the covariance function can be thought of as a function of $u - w$ alone. To indicate this, let $h = u - w$ and write

$$\begin{aligned}
\sigma(u, w) &= \sigma(u - w) \\
&= \sigma(h).
\end{aligned} \tag{3}$$

By definition, $\sigma(u, w) = \operatorname{Cov}\left(y(u), y(w)\right) = \operatorname{Cov}\left(y(w), y(u)\right) = \sigma(w, u)$. In particular, for a stationary process

$$\sigma(u - w) = \sigma(w - u).$$

As we have discussed, when variances exist property (1) implies properties (2) and (3). A *second-order* (weak) *stationary process* is any process that satisfies (2) and (3). A second-order stationary process may be strictly stationary in the sense that (1) holds, but it need not be.

We define a process to be *increment stationary* if it satisfies (2) and for any integer k, locations $u_1, \ldots, u_k$, sets $C_1, \ldots, C_{k-1}$, and vector h

$$\Pr[y(u_2) - y(u_1) \in C_1, \ldots, y(u_k) - y(u_{k-1}) \in C_{k-1}] =$$
$$\Pr[y(u_2 + h) - y(u_1 + h) \in C_1, \ldots, y(u_k + h) - y(u_{k-1} + h) \in C_{k-1}].$$

It is immediate that stationary processes are increment stationary, but the converse need not be true. Brownian motion (cf. Breiman, 1968) is a well-known process with $u \in [0, \infty)$ that is increment stationary, but not stationary.

In the geostatistics literature, second-order properties are not typically characterized using the covariance function. Instead, they are represented by either the variogram or the semivariogram. These functions are similar to the covariance function but are more appropriate for use with increment stationary processes because they are defined directly on the increments. For a process satisfying (2), the *semivariogram* is defined as

$$\begin{aligned}
\gamma(u, w) &= \frac{1}{2}\mathrm{E}[y(u) - y(w)]^2 \\
&= \frac{1}{2}\mathrm{Var}[y(u) - y(w)].
\end{aligned}$$

The *variogram* is twice the semivariogram. Clearly, the two functions contain equivalent information. The variogram has a more natural definition but, as will be seen below, for second-order stationary processes the semivariogram has advantages. Our discussion will use the semivariogram exclusively.

For an increment stationary process, $\gamma(u, w) = \gamma(u + h, w + h)$ for any h. Letting $h = -w$

$$\gamma(u, w) = \gamma(u - w, 0)$$

and we write

$$\gamma(u, w) = \gamma(u - w). \tag{4}$$

A process is said to be *intrinsically stationary* if it satisfies both (4) and (2). Intrinsic stationary processes need not be increment stationary but increment stationary processes are intrinsically stationary. The relationship between increment stationarity and intrinsic stationarity is similar to the relationship between stationarity and second-order stationarity. Again, if variances do not exist, a process can be increment stationary without being intrinsically stationary. Second-order stationary processes are intrinsically stationary. For any second-order stationary process

$$\gamma(u, w) = \frac{1}{2}\mathrm{Var}[y(u) - y(w)]$$

$$
\begin{aligned}
&= \frac{1}{2}\left[\operatorname{Var}\left(y(u)\right) + \operatorname{Var}\left(y(w)\right) - 2\operatorname{Cov}\left(y(w), y(u)\right)\right] \\
&= \frac{1}{2}\left[\sigma(0) + \sigma(0) - 2\sigma(w - u)\right] \\
&= \sigma(0) - \sigma(w - u) \\
&= \sigma(0) - \sigma(u - w),
\end{aligned}
$$

this is a function of $u - w$ so second-order stationary processes are also intrinsically stationary. In particular if $h = u - w$,

$$
\gamma(h) = \sigma(0) - \sigma(h). \tag{5}
$$

This simple relationship between the semivariogram and the covariance function is the reason for our use of the semivariogram in the remainder of the chapter.

Another generalization of stationarity is contained in the ideas of *generalized covariance functions* and *intrinsic random functions of order k*. Just as an intrinsically stationary process with a stationary semivariogram can be used to model a process with a nonstationary covariance function, an intrinsic random function of order k with a stationary generalized covariance function can be used to model a processes with a nonstationary covariance function. Knowledge of the generalized covariance function does not completely specify the covariance structure of a process but it can be used to obtain best linear unbiased predictions for universal kriging models. The equivalence of predictions based on the covariance function and the generalized covariance function can be established by modifying the results given in Section 3, see Christensen (1990b). Generalized covariance functions and intrinsic random functions were originally introduced by Matheron (1973) to avoid problems encountered with using residuals to estimate second-order properties in universal kriging. For data analysis, the fundamental idea is the same as in restricted maximum likelihood, cf. Patterson and Thompson (1974) and Christensen (1987, Section XII.6), but the methods are not necessarily based on likelihood analysis. Delfiner (1976) gives an introduction to these topics; they will not be discussed further in this chapter.

Often the second-order properties of a process can be assumed to depend only on the distance between two points and not on the direction between them. A second-order stationary process is *isotropic* if

$$
\sigma(u - w) = \sigma(\|u - w\|).
$$

An intrinsically stationary process is isotropic if

$$
\gamma(u - w) = \gamma(\|u - w\|).
$$

A process that is not isotropic is said to be *anisotropic*.

Finally, a characterization of processes that leads to computational advantages is separability. Let $h' = (h_1, \ldots, h_d)$. A second-order stationary process is said to be *separable* if its covariance function can be written as

$$\sigma(h) = \prod_{i=1}^{d} \sigma_i(h_i)$$

for some one-dimensional covariance functions $\sigma_1(\cdot), \ldots, \sigma_d(\cdot)$. A similar property for the semivariogram defines separability for intrinsically stationary processes. See Zimmerman (1989) and Zimmerman and Harville (1990) for applications of separability.

In this subsection, concepts of stationarity have been discussed for an arbitrary process $y(u)$ for which variances exist. In the remainder of the chapter, these variants of stationarity will be incorporated into divers models for the error process $e(u)$. In the universal kriging model, $\mathrm{E}[y(u)]$ is not constant so the process $y(u)$ fails to satisfy any of the definitions related to stationarity. On the other hand, the error process has a constant mean of zero so it may be modeled with some sort of stationarity assumption. As will be seen in the next two sections, the actual process of best linear unbiased prediction does not require any version of stationarity. Nonetheless, modeling the covariance structure is vital to the analysis of spatial data and stationarity assumptions are an important aspect of covariance modeling.

VI.2 Best Linear Unbiased Prediction of Spatial Data: Kriging

The purpose of this section is to illustrate the relationship between kriging and best linear unbiased prediction. The object of kriging is the prediction of unobserved spatial random variables based on the values of observed random variables. There is little to do except establish that we are working with a linear model. Given that fact, the BLUP has been found in Christensen (1987, Section XII.2). Kriging is often presented as either point kriging or block kriging. We give a detailed discussion of point kriging and mention the variations needed for block kriging.

Assume that the universal kriging model

$$m(u) = \sum_{j=1}^{p} \beta_j x_j(u)$$

holds, that observations have been taken at locations $u_1, \ldots, u_n$, and that we wish to predict the value of $y(u_0)$. Set the following notation:

$$y_i \;\; = \;\; y(u_i), \; i = 0, \ldots, n,$$

$$
\begin{aligned}
Y &= (y_1, \ldots, y_n)', \\
x_{ij} &= x_j(u_i), \ i = 0, \ldots, n, \\
x_i' &= (x_{i1}, \ldots, x_{ip}), \ i = 0, \ldots, n
\end{aligned}
$$

$$
X = \begin{bmatrix} x_1' \\ \vdots \\ x_n' \end{bmatrix},
$$

$$
\begin{aligned}
\beta &= (\beta_1, \ldots, \beta_p)', \\
e_i &= e(u_i)
\end{aligned}
$$

and

$$
e = (e_1, \ldots, e_n)'.
$$

The universal kriging model, as applied to the observations, can be written as

$$
Y = X\beta + e, \ \mathrm{E}(e) = 0, \ \mathrm{Cov}(e) = \Sigma \tag{1}
$$

where

$$
\Sigma = [\sigma_{ij}]
$$

and

$$
\sigma_{ij} = \sigma(u_i, u_j).
$$

Generally, Σ is nonsingular and the functions $x_j(u)$ and locations are taken so that X has full column rank with $J \in C(X)$. Under the full column rank assumption, β is estimable so, for any location u_0, $x_0'\beta$ is estimable. Let

$$
\Sigma_{Y0} = \begin{bmatrix} \sigma(u_1, u_0) \\ \vdots \\ \sigma(u_n, u_0) \end{bmatrix}.
$$

Applying the results of Christensen (1987, Section XII.2), the best linear unbiased predictor of y_0 is

$$
\hat{y}_0 = x_0'\hat{\beta} + \delta'(Y - X\hat{\beta}) \tag{2}
$$

where

$$
\hat{\beta} = (X'\Sigma^{-1}X)^{-1}X'\Sigma^{-1}Y
$$

and

$$
\delta = \Sigma^{-1}\Sigma_{Y0}.
$$

As in Christensen (1987, Section XII.2), we can write

$$
\hat{y}_0 = b'Y
$$

where

$$
b' = x_0'(X'\Sigma^{-1}X)^{-1}X'\Sigma^{-1} + \delta'\left(I - X(X'\Sigma^{-1}X)^{-1}X'\Sigma^{-1}\right). \tag{3}
$$

The mean squared prediction error (prediction variance) is

$$\begin{aligned}
\mathrm{Var}(y_0 - \hat{y}_0) &= \sigma(u_0, u_0) - \Sigma'_{Y0}\Sigma^{-1}\Sigma_{Y0} \\
&\quad + [x_0 - X'\delta]'(X'\Sigma^{-1}X)^{-1}[x_0 - X'\delta] \qquad (4) \\
&= \sigma(u_0, u_0) - 2b'\Sigma_{Y0} + b'\Sigma b\,.
\end{aligned}$$

Often the object of kriging is to produce a 2- or 3-dimensional map of the variable $y(u)$ over the region D. To do this, a large number of predictions for various locations u_0 are needed. Rewrite (2) as

$$\hat{y}_0 = x'_0\hat{\beta} + \Sigma'_{Y0}\Sigma^{-1}(Y - X\hat{\beta})\,.$$

Given $\hat{\beta}$ and $(Y - X\hat{\beta})'\Sigma^{-1}$, computation of different predictions is very inexpensive. It requires only the computation of inner products between x_0 and $\hat{\beta}$ and between $\Sigma^{-1}(Y - X\hat{\beta})$ and Σ_{Y0} for the various values of x_0 and Σ_{Y0}.

Kitanidis (1986) gives a Bayesian analysis of the universal kriging model. In particular, he relates Bayes point predictions to the best linear unbiased predictors and the variance of the predictive distribution to the prediction variance. Cressie (1986) presents prediction methods based on exploratory data analysis.

BLOCK KRIGING

So far we have assumed that our observations have been taken at point locations $u_1, \ldots, u_n$. In fact, it is physically impossible to take observations at points. An observation taken at u_i must actually be an observation on some neighborhood of u_i. If the neighborhood is small, the approximation to point observations should be good. However, in many applications, the neighborhood is sufficiently large that the properties of the neighborhood need to be incorporated into the analysis. In the geostatistical literature, the neighborhoods are referred to as blocks. The theory presented above carries through with almost no change. We need only to redefine the terms of the linear model.

Let B_i be the block associated with the i'th observation. Let $|B_i|$ be the volume (area) of the block. The observations and value to be predicted are now

$$y_i = \frac{1}{|B_i|} \int_{B_i} y(u)du$$

for $i = 0, 1, \ldots, n$. The elements of the design matrix X are

$$x_{ij} = \frac{1}{|B_i|} \int_{B_i} x_j(u)du\,.$$

Note that under the universal kriging model

$$\mathrm{E}(y_i) = \frac{1}{|B_i|} \int_{B_i} \sum_{j=1}^{p} \beta_j x_j(u)du$$

$$\begin{aligned} &= \sum_{j=1}^{p} \beta_j \left\{ \frac{1}{|B_i|} \int_{B_i} x_j(u)du \right\} \\ &= \sum_{j=1}^{p} \beta_j x_{ij} \,. \end{aligned}$$

Thus, the y_i's follow a linear model.

Note that $y_i - \mathrm{E}(y_i) = (1/|B_i|) \int_{B_i} e(u)du$ so the covariance of two observations is

$$\begin{aligned} \mathrm{Cov}(y_i, y_j) &= \mathrm{E}\left\{ \frac{1}{|B_i|}\frac{1}{|B_j|} \int_{B_i} e(u)du \int_{B_j} e(v)dv \right\} \\ &= \frac{1}{|B_i|}\frac{1}{|B_j|} \int_{B_i}\int_{B_j} \mathrm{E}[e(u)e(v)]dudv \\ &= \frac{1}{|B_i|}\frac{1}{|B_j|} \int_{B_i}\int_{B_j} \sigma(u,v)dudv \,. \end{aligned}$$

Given this linear model with known covariance structure, the BLUP can be obtained as usual.

VI.3   Prediction Based on the Semivariogram: Geostatistical Kriging

In much of the geostatistics literature, kriging is presented as a procedure that uses the semivariogram of the error process to determine optimal predictions. In general, given the semivariogram one cannot reproduce the covariance function, so it is not clear that kriging can be performed using only the semivariogram. A special case in which the semivariogram and the covariance function are equivalent is that of a second-order stationary process with

$$\lim_{\|u\| \to \infty} \sigma(u) = 0 \,. \tag{1}$$

In this case, by (6.1.5) $\gamma(h) = \sigma(0) - \sigma(h)$ so

$$\lim_{\|u\| \to \infty} \gamma(u) = \sigma(0)$$

and

$$\sigma(h) = \lim_{\|u\| \to \infty} \gamma(u) - \gamma(h) \,.$$

But generally, for intrinsically stationary processes and for second-order stationary processes that do not have property (1), we cannot expect to reproduce the covariance function from the semivariogram.

In this section, we give a condition under which the BLUP can be found as a function of the semivariogram. Interestingly, the condition does not involve the variability of the error process; the condition is essentially that the linear model (6.2.1) contains an intercept and that for predicting $y(u_0)$, the mean $m(u_0) = x_0'\beta$ also contains an intercept. In practice, this condition can be specified as

$$x_1(u) = 1 \qquad \text{all } u.$$

More generally, we can assume that

$$J_{n+1} \in C\left(\begin{bmatrix} X \\ x_0' \end{bmatrix}\right)$$

or equivalently that for some vector d

$$J = Xd$$

and

$$1 = x_0'd.$$

Our proof is based on the result of Christensen (1987, Exercise 12.1c), i.e., that if $b'Y$ is the BLUP, there exists a vector, that we will call λ, such that

$$\begin{bmatrix} \Sigma & X \\ X' & 0 \end{bmatrix}\begin{bmatrix} b \\ \lambda \end{bmatrix} = \begin{bmatrix} \Sigma_{Y0} \\ x_0 \end{bmatrix}. \tag{2}$$

There are two simple ways to prove this. First, b' as given in (6.2.3) can be substituted into (2) and a vector λ can be identified that satisfies the equality. Alternatively, if the inverses exist, one can establish that

$$\begin{bmatrix} \Sigma & X \\ X' & 0 \end{bmatrix}^{-1} = \begin{bmatrix} \Sigma^{-1}(I - A) & \Sigma^{-1}X(X'\Sigma^{-1}X)^{-1} \\ (X'\Sigma^{-1}X)^{-1}X'\Sigma^{-1} & -(X'\Sigma^{-1}X)^{-1} \end{bmatrix}$$

where $A = X(X'\Sigma^{-1}X)^{-1}X'\Sigma^{-1}$. This leads to

$$\begin{bmatrix} b \\ \lambda \end{bmatrix} = \begin{bmatrix} \Sigma & X \\ X' & 0 \end{bmatrix}^{-1}\begin{bmatrix} \Sigma_{Y0} \\ x_0 \end{bmatrix}.$$

It is easily seen that the solution b is the same as (6.2.3). In any case, if (2) has a unique solution, b must be as in (6.2.3).

Having established (2), we show that b satisfies a similar equation that involves only the semivariogram. If this equation has a unique solution, the solution must determine the BLUP.

In particular, define semivariogram matrices similar to Σ and Σ_{Y0}, say

$$\Gamma = [\gamma(u_i, u_j)]_{n \times n}$$

and

$$\Gamma_{Y0} = [\gamma(u_i, u_0)]_{n \times 1}.$$

The equation analogous to (2) is

$$\begin{bmatrix} -\Gamma & X \\ X' & 0 \end{bmatrix} \begin{bmatrix} b \\ \xi \end{bmatrix} = \begin{bmatrix} -\Gamma_{Y0} \\ x_0 \end{bmatrix}. \tag{3}$$

We now establish the equivalence of equations (2) and (3).

Proposition 6.3.1.   Suppose there exists a vector d such that $J = Xd$ and $1 = x_0'd$. Any solution $[b', \xi']'$ to (3) determines a solution $[b', \lambda']'$ to (2) and conversely.

PROOF.   Let $\sigma_y' = [\sigma_{11}, \ldots, \sigma_{nn}]$. It is easily seen that

$$\Sigma = \frac{1}{2}[\sigma_y J' + J\sigma_y'] - \Gamma \tag{4}$$

and

$$\Sigma_{Y0} = \frac{1}{2}[\sigma_y + \sigma(u_0, u_0)J] - \Gamma_{Y0}. \tag{5}$$

Both equations (2) and (3) imply that $X'b = x_0$ so a solution to either equation satisfies

$$J'b = d'X'b = d'x_0 = 1.$$

Suppose $[b', \xi']'$ is a solution to (3). We need to show that this determines a solution $[b', \lambda']'$ for (2). From equation (3)

$$-\Gamma b + X\xi = -\Gamma_{Y0}$$

Substituting for $-\Gamma$ and $-\Gamma_{Y0}$ from (4) and (5) gives

$$\Sigma b - \frac{1}{2}\sigma_y J'b - \frac{1}{2}J\sigma_y'b + X\xi = \Sigma_{Y0} - \frac{1}{2}\sigma_y - \frac{1}{2}\sigma(u_0, u_0)J.$$

Using the fact that $J'b = 1$ and rearranging terms gives

$$\Sigma b - \frac{1}{2}J\sigma_y'b + X\xi + \frac{1}{2}\sigma(u_0, u_0)J = \Sigma_{Y0}.$$

However $J = Xd$, so

$$\Sigma b + X\left(-\frac{1}{2}d\sigma_y'b + \xi + \frac{1}{2}\sigma(u_0, u_0)d\right) = \Sigma_{Y0}.$$

Pick $\lambda = -\frac{1}{2}d\sigma_y'b + \xi + \frac{1}{2}\sigma(u_0, u_0)d$ and $[b', \lambda']'$ is a solution to (2).

To show that any solution $[b', \lambda']'$ for (2) yields a solution to (3) use a similar argument. In particular, use (4) and (5) to substitute for Σ and Σ_{Y0} in (2).   □

Based on Proposition 6.3.1, if (2) and (3) have unique solutions then a solution to (3) determines a solution to (2) which determines the BLUP.

The traditional method for finding the BLUP in geostatistics is to find a solution to (3). We have established mild conditions for the validity of that approach.

The prediction variance can be written in terms of the semivariogram. By (6.2.4), the fact that $J'b = 1$ and equations (4) and (5)

$$
\begin{aligned}
\mathrm{E}(y_0 - \hat{y}_0)^2 &= \sigma(u_0, u_0) - 2b'\Sigma_{Y0} + b'\Sigma b \qquad (6) \\
&= \sigma(u_0, u_0) \\
&\quad - 2b'\left\{\frac{1}{2}\sigma_y + \frac{1}{2}\sigma(u_0, u_0)J - \Gamma_{Y0}\right\} \\
&\quad + b'\left\{\frac{1}{2}\sigma_y J' + \frac{1}{2}J\sigma'_y - \Gamma\right\}b \\
&= \sigma(u_0, u_0) \\
&\quad - b'\sigma_y - \sigma(u_0, u_0)b'J + 2b'\Gamma_{Y0} \\
&\quad + \frac{1}{2}b'\sigma_y J'b + \frac{1}{2}b'J\sigma'_y b - b'\Gamma b \\
&= 2b'\Gamma_{Y0} - b'\Gamma b.
\end{aligned}
$$

VI.4   Measurement Error and the Nugget Effect

One aspect of best linear (unbiased) prediction that is unappealing for some purposes is the fact, illustrated in Christensen (1987, Exercise 12.1b), that the predictor of a point that has been observed is just the point itself, i.e., $\hat{y}_i = y_i$. This phenomenon occurs because in predicting $y_0 \equiv y_i$,

$$
\Sigma_{Y0} = \begin{bmatrix} \sigma_{1i} \\ \vdots \\ \sigma_{ni} \end{bmatrix}
$$

which is just the i'th column of Σ. Clearly, a solution to $\Sigma\delta = \Sigma_{Y0}$ is given by the vector $\delta = (0, \ldots, 0, 1, 0, \ldots, 0)'$ where the 1 is in the i'th place. Thus

$$
\begin{aligned}
\hat{y}_i &= x'_i\hat{\beta} + \delta'(Y - X\hat{\beta}) \\
&= x'_i\hat{\beta} + (y_i - x'_i\hat{\beta}) \\
&= y_i.
\end{aligned}
$$

Given our current model, this is only reasonable. We have available only one realization of the stochastic process, so there is only one value observable at u_i and we have observed it. If we wish to predict at u_i the predictor must be y_i.

For many purposes, more smoothing is desired in the predictor. This can be accomplished by imagining that subsequent measurements taken at u_i

could be different from our original observation y_i. For this to happen, our observations must be subject to measurement error. Measurement errors are generally modeled as being uncorrelated with constant variance. In particular, we assume that observations follow the model

$$y(u) = m(u) + e(u) + e_M(u)$$

where all terms are defined as in Section 1 except that $e_M(u)$ is a second order stationary measurement error process, i.e.,

$$\begin{aligned} \mathrm{E}(e_M(u)) &= 0 \\ \mathrm{Var}(e_M(u)) &= \sigma_M^2 \\ \mathrm{Cov}(e_M(u), e_M(w)) &= 0 \qquad u \neq w \end{aligned}$$

and

$$\mathrm{Cov}(e_M(u), e(w)) = 0 \qquad \text{any } u, w\,.$$

Define covariance functions $\sigma_e(u, w)$ for $e(u)$, $\sigma_M(u, w)$ for $e_M(u)$, and $\sigma_\varepsilon(u, w)$ for $\varepsilon(u) = e(u) + e_M(u)$. Note that $\sigma_\varepsilon(u, w) = \sigma_e(u, w) + \sigma_M(u, w)$ and $\sigma_M(u, w) = \sigma_M^2 \delta_{u,w}$ where $\delta_{u,w}$ is the Kronecker delta; 1 if $u = w$ and 0 otherwise.

Letting $e_M = (e_M(u_1), \ldots, e_M(u_n))'$, the spatial linear model is

$$Y = X\beta + \varepsilon$$

where

$$\begin{aligned} \varepsilon &= e + e_M \\ \mathrm{E}(\varepsilon) &= 0 \end{aligned}$$

and writing $\mathrm{Cov}(\varepsilon) = V$ gives

$$V = \Sigma + \sigma_M^2 I\,.$$

With measurement error, the covariance matrix of Y is now V but the covariance between Y and a future observation at u_i is still Σ_{Y0}. The covariance is based entirely on $e(u)$. The measurement error process $e_M(u)$ contributes nothing. The prediction is

$$\hat{y}_i = x_i'\hat{\beta} + \Sigma_{Y0}' V^{-1}(Y - X\hat{\beta})$$

where

$$\hat{\beta} = (X'V^{-1}X)^{-1}X'V^{-1}Y\,.$$

Typically, the prediction does not simplify to y_i.

A well-known concept in geostatistics is that of the *nugget effect*. A nugget effect is said to occur for an intrinsically stationary process if

$$\lim_{\|u\| \to 0} \gamma(u) = \gamma_0 > 0\,.$$

Note that by definition, $\gamma(0) = 0$ (cf. Section 1). The existence of a nugget effect is in some sense equivalent to measurement error. To show this, we show that for second-order stationary errors, measurement error induces a nugget effect and that any semivariogram with a nugget effect can be produced by a measurement error model.

First, assume an arbitrary measurement error model. The total error process is

$$\varepsilon(u) = e(u) + e_M(u).$$

The measurement error process $e_M(u)$ is second-order stationary by definition. Assume that $e(u)$ is second-order stationary so that $\varepsilon(u)$ must also be second-order stationary. Also assume that $\lim_{\|u\|\to 0} \sigma_e(u)$ exists. The covariance functions of the processes satisfy

$$\sigma_\varepsilon(u) = \sigma_e(u) + \sigma_M(u)$$

because $e_M(u)$ is uncorrelated with $e(w)$ for any u, w. Moreover, by definition

$$\sigma_M(u) = \begin{cases} \sigma_M^2 & u = 0 \\ 0 & u \neq 0 \end{cases}.$$

As illustrated in (6.1.5)

$$\begin{aligned} \gamma_\varepsilon(u) &= \sigma_\varepsilon(0) - \sigma_\varepsilon(u) \\ &= \sigma_e(0) - \sigma_e(u) + \sigma_M^2. \end{aligned}$$

By the Cauchy-Schwartz inequality, for any u, w

$$\sigma_e(u - w) = \mathrm{Cov}\,(e(u), e(w)) \leq \sqrt{\mathrm{Var}\,(e(u))\,\mathrm{Var}\,(e(w))} = \sigma_e(0),$$

hence

$$\sigma_e(u) \leq \sigma_e(0)$$

and

$$\lim_{\|u\|\to 0} \gamma_\varepsilon(u) = \sigma_e(0) - \lim_{\|u\|\to 0} \sigma_e(u) + \sigma_M^2 > 0.$$

Thus measurement error induces a nugget effect.

Conversely, if $\varepsilon(u)$ is a second-order stationary error process with a nugget effect, then there exist second-order stationary error processes $e(u)$ and $e_M(u)$ such that $e_M(u)$ is a measurement error process and the process $\eta(u) = e(u) + e_M(u)$ has the same semivariogram as $\varepsilon(u)$. Simply define $e(u)$ to be a second-order stationary process with

$$\sigma_e(u) = \begin{cases} \sigma_\varepsilon(u) & u \neq 0 \\ \sigma_\varepsilon(0) - \gamma_0 & u = 0 \end{cases}$$

and $e_M(u)$ to be the measurement error process with

$$\sigma_M(u) = \begin{cases} 0 & u \neq 0 \\ \gamma_0 & u = 0 \end{cases}.$$

Because the measurement error process is taken to be uncorrelated with $e(u)$,

$$
\begin{aligned}
\gamma_\eta(u) &= \sigma_\eta(0) - \sigma_\eta(u) \\
&= \sigma_e(0) - \sigma_e(u) + \sigma_M(0) - \sigma_M(u) \\
&= \sigma_\varepsilon(0) - \gamma_0 - \sigma_\varepsilon(u) + \gamma_0 - 0 \\
&= \sigma_\varepsilon(0) - \sigma_\varepsilon(u) \\
&= \gamma_\varepsilon(u) .
\end{aligned}
$$

For prediction, the difference between measurement error and a nugget effect without measurement error exists only in Σ_{Y0} and then only when $y_0 = y_i$ for some $i = 1, \ldots, n$. With measurement error, a new observation at location u_i can be different from y_i. The covariance between Y and a new observation at u_i is

$$
\Sigma_{YiM} = \begin{bmatrix} \sigma_e(u_1, u_i) \\ \vdots \\ \sigma_e(u_n, u_i) \end{bmatrix} .
$$

With a pure nugget effect, the only observation has already been taken so

$$
\Sigma_{YiN} = \begin{bmatrix} \sigma_\varepsilon(u_1, u_i) \\ \vdots \\ \sigma_\varepsilon(u_n, u_i) \end{bmatrix} = \Sigma_{YiM} + \sigma_M^2 \delta
$$

where $\delta' = (0, \ldots, 0, 1, 0, \ldots, 0)$ with the 1 in the i'th place. For predictions not at a data point, either model gives

$$
\Sigma_{Y0} = \begin{bmatrix} \sigma_e(u_1, u_0) \\ \vdots \\ \sigma_e(u_n, u_0) \end{bmatrix} .
$$

If $\sigma_e(u, w)$ is continuous as $u_0 \to u_i$, the measurement error model gives continuous predictions because $\Sigma_{Y0} \to \Sigma_{YiM}$. In the pure nugget effect model, Σ_{Y0} does not converge to Σ_{YiM} so the predictions are discontinuous at the data points. The discontinuities exist so that data points will be predicted as themselves.

VI.5   The Effect of Estimated Covariances on Prediction

In many problems, including the examination of spatial data, the analysis is based on the linear model

$$
Y = X\beta + e, \ \ E(e) = 0, \ \ \mathrm{Cov}(e) = \Sigma .
$$

For a given Σ, the BLUE of $X\beta$ is

$$X\hat{\beta} = AY$$

where

$$A = X(X'\Sigma^{-1}X)^{-}X'\Sigma^{-1}$$

is an oblique projection operator onto $C(X)$. Note that the BLUP (6.2.2) involves both $X\hat{\beta}$ and the residual vector

$$Y - X\hat{\beta} = (I - A)Y.$$

In practice, the covariance matrix Σ is rarely known. In some applications Σ can be assumed to have a special form that allows a simple analysis without knowing the exact value of Σ, cf. Christensen (1987, Chapters II, III, XI) and Chapter I. (In particular, it suffices to know Σ up to a constant multiple.) However, in the analysis of spatial data and many other problems (cf. Carroll and Ruppert, 1988) Σ is not known nor does it have a special form that allows a simple analysis. In practice, Σ is estimated with a function of Y say $\tilde{\Sigma}(Y) = \tilde{\Sigma}$, that is nonnegative definite. Estimation of β is performed by plugging $\tilde{\Sigma}$ in for Σ, i.e., an estimate is $\tilde{\beta}$ where

$$X\tilde{\beta} = \tilde{A}Y$$

and $\tilde{A}$ is the random projection operator

$$\tilde{A} = X(X'\tilde{\Sigma}^{-1}X)^{-}X'\tilde{\Sigma}^{-1}.$$

$\tilde{A}$ is random because $\tilde{\Sigma}$ is a function of Y, hence random. For simplicity, in the discussion below we will assume that Σ and $\tilde{\Sigma}$ are positive definite.

For the prediction problem, estimates $\tilde{\Sigma}(Y)$ and $\tilde{\Sigma}_{Y0}(Y)$ are required. The plug-in predictor is taken by analogy with (6.2.2) as

$$\tilde{y}_0 = x_0'\tilde{\beta} + \tilde{\delta}'(Y - X\tilde{\beta})$$

where $\tilde{\beta}$ is defined as above,

$$\tilde{\delta} = \tilde{\Sigma}^{-1}\tilde{\Sigma}_{Y0},$$

and $x_0'\beta$ is assumed to be estimable.

Our discussion consists of two parts. We begin with some comments on estimating covariances for spatial data problems. The section concludes with some general mathematical results originally from Eaton (1985) and Harville (1985).

SPATIAL DATA

In practice, neither the covariance function nor the semivariogram are known. These must be estimated.

For a stationary process, given an estimate $\tilde{\sigma}(u)$, estimates $\tilde{\Sigma}$ and $\tilde{\Sigma}_{Y0}$ can be generated for use in (6.2.2) to obtain estimated optimal predictions and in (6.2.4) to obtain estimated prediction variances. For an intrinsically stationary process, an estimate $\tilde{\gamma}(u)$ leads to estimates $\tilde{\Gamma}$ and $\tilde{\Gamma}_{Y0}$ for use in (6.3.3) and thus to an estimate of b, say $\tilde{b}$, which leads to predictions and can also be used in (6.3.6) to obtain estimated prediction variances. These estimated prediction variances are notorious for underestimating the true prediction variance.

Cressie (1988) observes that to ensure consistency of estimators in ordinary kriging one must assume not only that the process is strictly stationary but also *ergodic* (cf. Adler, 1981). Recall that strict stationarity implies second order, increment, and intrinsic stationarity. None of these three imply stationarity. For Gaussian processes, ergodicity is implied by

$$\lim_{\|h\| \to \infty} \sigma(h) = 0 .$$

Because ordinary kriging is a special case of universal kriging, even stronger assumptions may be necessary to ensure consistency in the more general model. In particular, some assumptions about the asymptotic behavior of the design matrix are probably needed.

It should be emphasized that we are not discussing consistency of predictors. Even if the joint distributions (first and second moments) were known, the best predictor (best linear predictor) would not give perfect predictions. Best linear unbiased estimates depend on estimating the mean function. Consistency is concerned with the estimated mean converging to the true mean so that the BLUP converges to the BLP. When the covariances are also estimated, we obtain only an estimated BLUP. In this case again, consistency refers to the estimated BLUP converging to the BLP. Stein (1988) discusses asymptotically efficient prediction. Diamond and Armstrong (1984) indicate that prediction is reasonably robust to the choice of different covariance functions.

MATHEMATICAL RESULTS

Eaton (1985) and Harville (1985) have given conditions under which the plug-in estimates and predictors discussed above are unbiased and have variances at least as great as the corresponding BLUE's and BLUP's. This is one of those cases in which two people have developed very similar ideas simultaneously and independently. We follow Eaton's development. As discussed in the previous subsection, to estimate the variance of a plug-in estimate or predictor, the variance formula for the BLUE or BLUP is frequently used with Σ replaced by $\tilde{\Sigma}$. Under mild conditions, when $\tilde{\Sigma}$ is unbiased for

Σ, the expected value of this estimated variance is less than or equal to the variance of the BLUE or BLUP (which in turn is less than or equal to the true variance of the plug-in estimator or predictor). These results establish a theoretical basis for the often observed phenomenon that these estimated variances for plug-in estimators (predictors) are often misleadingly small. Although Eaton's results do not explicitly use any parameterization for Σ, it is typically the case that the covariance matrix depends on a parameter vector θ, i.e., $\Sigma = \Sigma(\theta)$, and that the estimate of $\Sigma(\theta)$ is $\tilde{\Sigma} = \Sigma(\tilde{\theta})$ where $\tilde{\theta}$ is an estimate of θ.

The first results on the unbiasedness of plug-in estimates and predictors are apparently from Kackar and Harville (1981). Other results on improved variance estimation for plug-in predictors are given by Kackar and Harville (1984), Jeske and Harville (1986), and Zimmerman and Cressie (1989).

Definition 6.5.1. $\tilde{B}(Y)$ is a residual type statistic if

$$\tilde{B}(Y) = \tilde{B}(Y - X\beta) \qquad \text{for any} \quad \beta \tag{1}$$

and

$$\tilde{B}(Y) = \tilde{B}(-Y). \tag{2}$$

Note that any residual type statistic has $\tilde{B}(Y) = \tilde{B}(Y - X\hat{\beta}) = \tilde{B}((I - A)Y)$, so residual type statistics can be viewed as functions of the residual vector.

In the discussion below, we will assume that $\tilde{\Sigma}$ and $\tilde{\Sigma}_{Y0}$ are residual type statistics. Most standard methods for estimating covariance matrices satisfy the conditions of Definition 6.5.1. Clearly, if $\tilde{\Sigma}$ and $\tilde{\Sigma}_{Y0}$ are residual type statistics, any functions of them are also of the residual type. In particular, functions such as $\tilde{\Sigma}^{-1}$,

$$\tilde{A} = X(X'\tilde{\Sigma}^{-1}X)^{-}X'\tilde{\Sigma}^{-1},$$

and

$$\tilde{\delta} = \tilde{\Sigma}^{-1}\tilde{\Sigma}_{Y0}$$

are residual type statistics.

EXERCISE 6.1.   Show that all of the variance component estimation procedures in Christensen (1987, Chapter XII) yield covariance matrix estimates that are residual type statistics.

The key result in establishing that plug-in estimators are unbiased is the following proposition.

Proposition 6.5.2.   If e and $-e$ have the same distribution and if

$\tilde{B}(Y)$ is a residual type statistic of order $r \times n$, then

$$\mathrm{E}[\tilde{B}(Y)Y] = \mathrm{E}[\tilde{B}(Y)X\beta].$$

PROOF.
$$\mathrm{E}[\tilde{B}(Y)Y] = \mathrm{E}[\tilde{B}(Y)X\beta] + \mathrm{E}[\tilde{B}(Y)e].$$

It suffices to show that $\mathrm{E}[\tilde{B}(Y)e] = 0$. By Definition 6.5.1

$$\tilde{B}(Y) = \tilde{B}(Y - X\beta) = \tilde{B}(e)$$

and by the symmetry property of e assumed in the proposition

$$\mathrm{E}[\tilde{B}(e)e] = -\mathrm{E}[\tilde{B}(-e)e] = -\mathrm{E}[\tilde{B}(e)e].$$

The only way a real vector can equal its negative is if the vector is zero, thus completing the proof. $\square$

Henceforth assume that e and $-e$ have the same distribution.

Proposition 6.5.2 leads immediately to two results on unbiased estimation.

Corollary 6.5.3. $\mathrm{E}[X\tilde{\beta}] = X\beta.$

PROOF.   By definition, $X\tilde{\beta} = \tilde{A}Y$ where $\tilde{A}$ is a residual type statistic so by Proposition 6.5.2, $\mathrm{E}[X\tilde{\beta}] = \mathrm{E}[\tilde{A}X\beta]$. Because $\tilde{A}$ is a projection operator onto $C(X)$ for any $\tilde{\Sigma}$, $\mathrm{E}[\tilde{A}X\beta] = \mathrm{E}[X\beta] = X\beta$. $\square$

Corollary 6.5.4.   If $\lambda'\beta$ is estimable, then

$$\mathrm{E}[\lambda'\tilde{\beta}] = \lambda'\beta.$$

PROOF.   By estimability, $\lambda' = \rho'X$

$$\mathrm{E}[\lambda'\tilde{\beta}] = \mathrm{E}[\rho'\tilde{A}Y] = \rho'\mathrm{E}[\tilde{A}Y] = \rho'X\beta.$$

$\square$

Now consider the prediction problem. We seek to predict y_0 where $\mathrm{E}(y_0) = x_0'\beta$ and $x_0'\beta$ is estimable. The plug-in predictor is

$$\begin{aligned}
\tilde{y}_0 &= x_0'\tilde{\beta} + \tilde{\delta}'(Y - X\tilde{\beta}) \\
&= x_0'\tilde{\beta} + \tilde{\delta}'(I - \tilde{A})Y
\end{aligned}$$

where $\tilde{\delta} = \tilde{\Sigma}^{-1}\tilde{\Sigma}_{Y0}$ is a residual type statistic. Before proving that the plug-in predictor is unbiased, we establish another result.

Lemma 6.5.5. $\mathrm{E}[\tilde{\delta}'(I - \tilde{A})Y] = 0.$

PROOF.   Because $\tilde{\delta}$ and $(I - \tilde{A})$ are residual type statistics, $\tilde{\delta}'(I - \tilde{A})$ is also of residual type. Applying Proposition 6.5.2 and using the fact that for each Y, $\tilde{A}$ is a projection operator onto $C(X)$,

$$\mathrm{E}[\tilde{\delta}'(I - \tilde{A})Y] = \mathrm{E}[\tilde{\delta}'(I - \tilde{A})X\beta] = 0.$$

$\square$

The plug-in predictor is unbiased.

Proposition 6.5.6. $\mathrm{E}[\tilde{y}_0] = x_0'\beta = \mathrm{E}[y_0].$

PROOF.   By Corollary 6.5.4 and Lemma 6.5.5

$$\begin{aligned}
\mathrm{E}[\tilde{y}_0] &= \mathrm{E}[x_0'\tilde{\beta} + \tilde{\delta}'(I - \tilde{A})Y] \\
&= \mathrm{E}[x_0'\tilde{\beta}] + \mathrm{E}[\tilde{\delta}(I - \tilde{A})Y] \\
&= x_0'\beta.
\end{aligned}$$

$\square$

The next two propositions establish conditions under which the variance of the plug-in estimate and the prediction variance of the plug-in predictor are known to be no less than the variance of the BLUE and the BLUP. After proving the results, a brief discussion of the conditions necessary for the results will be given.

Proposition 6.5.7.   If $\mathrm{E}[Ae|(I - A)e] = 0$ and $\lambda' = \rho'X$, then

$$\mathrm{Var}[\lambda'\tilde{\beta}] = \mathrm{Var}[\lambda'\hat{\beta}] + \mathrm{Var}[\lambda'\tilde{\beta} - \lambda'\hat{\beta}].$$

PROOF.

$$\begin{aligned}
\mathrm{Var}(\lambda'\tilde{\beta}) &= \mathrm{Var}(\lambda'\tilde{\beta} - \lambda'\hat{\beta} + \lambda'\hat{\beta}) \\
&= \mathrm{Var}(\lambda'\hat{\beta}) + \mathrm{Var}(\lambda'\tilde{\beta} - \lambda'\hat{\beta}) \\
&\quad + 2\mathrm{Cov}\left(\lambda'\hat{\beta}, \lambda'(\tilde{\beta} - \hat{\beta})\right),
\end{aligned}$$

so it suffices to show that

$$\mathrm{Cov}\left(\lambda'\hat{\beta}, \lambda'(\tilde{\beta} - \hat{\beta})\right) = 0.$$

Because $\mathrm{E}[\lambda'\tilde{\beta} - \lambda'\hat{\beta}] = 0$, it is enough to show that

$$\mathrm{E}[\lambda'\hat{\beta}\{\lambda'\tilde{\beta} - \lambda'\hat{\beta}\}] = 0. \tag{1}$$

Now observe that because $AY \in C(X)$ and $\tilde{A}$ is a projection operator onto $C(X)$

$$
\begin{aligned}
X\tilde{\beta} &= \tilde{A}Y = \tilde{A}[AY + (I - A)Y] \\
&= AY + \tilde{A}(I - A)Y.
\end{aligned}
$$

Hence

$$
\begin{aligned}
\lambda'\tilde{\beta} = \rho'X\tilde{\beta} &= \rho'AY + \rho'\tilde{A}(I - A)Y \\
&= \lambda'\hat{\beta} + \rho'\tilde{A}(I - A)Y
\end{aligned}
$$

and

$$\lambda'\tilde{\beta} - \lambda'\hat{\beta} = \rho'\tilde{A}(I - A)Y.$$

Thus, (1) is equivalent to

$$\mathrm{E}[\lambda'\hat{\beta}\{\rho'\tilde{A}(I - A)Y\}] = 0.$$

Before proceeding, note two things: first, $(I - A)Y = (I - A)e$ and second, because $\tilde{A}$ is a residual-type function of Y, $\tilde{A}$ is also a function of $(I - A)Y$. Now consider the conditional expectation

$$
\begin{aligned}
\mathrm{E}\big[\lambda'\hat{\beta}\{\rho'\tilde{A}(I - A)Y\}\big|(I - A)Y\big] \\
= \{\rho'\tilde{A}(I - A)Y\}\mathrm{E}\big[\lambda'\hat{\beta}\big|(I - A)Y\big] \\
= \{\rho'\tilde{A}(I - A)Y\}\mathrm{E}\big[\rho'X\beta + \rho'Ae\big|(I - A)Y\big] \tag{2} \\
= \{\rho'\tilde{A}(I - A)Y\}\rho'X\beta.
\end{aligned}
$$

The last equality follows because $\rho'X\beta$ is a constant and $\mathrm{E}[\rho'Ae|(I-A)Y] = \rho'\mathrm{E}[Ae|(I - A)e] = 0$ by assumption. The statistic $\rho'\tilde{A}(I - A)$ is of residual type so using (2) and Proposition 6.5.2

$$
\begin{aligned}
\mathrm{E}\big[\lambda'\hat{\beta}\{\rho'\tilde{A}(I - A)Y\}\big] &= \mathrm{E}\Big(\mathrm{E}\big[\lambda'\hat{\beta}\{\rho'\tilde{A}(I - A)Y\}\big|(I - A)Y\big]\Big) \\
&= \mathrm{E}\Big(\{\rho'\tilde{A}(I - A)Y\}\rho'X\beta\Big) \\
&= \rho'X\beta\,\mathrm{E}\Big(\rho'\tilde{A}(I - A)Y\Big) \\
&= \rho'X\beta\,\mathrm{E}[\rho'\tilde{A}(I - A)X\beta] \\
&= 0
\end{aligned}
$$

which proves the result. $\square$

Proposition 6.5.8.   If $\mathrm{E}[y_0 - \hat{y}_0 | (I - A)e] = 0$, then

$$\mathrm{Var}(y_0 - \tilde{y}_0) = \mathrm{Var}(y_0 - \hat{y}_0) + \mathrm{Var}(\hat{y}_0 - \tilde{y}_0).$$

PROOF.   The proof is very similar to the proof of Proposition 6.5.7.

$$\begin{aligned}
\mathrm{Var}(y_0 - \tilde{y}_0) &= \mathrm{Var}(y_0 - \hat{y}_0 + \hat{y}_0 - \tilde{y}_0) \\
&= \mathrm{Var}(y_0 - \hat{y}_0) + \mathrm{Var}(\hat{y}_0 - \tilde{y}_0) \\
&\quad + 2\mathrm{Cov}(y_0 - \hat{y}_0, \hat{y}_0 - \tilde{y}_0).
\end{aligned}$$

We show that $\mathrm{Cov}(y_0 - \hat{y}_0, \hat{y}_0 - \tilde{y}_0) = 0$ or equivalently, because $\mathrm{E}[\hat{y}_0 - \tilde{y}_0] = 0$, that

$$\mathrm{E}[(y_0 - \hat{y}_0)(\hat{y}_0 - \tilde{y}_0)] = 0.$$

Recall that $\tilde{b}$ satisfies

$$\tilde{y}_0 = \tilde{b}'Y = \rho'\tilde{A}Y + \tilde{\delta}'(I - \tilde{A})Y. \tag{3}$$

Substituting AY for Y in (3) and using the facts that $AY \in C(X)$ and $\tilde{A}$ is a projection operator onto $C(X)$, we have $\tilde{b}'AY = \rho'\tilde{A}AY = \rho'AY = x_0'\hat{\beta}$. Moreover,

$$\begin{aligned}
\tilde{y}_0 &= \tilde{b}'AY + \tilde{b}'(I - A)Y \\
&= x_0'\hat{\beta} + \tilde{b}'(I - A)Y \\
&= x_0'\hat{\beta} + \delta'(I - A)Y - \delta'(I - A)Y + \tilde{b}'(I - A)Y \\
&= \hat{y}_0 + (\tilde{b} - \delta)'(I - A)Y,
\end{aligned}$$

and

$$\hat{y}_0 - \tilde{y}_0 = (\delta - \tilde{b})'(I - A)Y$$

which is a function of $(I - A)Y$ because $\tilde{b}$ is a residual type statistic.

As in the previous proof, evaluate the conditional expectation. This gives

$$\begin{aligned}
\mathrm{E}[(y_0 - \hat{y}_0)(\hat{y}_0 - \tilde{y}_0)|(I - A)Y] &= (\hat{y}_0 - \tilde{y}_0)\mathrm{E}[(y_0 - \hat{y}_0)|(I - A)Y] \\
&= 0
\end{aligned}$$

by assumption and the fact that $(I - A)Y = (I - A)e$. Because the conditional expectation is zero for all $(I - A)Y$, the unconditional expectation is zero and the result is proven. $\square$

Proposition 6.5.7 shows that the plug-in variance equals the BLUE variance plus a nonnegative quantity. Thus, the plug-in variance is at least as large as the BLUE variance. Proposition 6.5.8 gives a similar result for prediction.

Eaton (1985) discusses situations under which the conditions $\mathrm{E}[Ae|(I - A)e] = 0$ and $\mathrm{E}[y_0 - \hat{y}_0|(I - A)e] = 0$ hold. In particular, the first condition holds if e has an elliptical distribution and the second condition holds if (e_0, e') has an elliptical distribution. Elliptical distributions can be generated as follows. Y has an elliptical distribution if Y has a density that can be written as $\varphi(y'By)$ where B is a positive definite matrix and $\varphi(u)$ is a density on the nonnegative real numbers u. From Christensen (1987, Section I.2), the multivariate normal distribution is a special case of elliptical distributions.

Eaton's last results involve concave functions and Jensen's inequality. A set ζ is convex if for any $\alpha \in [0,1]$ and $s_1, s_2 \in \zeta$, the point $\alpha s_1 + (1 - \alpha)s_2 \in \zeta$. A function $\Psi : \zeta \to \mathbf{R}$ is concave if $\Psi(\alpha s_1 + (1 - \alpha)s_2) \geq \alpha \Psi(s_1) + (1 - \alpha)\Psi(s_2)$. Finally, if s is random and defined on ζ, Jensen's inequality states that $\Psi(\mathrm{E}[s]) \geq \mathrm{E}[\Psi(s)]$. See Ferguson (1967) for a more complete discussion of convexity and a proof of Jensen's inequality.

Let ζ be the set of all positive definite matrices Σ and observe that ζ is a convex set. Let $\mathcal{P}$ be the set of all matrices that are projection operators onto spaces that contain $C(X)$, i.e.,

$$\mathcal{P} = \{P|PP = P \text{ and } C(X) \subset C(P)\}.$$

For any $\rho \in \mathbf{R}^n$ and each $P \in \mathcal{P}$ define

$$\Psi_P(\Sigma) = \rho' P\Sigma P'\rho.$$

Observe that $\Psi_P(\Sigma)$ is a concave function. To see this, note that

$$\begin{aligned}
\Psi_P(\alpha\Sigma_1 + (1 - \alpha)\Sigma_2) &= \rho' P\{\alpha\Sigma_1 + (1 - \alpha)\Sigma_2\}P'\rho \\
&= \alpha\rho' P\Sigma_1 P'\rho + (1 - \alpha)\rho' P\Sigma_2 P'\rho \\
&= \alpha\Psi_P(\Sigma_1) + (1 - \alpha)\Psi_P(\Sigma_2).
\end{aligned}$$

For any P, $\rho'PY$ is a linear unbiased estimate of $\rho'X\beta$ so the variance of the BLUE, $\rho'AY$, is at least as small as the variance of $\rho'PY$, i.e.,

$$\rho' A\Sigma A'\rho = \inf_{P\in\mathcal{P}} \rho' P\Sigma P'\rho.$$

Define

$$\Psi(\Sigma) = \rho' A\Sigma A'\rho$$

so we see that

$$\Psi(\Sigma) = \inf_{P\in\mathcal{P}} \Psi_P(\Sigma).$$

The infinum of a set of concave functions is also concave so $\rho' A\Sigma A'\rho$ is concave. (Note that a direct proof of concavity is difficult because A is a function of Σ.)

EXERCISE 6.2.   Show that if $f_\lambda(x)$ is concave for any $\lambda \in \Lambda$, then $f(x) = \inf_{\lambda \in \Lambda} f_\lambda(x)$ is also concave.

Hint: By definition, for any point x_0 and any $\varepsilon > 0$ there exists λ such that $f(x_0) \geq f_\lambda(x_0) - \varepsilon$.

The estimated variance of the plug-in estimator is $\rho' \tilde{A} \tilde{\Sigma} \tilde{A}' \rho$. Define

$$\widetilde{\mathrm{Var}}(\rho' X \tilde{\beta}) = \rho' \tilde{A} \tilde{\Sigma} \tilde{A}' \rho \,.$$

Recalling that

$$\mathrm{Var}(\rho' X \hat{\beta}) = \rho' A \Sigma A' \rho$$

and that by Proposition 6.5.7

$$\mathrm{Var}(\rho' X \hat{\beta}) \leq \mathrm{Var}(\rho' X \tilde{\beta}) \,,$$

we can prove the following.

Proposition 6.5.9.   If $\tilde{\Sigma}$ is unbiased for Σ and if Proposition 6.5.7 holds, then

$$\mathrm{E}[\widetilde{\mathrm{Var}}(\rho' X \tilde{\beta})] \leq \mathrm{Var}(\rho' X \hat{\beta}) \leq \mathrm{Var}(\rho' X \tilde{\beta}) \,.$$

PROOF.   We need only prove that

$$\mathrm{E}[\rho' \tilde{A} \tilde{\Sigma} \tilde{A}' \rho] \leq \rho' A \Sigma A' \rho \,.$$

Because $\Psi(\Sigma)$ is concave, Jensen's inequality gives

$$\begin{aligned}
\rho' A \Sigma A' \rho &= \Psi(\Sigma) \\
&= \Psi(\mathrm{E}[\tilde{\Sigma}]) \\
&\geq \mathrm{E}[\Psi(\tilde{\Sigma})] \\
&= \mathrm{E}[\rho' \tilde{A} \tilde{\Sigma} \tilde{A}' \rho] \,.
\end{aligned}$$

$\square$

Thus, for an unbiased covariance matrix estimate $\tilde{\Sigma}$, the expected value of the estimated variance of $\rho' X \tilde{\beta}$ is no greater than the variance of $\rho' X \hat{\beta}$ while the true variance of $\rho' X \tilde{\beta}$ is no less than the variance of $\rho' X \hat{\beta}$. This establishes that there is a tendency for the estimated variance of $\rho' X \tilde{\beta}$ to underestimate the true variance and illustrates how the underestimation could be very substantial.

The result for prediction follows similarly. Consider the set

$$\mathcal{D} = \{d | d' X = x_0'\} \,.$$

Then any linear unbiased predictor can be written as $d'Y$ for some $d \in \mathcal{D}$. Let $\sigma_{00} = \text{Var}(y_0)$,

$$V = \begin{bmatrix} \sigma_{00} & \Sigma_{0Y} \\ \Sigma_{Y0} & \Sigma \end{bmatrix}$$

and take ζ as the set of all positive definite V. Define

$$\begin{aligned} \Psi_d(V) &= \text{Var}(y_0 - d'Y) \\ &= \sigma_{00} - 2d'\Sigma_{Y0} + d'\Sigma d. \end{aligned}$$

It is easily seen that

$$\Psi_d(\alpha V_1 + (1-\alpha)V_2) = \alpha\Psi_d(V_1) + (1-\alpha)\Psi_d(V_2)$$

so $\Psi_d(V)$ is concave. Define

$$\Psi(V) = \inf_{d \in \mathcal{D}} \Psi_d(V).$$

Note that $\Psi(V)$ is concave and

$$\begin{aligned} \Psi(V) &= \text{Var}(y_0 - b'Y) \\ &= \text{Var}(y_0 - \hat{y}_0). \end{aligned}$$

Once again, for $\tilde{V}$ unbiased and writing (6.2.4) with $\tilde{V}$ substituted for V as $\widetilde{\text{Var}}(y_0 - \tilde{y}_0)$, Jensen's inequality gives

$$\begin{aligned} \text{E}[\widetilde{\text{Var}}(y_0 - \tilde{y}_0)] &= \text{E}[\Psi(\tilde{V})] \\ &\leq \Psi(\text{E}[\tilde{V}]) \\ &= \Psi(V) \\ &= \text{Var}(y_0 - \hat{y}_0). \end{aligned}$$

We have proved the following proposition.

Proposition 6.5.10.   If $\tilde{V}$ is unbiased for V and if Proposition 6.5.8 holds then

$$\text{E}[\widetilde{\text{Var}}(y_0 - \tilde{y}_0)] \leq \text{Var}(y_0 - \hat{y}_0) \leq \text{Var}(y_0 - \tilde{y}_0).$$

Note that these results depend on the assumption that the procedure used for estimating Σ does not depend on the true value of β. Residual type estimators take on the same value when based on Y or on $e = Y - X\beta$. (The fact that e is unobservable is irrelevant to our argument.) Because the distribution of e does not depend on β, neither does the distribution of $\Sigma(Y)$. For example, if $\Sigma(\theta)$ is the covariance matrix for observations from a process with a stationary covariance function then clearly, because the covariance function does not depend on β, residual type estimators for $\Sigma(\theta)$ are reasonable. However situations exist in which residual type estimators are not reasonable.

A common problem in regression analysis is the presence of heteroscedastic errors, cf. Christensen (1987, Section XIII.4). To deal with this problem one often assumes a model

$$\Sigma(\theta) = [\sigma_{ij}(\theta)]$$

where

$$\sigma_{ij}(\theta) = 0 \quad i \neq j$$

and

$$\sigma_{ii}(\theta) = h_i(\alpha, \beta).$$

Here

$$\theta' = (\alpha', \beta').$$

Two common choices for the function h_i take α as a scalar and

$$h_i(\alpha, \beta) = \alpha(x_i'\beta)^2$$

or

$$h_i(\alpha, \beta) = \alpha\, x_i'\beta\,,$$

cf. Carroll and Ruppert (1988).

When the variance function depends on β, it is counter intuitive to use residual type estimation procedures. In particular, MLE's will not be residual type estimates for these variance functions. For normal errors, van Houwelingen (1988) has established that the variance of the optimal weighted least squares estimate $\hat{\beta}$ (based on known variances) is at least as great as the asymptotic variance of $\beta^* - \beta$ where β^* is the MLE of β. (Actually, the proper comparison is between $\sqrt{n}\hat{\beta}$ and $\sqrt{n}(\beta^* - \beta)$ so that both quantities have nontrivial asymptotic distributions.) This remarkable result occurs because there may be extra information to be gained about β from the variability of the observations. Moreover, van Houwelingen also established that for such variance models the MLE β^* *may* not even be consistent.

VI.6 Models for Covariance Functions and Semivariograms

In practice, the covariance function $\sigma(u, w)$ is rarely known. To obtain an estimate of the covariance matrix of Y, i.e., Σ, some method of estimating $\sigma(u, w)$ is needed. Recalling that Σ is an $n \times n$ matrix with $n(n+1)/2$ distinct elements, there is little hope of estimating Σ from the n observations in

Y without making assumptions about Σ or equivalently assumptions about $\sigma(\cdot,\cdot)$. In particular, we assume that the covariance function depends on a vector of parameters θ. Write the covariance function as

$$\sigma(u,w;\theta)$$

which is a known function for given θ. We can also write

$$\mathrm{Cov}(Y) = \Sigma(\theta) = [\sigma(u_i,u_j;\theta)]$$

and

$$\mathrm{Cov}(Y,y_0) = \Sigma_{Y0}(\theta) = [\sigma(u_i,u_0;\theta)]\,.$$

Alternatively, we can assume that the semivariogram depends on a vector of parameters θ and write

$$\gamma(u,w;\theta)\,,$$
$$\Gamma(\theta) = [\gamma(u_i,u_j;\theta)]$$

and

$$\Gamma_{Y0}(\theta) = [\gamma(u_i,u_0;\theta)]\,.$$

In this section, we consider some of the standard models for $\sigma(u,w;\theta)$ and $\gamma(u,w;\theta)$. In Section 7, methods of estimating θ are discussed.

THE LINEAR COVARIANCE MODEL

This model assumes that $\sigma(u,w;\theta)$ is linear in the components of θ. In particular, write $\theta = (\theta_0,\theta_1,\ldots,\theta_r)$ and for $s = 0,1,\ldots,r$ let $\sigma_s(u,w)$ be a known covariance function. The linear covariance model is

$$\sigma(u,w;\theta) = \sum_{s=0}^{r} \theta_s \sigma_s(u,w)\,.$$

To ensure that $\sigma(u,w;\theta)$ is a legitimate covariance function, we assume that $\theta_s \geq 0$ for all s. Writing

$$\Sigma_s = [\sigma_s(u_i,u_j)]\,,$$

we have

$$\Sigma(\theta) = \sum_{s=0}^{r} \theta_s \Sigma_s\,.$$

The linear semivariogram model is defined similarly

$$\gamma(u,w;\theta) = \sum_{s=0}^{r} \theta_s \gamma_s(u,w)$$

for known semivariograms $\gamma_s(\cdot,\cdot)$.

A commonly used linear semivariogram is the isotropic function

$$\gamma(\|h\|\,;\theta) = \theta_0 + \theta_1\|h\|\,,\tag{1}$$

where θ_0 and θ_1 are nonnegative and $\theta_0 = \sigma_M^2$. In fact, this is often referred to as *the* linear semivariogram model. This semivariogram cannot correspond to a second-order stationary process because

$$\lim_{\|h\|\to\infty}\{\theta_0 + \theta_1\|h\|\} = \infty\,.$$

Recall that for a second-order stationary process $\gamma(\|h\|) = \sigma(0) - \sigma(\|h\|)$ and, by Cauchy-Schwartz, $|\sigma(\|h\|)| \le \sigma(0)$ thus the variance would have to be infinite. Brownian motion is a process in $\mathbf{R}$ that has a linear semivariogram. The linear semivariogram model would seem to be most appropriate for data that has a logical origin, e.g., data collected around a smelter, and that resembles a random walk in that the variability increases as one gets further from the origin.

There is a temptation to modify (1) so that

$$\lim_{\|h\|\to\infty}\gamma(\|h\|\,;\theta) \neq \infty\,.$$

The idea is that the linear semivariogram may be a reasonable approximation up to a point, but that the variability of real data would not go on increasing indefinitely. It is sometimes suggested that the function

$$g(\|h\|\,;\theta) = \begin{cases} \theta_0 + \theta_1\|h\| & \|h\| \le \theta_2 \\ \theta_0 + \theta_1\theta_2 & \|h\| > \theta_2 \end{cases}$$

could be used. This would correspond to a second-order stationary process with measurement error $\sigma_M^2 = \theta_0$ and a variance for each observation of $K = \theta_0 + \theta_1\theta_2$. Unfortunately, the corresponding "covariance" function

$$s(\|h\|\,;\theta) = K - g(\|h\|\,;\theta)$$

is not a legitimate covariance function. It is not nonnegative definite. One can find locations $u_1,\ldots,u_k$ such that the $k \times k$ matrix $[s(\|u_i - u_j\|\,;\theta)]$ is not nonnegative definite. Moreover, $g(\cdot\,;\theta)$ does not satisfy a property similar to nonnegative definiteness that is necessary for all semivariograms, cf. Journel and Hüijbregts (1978).

Nonlinear Isotropic Covariance Models

We now present some of the standard isotropic covariance models that are nonlinear. In the next subsection, some methods of dealing with non-isotropic [anisotropic] covariances will be considered.

The *spherical covariance function* is

$$\sigma(\|h\|;\theta) = \begin{cases} \theta_1 \left[1 - \frac{3\|h\|}{2\theta_2} + \frac{\|h\|^3}{2\theta_2^3}\right] & 0 < \|h\| \le \theta_2 \\ \theta_0 + \theta_1 & \|h\| = 0 \\ 0 & \|h\| > \theta_2 \end{cases}$$

for θ_0, θ_1, θ_2 nonnegative. This covariance function arises naturally on $\mathbf{R}^3$ (cf. Matern, 1986, Section 3.2) and also defines a covariance function on $\mathbf{R}^2$. The measurement error variance is $\sigma_M^2 = \theta_0$. The total variance is $\theta_0 + \theta_1$. The *range* of a covariance function is the distance after which observations become uncorrelated. For the spherical model, observations more than θ_2 units apart are uncorrelated so the range is θ_2.

The *exponential covariance function* is

$$\sigma(\|h\|;\theta) = \begin{cases} \theta_1 \exp(-\theta_2\|h\|) & \|h\| > 0 \\ \theta_0 + \theta_1 & \|h\| = 0 \end{cases}$$

for θ_0, θ_1, θ_2 nonnegative. The measurement error variance is θ_0, the total variance is $\theta_0 + \theta_1$, and the range is infinite. While the range is infinite, correlations decrease very rapidly as $\|h\|$ increases. Of course, this phenomenon depends on the value of θ_2.

Another common covariance model, often called the *Gaussian covariance function*, is

$$\sigma(\|h\|;\theta) = \begin{cases} \theta_1 \exp(-\theta_2\|h\|^2) & \|h\| > 0 \\ \theta_0 + \theta_1 & \|h\| = 0 \end{cases}$$

for θ_0, θ_1, θ_2 nonnegative. Ripley (1981, p. 56) takes exception to the name Gaussian. The name originates from the similarity of the function to the characteristic function of a normal distribution, cf. Matern (1986, Section 2.4). The behavior of the Gaussian model is similar to that of the exponential model. However, with the term $\|h\|^2$ in the numerator, covariances at distances greater than 1 approach zero much more rapidly than in the exponential model. Also, for small distances $\|h\|$, the covariance approaches the value θ_1 much more rapidly than does the exponential. Note that the Gaussian model is separable in the sense of Section 1.

Whittle (1954) has shown that a covariance function that depends on $K_1(\cdot)$, the first-order modified Bessel function of the second kind, arises naturally in $\mathbf{R}^2$. In particular, for $\theta_1, \theta_2 > 0$ the function is

$$\sigma(\|h\|;\theta) = \theta_1\|h\|\theta_2 K_1(\|h\|\theta_2).$$

This can be modified by adding a measurement error of variance θ_0 when $\|h\| = 0$. Whittle (1963) considers more general functions

$$\sigma(\|h\|;\theta) = [\theta_1/2^{\nu-1}\Gamma(\nu)](\theta_2\|h\|)^\nu K_\nu(\theta_2\|h\|)$$

where $\nu > 0$ is known and $K_\nu(\cdot)$ is the ν order modified Bessel function of the second kind. Ripley (1981, p. 56) gives some graphs of these functions and mentions that for $\nu = 1/2$, one gets the exponential model (without measurement error).

MODELING ANISOTROPIC COVARIANCE FUNCTIONS

Anisotropic covariance functions are simply covariance functions that are not isotropic. We mention only two possible approaches to modeling such functions. Suppose that $h = (h_1, h_2, h_3)'$ and that we suspect the variability in the direction $(0, 0, 1)'$ is causing the anisotropicity. (Isn't anisotropicity a wonderful word?) For example, h_1 and h_2 could determine a surface location (e.g., longitude and latitude), while h_3 determines depth. For fixed h_3, variability might very well be isotropic in h_1 and h_2; however, the variability in depth may not behave like that in the other two directions.

Ripley (1981) suggests modifying isotropic models. Rather than using $\sigma(\|h\|)$ where $\|h\| = \sqrt{h_1^2 + h_2^2 + h_3^2}$, use $\sigma(\sqrt{h_1^2 + h_2^2 + \lambda h_3^2})$ where λ is an additional parameter to be estimated. For example, the exponential model becomes

$$\sigma(h; \theta, \lambda) = \begin{cases} \theta_1 \exp\left(-\theta_2 \sqrt{h_1^2 + h_2^2 + \lambda h_3^2}\right) & \|h\| > 0 \\ \theta_0 + \theta_1 & \|h\| = 0 \end{cases}.$$

This is a special case of the elliptical covariance functions discussed by Matern (1986). Elliptical covariance functions are isotropic functions $\sigma(\cdot)$ evaluated at $\sigma(\sqrt{h' A h})$. Here A can be taken as any nonnegative definite matrix.

Hüijbregts (1975) suggests adding different isotropic models, e.g.,

$$\sigma(h; \theta) = \sigma_1(\|h\|; \theta_1) + \sigma_2(|h_3|; \theta_2),$$

where $\sigma_1(\cdot; \theta_1)$ is an isotropic covariance on the entire vector h that depends on a parameter vector θ_1. Similarly, $\sigma_2(\cdot; \theta_2)$ is isotropic in the component h_3 and depends on the parameter vector θ_2.

NONLINEAR SEMIVARIOGRAMS

In geostatistics, if the semivariogram $\gamma(\cdot)$ has the property that

$$\lim_{\|h\| \to \infty} \gamma(h) = \gamma_\infty < \infty,$$

then γ_∞ is called the *sill* of the semivariogram. Any semivariogram with a sill can be obtained from a second-order stationary process with the property that

$$\lim_{\|h\| \to \infty} \sigma(h) = 0.$$

In particular, the stationary covariance function is

$$\begin{aligned} \sigma(0) &= \gamma_\infty \\ \sigma(h) &= \gamma_\infty - \gamma(h). \end{aligned}$$

Conversely, any stationary covariance function with

$$\lim_{\|h\|\to\infty} \sigma(h) = 0 \tag{2}$$

determines a semivariogram with a sill. This follows from the fact that for second-order stationary processes

$$\gamma(h) = \sigma(0) - \sigma(h). \tag{3}$$

All of the nonlinear covariance functions we have considered satisfy (2). It is a simple matter to convert them to semivariogram models using (3).

VI.7   Estimation of Covariance Functions and Semivariograms

In Christensen (1987, Chapter XII), we considered the linear model

$$Y = X\beta + \varepsilon, \ \mathrm{E}(\varepsilon) = 0, \ \mathrm{Cov}(\varepsilon) = \Sigma(\theta)$$

where $\theta = (\sigma_0^2, \ldots, \sigma_r^2)$ and

$$\Sigma(\theta) = \sigma_0^2 I + \sum_{s=1}^{r} \sigma_s^2 Z_s Z_s' \tag{1}$$

for known matrices $Z_1, \ldots, Z_r$ Equation (1) is a linear model for $\Sigma(\theta)$ and, as will be seen below, is equivalent to the linear covariance model discussed in the previous section. For the linear covariance model, the estimation of the parameters can be executed as in Christensen (1987, Chapter XII).

In general, we assume that

$$Y = X\beta + \varepsilon, \ \mathrm{E}(\varepsilon) = 0, \ \mathrm{Cov}(\varepsilon) = \Sigma(\theta) \tag{2}$$

where $\theta = (\theta_0, \ldots, \theta_r)'$ and $\Sigma(\theta)$ is generated by a model such as those discussed in the previous section. For normal errors ε, the maximum likelihood and restricted maximum likelihood methods can be modified to find estimates of θ even for nonlinear covariance functions. The geostatistics literature also includes various ad hoc model-fitting procedures (cf. Cressie, 1985). These are usually based on empirical estimates $\tilde{\sigma}(u_i, u_j)$ of the elements of $\Sigma(\theta)$ combined with an ad hoc method of choosing $\hat{\theta}$ so that the values $\sigma(u_i, u_j; \hat{\theta})$ are in some sense close to the values $\tilde{\sigma}(u_i, u_j)$.

Of course, the observed data is Y so estimation must be based on Y and the model for Y. It is a subtle point, but worth mentioning, that we must estimate the parameter θ in $\Sigma(\theta)$ rather than the parameter θ in $\sigma(u, w; \theta)$. Of course, given an estimate $\hat{\theta}$, not only does $\Sigma(\hat{\theta})$ estimate the covariance matrix but also $\sigma(u, w; \hat{\theta})$ estimates the covariance function. Nonetheless, our observations only give direct information about $\Sigma(\theta)$.

ESTIMATION FOR LINEAR COVARIANCE FUNCTIONS

We begin by illustrating that the linear covariance function model,

$$\sigma(u, w; \theta) = \sum_{i=0}^{r} \theta_i \sigma_i(u, w)$$

for nonnegative θ_i's and known $\sigma_i(\cdot, \cdot)$'s, generates a mixed model as in Christensen (1987, Chapter XII). The matrix obtained by evaluating $\sigma_i(\cdot, \cdot)$ at the points u_j, u_k, $j, k = 1, \ldots, n$, say

$$\Sigma_i = [\sigma_i(u_j, u_k)]$$

is a known nonnegative definite matrix because each $\sigma_i(u, w)$ is a known covariance function. By Christensen (1987, Corollary B.25), for $q(i) \geq r(\Sigma_i)$ we can write

$$\Sigma_i = Z_i Z_i'$$

where Z_i is an $n \times q(i)$ matrix. For $i = 0, \ldots, r$, let ξ_i be a $q(i)$-dimensional random vector with

$$
\begin{aligned}
\mathrm{E}(\xi_i) &= 0 \\
\mathrm{Cov}(\xi_i) &= \theta_i I_{q(i)}
\end{aligned}
$$

and

$$\mathrm{Cov}(\xi_i, \xi_j) = 0 \qquad i \neq j \,.$$

The ξ_i's have been defined so that the mixed model

$$Y = X\beta + \sum_{i=0}^{r} Z_i \xi_i$$

is identical to model (2) where

$$
\begin{aligned}
\Sigma(\theta) &= \sum_{i=0}^{r} \theta_i Z_i Z_i' \\
&= \sum_{i=0}^{r} \theta_i \Sigma_i \,.
\end{aligned}
$$

By analogy with Christensen (1987, Chapter XII), we will typically assume that $\theta_0 \sigma_0(u, w)$ is the covariance function for a measurement error process, i.e., $\theta_0 = \sigma_M^2$ and

$$\sigma_0(u, w) = \begin{cases} 1 & u = w \\ 0 & u \neq w \end{cases} \,.$$

This leads to

$$\Sigma_0 = I_n \,.$$

With this assumption, $\xi_0 = e$ where e is the error vector from Christensen (1987, Chapter XII). If a model without measurement error is desired, the terms γ_0, Z_0, and Σ_0 can be dropped.

Note that all of the estimation methods in Christensen (1987, Chapter XII) depend on Z_i only through $Z_i Z_i'$ which in the current model is just Σ_i. The estimation procedures can be found by substitution.

For normal errors, the maximum likelihood estimates $\hat{\beta}$, $\hat{\theta}$ satisfy

$$X'\Sigma^{-1}(\hat{\theta})X\hat{\beta} = X'\Sigma^{-1}(\hat{\theta})Y$$

and

$$\mathrm{tr}[\Sigma^{-1}(\hat{\theta})\Sigma_i] = (Y - X\hat{\beta})'\Sigma^{-1}(\hat{\theta})\Sigma_i\Sigma^{-1}(\hat{\theta})(Y - X\hat{\beta})$$

for $i = 0, \ldots, r$. The restricted maximum likelihood estimate $\hat{\theta}$ satisfies

$$\sum_{j=0}^{r} \hat{\theta}_j \mathrm{tr}[\Sigma_j\Sigma^{-1}(\hat{\theta})(I - A)\Sigma_i\Sigma^{-1}(\hat{\theta})(I - A)] =$$

$$Y'(I - A)'\Sigma^{-1}(\hat{\theta})\Sigma_i\Sigma^{-1}(\hat{\theta})(I - A)Y$$

for $i = 0, \ldots, r$ where

$$A = X(X'\Sigma^{-1}(\hat{\theta})X)^- X'\Sigma^{-1}(\hat{\theta}) \,.$$

Both sets of equations require iterative computing methods to obtain solutions.

Without assuming normality, minimum norm quadratic unbiased estimates (MINQUE's) can be obtained. Select weights $w_0, w_1, \ldots, w_r$ and define

$$\Sigma_w = \sum_{i=0}^{r} w_i\Sigma_i$$

and

$$A_w = X(X'\Sigma_w^{-1}X)^- X'\Sigma_w^{-1} \,.$$

The MINQUE estimates satisfy

$$\sum_{j=0}^{r} \hat{\theta}_j \mathrm{tr}[\Sigma_j\Sigma_w^{-1}(I - A_w)\Sigma_i\Sigma_w^{-1}(I - A_w)] = Y'(I - A_w)'\Sigma_w^{-1}\Sigma_i\Sigma_w^{-1}(I - A)Y$$

for $i = 0, \ldots, r$. Computationally, one can use a matrix package to compute the Z_i's and then use a standard mixed model package to estimate $\hat{\theta}$.

The equivalence of the linear covariance function model and random effects models was apparently first recognized by Kitanidis (1983, 1985). Marshall and Mardia (1985) also proposed MINQUE estimation. Stein (1987) gives asymptotic efficiency and consistency results for MINQUE estimates.

MAXIMUM LIKELIHOOD ESTIMATION

Maximum likelihood for nonlinear covariance functions requires only minor modifications to the procedure presented in Christensen (1987, Section XII.4). The partial derivatives of the log-likelihood are

$$\frac{\partial \ell}{\partial \beta} = -X'\Sigma^{-1}X\beta + X'\Sigma^{-1}Y$$

and

$$\frac{\partial \ell}{\partial \theta_i} = -\frac{1}{2}\mathrm{tr}\left(\Sigma^{-1}\frac{\partial \Sigma}{\partial \theta_i}\right) + \frac{1}{2}(Y - X\beta)'\Sigma^{-1}\frac{\partial \Sigma}{\partial \theta_i}\Sigma^{-1}(Y - X\beta).$$

Setting the partial derivatives equal to zero leads to solving the equations

$$X'\Sigma^{-1}X\beta = X'\Sigma^{-1}Y$$

and

$$\mathrm{tr}\left(\Sigma^{-1}\frac{\partial \Sigma}{\partial \theta_i}\right) = (Y - X\beta)'\Sigma^{-1}\frac{\partial \Sigma}{\partial \theta_i}\Sigma^{-1}(Y - X\beta)$$

for $i = 0, \ldots, r$. As has been discussed in other contexts, the likelihood is maximized for any Σ by taking

$$X\hat{\beta} = AY$$

where $A = X(X'\Sigma^{-1}X)^{-1}X'\Sigma^{-1}$. It follows that $\hat{\theta}$ satisfies

$$\mathrm{tr}\left(\hat{\Sigma}^{-1}\frac{\partial \hat{\Sigma}}{\partial \theta_i}\right) = Y'(I - \hat{A})'\hat{\Sigma}^{-1}\frac{\partial \hat{\Sigma}}{\partial \theta_i}\hat{\Sigma}^{-1}(I - \hat{A})Y, \tag{3}$$

$i = 0, \ldots, r$ where $\partial\hat{\Sigma}/\partial\theta_i$ indicates $\partial\Sigma/\partial\theta_i$ evaluated at $\hat{\theta}$.

The matrix

$$\frac{\partial \Sigma}{\partial \theta_i} = [\partial\sigma(u_j, u_k; \theta)/\partial\theta_i]$$

depends on the particular covariance model being used. For example, assuming the isotropic exponential model without measurement error gives

$$\sigma(u, w; \theta) = \theta_1 e^{-\theta_2\|u-w\|}.$$

Differentiation yields

$$\partial\sigma(u, w; \theta)/\partial\theta_1 = e^{-\theta_2\|u-w\|}$$

thus defining $\partial\Sigma/\partial\theta_1$. Also,

$$\partial\sigma(u, w; \theta)/\partial\theta_2 = -\theta_1\|u - w\|e^{-\theta_2\|u-w\|}$$

which defines $\partial \Sigma / \partial \theta_2$.

The standard covariance models are discontinuous at $\|h\| = 0$ when measurement error occurs. This might give one doubts about whether these methods for obtaining MLE's can be executed. There is no problem. Derivatives are taken with respect to the θ_i's and all of the standard models are continuous in θ.

The maximum likelihood approach for spatial data was apparently first proposed by Kitanidis (1983) for linear covariance functions. Mardia and Marshall (1984) independently proposed using MLE's for general covariance models. Kitanidis and Lane (1985) also extended Kitanidis (1983) to general covariance functions. All of these articles discuss computational procedures. In their article on analyzing field-plot experiments, Zimmerman and Harville (1990) present a nice general discussion of maximum likelihood and restricted maximum likelihood methods. They also point out that results in Zimmerman (1989) can be used to reduce the computational burden when many of the standard covariance models are used. Warnes and Ripley (1987) have pointed out that the likelihood function is often multimodal and that care must be taken to obtain the global rather than some local maximum of the likelihood function, cf. also Mardia and Watkins (1989). As always, high correlations between the parameter estimates can cause instability in the estimates.

RESTRICTED MAXIMUM LIKELIHOOD ESTIMATION

The restricted maximum likelihood (REML) estimation procedure of Patterson and Thompson (1974) can be modified to treat general covariance models. As in Christensen (1987, Section XII.6), the method maximizes the likelihood associated with

$$B'Y \sim N(0, B'\Sigma(\theta)B)$$

where B is a full column rank matrix with

$$C(B) = C(X)^{\perp}.$$

Setting the partial derivatives of the log-likelihood function equal to zero leads to solving

$$\text{tr}\left[(B'\Sigma B)^{-1}\frac{\partial B'\Sigma B}{\partial \theta_i}\right] = Y'B(B'\Sigma B)^{-1}\frac{\partial B'\Sigma B}{\partial \theta_i}(B'\Sigma B)^{-1}B'Y.$$

Noting that
$$\partial B'\Sigma B/\partial \theta_i = B'(\partial \Sigma/\partial \theta_i)B,$$

it is clearly equivalent to solve

$$\text{tr}[(B'\Sigma B)^{-1}B'(\partial \Sigma/\partial \theta_i)B] = Y'B(B'\Sigma B)^{-1}B'\frac{\partial \Sigma}{\partial \theta_i}B(B'\Sigma B)^{-1}B'Y.$$

Rewriting this as in Christensen (1987, Section XII.6), a REML estimate $\hat{\theta}$ is a solution to

$$\mathrm{tr}[\hat{\Sigma}^{-1}(I - \hat{A})\partial\hat{\Sigma}/\partial\theta_i] = Y'(I - \hat{A})'\hat{\Sigma}^{-1}(\partial\hat{\Sigma}/\partial\theta_i)\hat{\Sigma}^{-1}(I - \hat{A})Y \quad (4)$$

for $i = 0, \ldots, r$. Note that the only difference between the REML equations (4) and the MLE equations (3) is the existence of the term $(I - \hat{A})$ in the trace in (4).

TRADITIONAL GEOSTATISTICAL ESTIMATION

The traditional approach to covariance function or semivariogram estimation (cf. Journel and Hüijbregts, 1978 or David, 1977) is to obtain an "empirical" estimate and to fit a model to the empirical estimate. We concentrate on fitting covariance functions and discuss fitting semivariograms at the end of the subsection. We begin by discussing empirical estimation. In order to have enough data to perform estimation, we assume second-order stationarity, i.e., $\sigma(u, w) = \sigma(u - w)$. The empirical estimate is nonparametric in the sense that estimates are not based on any covariance model with a small number of parameters. To begin the procedure, choose a nonnegative definite weighting matrix, say Σ_0, and fit the model

$$Y = X\beta + e, \; \mathrm{E}(e) = 0, \; \mathrm{Cov}(e) = \Sigma_0$$

to obtain residuals

$$\hat{e}_0 = (I - A_0)Y - Y - X\hat{\beta}_0$$

where

$$A_0 = X(X'\Sigma_0^{-1}X)^- X'\Sigma_0^{-1}$$

and

$$X\hat{\beta}_0 = A_0 Y.$$

These residuals are the basis of empirical estimates of the covariance function. For any vector h, there is a finite number N_h of pairs of observations y_i and y_j for which $u_i - u_j = h$. For each of these pairs, list the corresponding residual pairs, say $(\hat{e}_{0i}, \hat{e}_{0i(h)})$, $i = 1, \ldots, N_h$. If $N_h \geq 1$, the traditional empirical estimator is

$$\hat{\sigma}(h) = \hat{\sigma}(-h) = \frac{1}{N_h}\sum_{i=1}^{N_h}\hat{e}_i\hat{e}_{i(h)}.$$

If N_h is zero, no empirical estimate is possible because no data has been collected with $u_i - u_j = h$. With a finite number of observations, there will be only a finite number of vectors, say $h(1), h(2), \ldots, h(q)$ that have $N_{h(i)} > 0$. In practice, if $N_{h(i)}$ is not substantially greater than 1, we may not wish to include $h(i)$ as a vector for which the covariance function will be estimated.

Given a parametric stationary covariance function, say $\sigma(h;\theta)$, a least squares estimate of θ can be obtained by minimizing

$$\sum_{i=1}^{q} [\hat{\sigma}(h(i)) - \sigma(h(i);\theta)]^2 \;.$$

Weighted least squares estimates can also be computed. If the covariances or asymptotic covariances of $\hat{\sigma}(h(i))$ and $\hat{\sigma}(h(j))$ can be computed, say $\mathrm{Cov}(\hat{\sigma}(h(i)), \hat{\sigma}(h(j))) = v_{ij}$, write

$$V = [v_{ij}]$$

and choose θ to minimize

$$S'V^{-1}S$$

where $S' = [\hat{\sigma}(h(1)) - \sigma(h(1);\theta), \ldots, \hat{\sigma}(h(q)) - \sigma(h(q);\theta)]$. In many cases, the covariances will be small relative to the variances so a reasonable estimate can be obtained by minimizing

$$\sum_{i=1}^{q} [\hat{\sigma}(h(i)) - \sigma(h(i);\theta)]^2 \,/v_{ii} \;.$$

The Gauss-Newton procedure described in Section V.4 can be used to find the estimate, say $\hat{\theta}_0$, using any of these criteria.

The best fit of the linear model and hence the best residuals are obtained by taking $\Sigma_0 = \Sigma$. Because Σ is unknown, it is reasonable to use $\hat{\theta}_0$ to estimate it. Let

$$\Sigma_1 = [\sigma(u_i, u_j; \hat{\theta}_0)]$$

and find residuals

$$\hat{e}_1 = (I - A_1)Y$$

where

$$A_1 = X(X'\Sigma_1^{-1}X)^- X'\Sigma_1^{-1} \;.$$

These residuals lead to pairs $(\hat{e}_{1i}, \hat{e}_{1i(h)})$, $i = 1, \ldots, N_h$ and estimates $\hat{\sigma}_1(h) = (1/N_h)\sum_{i=1}^{N_h} \hat{e}_{1i}\hat{e}_{1i(h)}$. The estimates can then be used to obtain $\hat{\theta}_1$ and define

$$\hat{\Sigma}_2 = [\sigma(u_i, u_j; \hat{\theta}_1)] \;.$$

This procedure can be iterated in the hope that the sequence $\hat{\theta}_t$ converges to some value $\hat{\theta}$. Armstrong (1984) presents criticisms of this method.

The idea of using weighted least squares as a criterion for fitting semivariograms was first presented by Cressie (1985). The presentation above is a covariance function version of Cressie's ideas. Cressie's discussion was restricted to the ordinary kriging model. For this model, he computed the necessary variances and covariances. Cressie also suggested using robust empirical semivariogram estimates, in particular those proposed by

Hawkins and Cressie (1984). Again, Cressie computed the variances and covariances necessary for weighted least squares. The traditional empirical semivariogram estimator in ordinary kriging is

$$\hat{\gamma}(h) = \frac{1}{2N_h} \sum_{i=1}^{N_h} (y_i - y_{i(h)})^2$$

where the pairs $(y_i, y_{i(h)})$ are the N_h pairs whose locations differ by h. The robust estimates from Hawkins and Cressie (1984) are

$$2\tilde{\gamma}(h) = \left[\frac{1}{N_h} \sum_{i=1}^{N_h} |y_i - y_{i(h)}|^{1/2} \right]^4 \bigg/ [0.457 + 0.494/N_h]$$

and

$$2\tilde{\gamma}(h) = [\text{median}\{|y_i - y_{i(h)}|^{1/2}\}]^4 \bigg/ B_h$$

where B_h is a bias correction factor.

Methods other than least squares and weighted least squares are often used to fit covariance functions $\sigma(h; \theta)$ and semivariograms $\gamma(h; \theta)$ to their empirical counterparts. Various methods have been devised for particular covariance and semivariogram models. Models are also frequently fit by visual inspection.

If an isotropic covariance function or semivariogram is assumed, the empirical estimates change slightly. For covariance functions

$$\hat{\sigma}(\|h\|) = \frac{1}{N_h} \sum_{i=1}^{N_h} \hat{e}_i \hat{e}_{i(\|h\|)}$$

where the pairs $(\hat{e}_i, \hat{e}_{i(\|h\|)})$, $i = 1, \ldots, N_h$ are all residual pairs with locations separated by the distance $\|h\|$. For the semivariogram in ordinary kriging

$$2\hat{\gamma}(\|h\|) = \frac{1}{N_h} \sum_{i=1}^{N_h} [y_i - y_{i(\|h\|)}]^2.$$

The pairs $(y_i, y_{i(\|h\|)})$, $i = 1, \ldots, N_h$ consist of all observations with locations separated by $\|h\|$.

Zimmerman and Zimmerman (1990) present results from a Monte Carlo experiment comparing various techniques of estimating the variogram in ordinary kriging. Cressie (1989) gives a very complete illustration of traditional methods for semivariogram estimation in ordinary kriging.

References

Adler, Robert J. (1981). *The Geometry of Random Fields*. John Wiley and Sons, New York.

Aitchison, J. (1975). Goodness of prediction fit. *Biometrika*, **62**, 547-554.

Aitchison, J. and Dunsmore, I.R. (1975). *Statistical Prediction Analysis*. Cambridge University Press, Cambridge.

Akaike, Hirotugu (1973). Information theory and an extension of the maximum likelihood principle. In *Proceedings of the 2nd International Symposium on Information*, edited by B.N. Petrov and F. Czaki. Akademiai Kiado, Budapest.

Anderson, T.W. (1971). *The Statistical Analysis of Time Series*. John Wiley and Sons, New York.

Anderson, T.W. (1984). *An Introduction to Multivariate Statistical Analysis*, Second Edition. John Wiley and Sons, New York.

Andrews, D.F., Gnanadesikan, R., and Warner, J.L. (1971). Transformations of multivariate data. *Biometrics*, **27**, 825-840.

Ansley, C.F. and Kohn, R. (1984). On the estimation of ARIMA models with missing values. In *Time Series Analysis of Irregularly Observed Data*, edited by E. Parzen. Springer-Verlag, New York.

Armstrong, M. (1984). Problems with universal kriging. *Journal of the International Association for Mathematical Geology*, **16**, 101-108.

Arnold, Stephen F. (1981). *The Theory of Linear Models and Multivariate Analysis*. John Wiley and Sons, New York.

Bartlett, M.S. (1946). On the theoretical specification of sampling properties of autocorrelated time series. *Journal of the Royal Statistical Society, Supplement*, **8**, 27-41.

Berger, James O. (1985). *Statistical Decision Theory and Bayesian Analysis*. Springer-Verlag, New York.

Bloomfied, Peter (1976). *Fourier Analysis of Time Series: An Introduction*. John Wiley and Sons, New York.

Box, George E.P. (1950). Problems in the analysis of growth and wear curves. *Biometrics*, **6**, 362-389.

Box, G.E.P. and Cox, D.R. (1964). An analysis of transformations (with discussion). *Journal of the Royal Statistical Society, Series B*, **26**, 211-246.

Box, George E.P. and Jenkins, Gwylem M. (1970). *Time Series Analysis: Forecasting and Control*, (Revised Edition, 1976). Holden Day, San Francisco.

Breiman, Leo (1968). *Probability*. Addison-Wesley, Reading, MA.

Brillinger, David R. (1981). *Time Series: Data Analysis and Theory*, Second Edition. Holden Day, San Francisco.

Brockwell, Peter J. and Davis, Richard A. (1987). *Time Series: Theory and Methods*. Springer-Verlag, New York.

Carroll, R.J. and Ruppert, D. (1988). *Transformations and Weighting in Regression*. Chapman and Hall, New York.

Christensen, Ronald (1987). *Plane Answers to Complex Questions: The Theory of Linear Models*. Springer-Verlag, New York.

Christensen, Ronald (1990). *Log-linear Models*. Springer-Verlag, New York.

Christensen, Ronald (1990b). The equivalence of predictions from universal kriging and intrinsic random function kriging. *Mathematical Geology*, **22**, 655-664.

Clayton, Murray K., Geisser, Seymour, and Jennings, Dennis E. (1986). A comparison of several model selection procedures. In *Bayesian Inference and Decision Techniques*, edited by P. Goel and A. Zellner. North Holland, Amsterdam.

Cliff, A. and Ord, J.K. (1981). *Spatial Processes: Models and Applications*. Pion, London.

Cox, D.R. (1972). Regression models and life tables (with discussion). *Journal of the Royal Statistical Society, Series B*, **34**, 187-220.

Cressie, Noel (1985). Fitting variogram models by weighted least squares. *Journal of the International Association for Mathematical Geology*, **17**, 563-586.

Cressie, Noel (1986). Kriging nonstationary data. *Journal of the American Statistical Association*, **81**, 625-634.

Cressie, Noel (1988). Spatial prediction and ordinary kriging. *Mathematical Geology*, **20**, 405-421.

Cressie, Noel (1989). Geostatistics. *The American Statistician*, **43**, 197-202.

Danford, M.B., Hughes, H.M., and McNee, R.C. (1960). On the analysis of repeated-measurements experiments. *Biometrics*, **16**, 547-565.

David, M. (1977). *Geostatistical Ore Reserve Estimations*. Elsevier, New York.

Deely, J.J. and Lindley, D.V. (1981). Bayes empirical Bayes. *Journal of the American Statistical Association*, **76**, 833-841.

Delfiner, P. (1976). Linear estimation of nonstationary spatial phenomena. In *Advanced Geostatistics in the Mining Industry*, edited by M. Guarascia, M. David, and C. Hüijbregts. Reidel, Dordrecht.

Dempster, A.P., Laird, N.M., and Rubin, D.B. (1977). Maximum likelihood from incomplete data via the EM algorithm. *Journal of the Royal Statistical Society, Series B*, **39**, 1-38.

Diamond, P. and Armstrong, M. (1984). Robustness of variograms and conditioning of kriging matrices. *Journal of the International Association for Mathematical Geology*, **16**, 809-822.

Diderrich, George T. (1985). The Kalman filter from the perspective of Goldberger–Theil estimators. *The American Statistician*, **39**, 193-198.

Dillon, Wm. R. and Goldstein, Matthew (1984). *Multivariate Analysis: Methods and Applications*. John Wiley and Sons, New York.

Dixon, W.J., Brown, M.B., Engelman, L., Hill, M.A., and Jennrich, R.I. (1988). *BMDP Statistical Software Manual*, Vol. 2. University of California Press, Berkeley.

Dixon, Wilfrid J. and Massey, Frank J., Jr. (1983). *Introduction to Statistical Analysis*. McGraw-Hill, New York.

Doob, J.L. (1953). *Stochastic Processes*. John Wiley and Sons, New York.

Draper, Norman and Smith, Harry (1981). *Applied Regression Analysis, Second Edition*. John Wiley and Sons, New York.

Eaton, Morris L. (1983). *Multivariate Statistics: A Vector Space Approach*. John Wiley and Sons, New York.

Eaton, Morris L. (1985). The Gauss-Markov theorem in multivariate analysis. In *Multivariate Analysis - VI*, edited by P.R. Krishnaiah. North Holland, Amsterdam.

Efron, Bradley (1983). Estimating the error rate of a prediction rule: Improvement on cross-validation. *Journal of the American Statistical Association*, **78**, 316-331.

Ferguson, Thomas S. (1967). *Mathematical Statistics: A Decision Theoretic Approach*. Academic Press, New York.

Fisher, Ronald A. (1936). The use of multiple measurements in taxonomic problems, *Annals of Eugenics*, **7**, 179-188.

Fisher, Ronald A. (1938). The statistical utilization of multiple measurements. *Annals of Eugenics*, **8**, 376-386.

Friedman, Jerome H. (1989). Regularized discriminant analysis. *Journal of the American Statistical Association*, **84**, 165-175.

Fuller, Wayne A. (1976). *Introduction to Statistical Time Series*. John Wiley and Sons, New York.

Geisser, Seymour (1971). The inferential use of predictive distributions. In *Foundations of Statistical Inference*, edited by V.P. Godambe and D.A. Sprott. Holt, Rinehart, and Winston, Toronto.

Geisser, Seymour (1977). Discrimination, allocatory and separatory, linear aspects. In *Classification and Clustering*, edited by J. Van Ryzin. Academic Press, New York.

Geweke, J.F. and Singleton, K.J. (1980). Interpreting the likelihood ratio statistic in factor models when sample size is small. *Journal of the American Statistical Association*, **75**, 133-137.

Gnanadesikan, R. (1977). *Methods for Statistical Data Analysis of Multivariate Observations*. John Wiley and Sons, New York.

Goldberger, Arthur S. (1962). Best linear unbiased prediction in the generalized linear regression model. *Journal of the American Statistical Association*, **57**, 369-375.

Greenhouse, S.W. and Geisser, S. (1959). On methods in the analysis of profile data. *Psychometrika*, **24**, 95-112.

Gupta, N.K. and Mehra, R.K. (1974). Computational aspects of maximum likelihood estimation and reduction in sensitivity function calculations. *IEEE Transactions on Auto. Control*, **AC-19**, 774-783.

Hand, D.J. (1981). *Discrimination and Classification*. John Wiley and Sons, New York.

Hand, D.J. (1983). A comparison of two methods of discriminant analysis applied to binary data. *Biometrics*, **39**, 683-694.

Hannan, Edward James (1970). *Multiple Time Series*. John Wiley and Sons, New York.

Harrison, P.J. and Stevens, C.F. (1971). A Bayesian approach to short-term forecasting. *Operations Research Quarterly*, **22**, 341-362.

Harrison, P.J. and Stevens, C.F. (1976). Bayesian forecasting. *Journal of the Royal Statistical Society, Series B*, **38**, 205-247.

Harville, David A. (1985). Decomposition of prediction error. *Journal of the American Statistical Association*, **80**, 132-138.

Hawkins, Douglas M. and Cressie, Noel (1984). Robust kriging – a proposal. *Journal of the International Association for Mathematical Geology*, **16**, 3-18.

Heck, D.L. (1960). Charts of some upper percentage points of the distribution of the largest characteristic root. *Annals of Mathematical Statistics*, **31**, 625-642.

Hotelling, Harold (1933). Analysis of a complex of statistical variables into principal components. *Journal of Educational Psychology*, **24**, 417-441, 498-520.

Hüijbregts, C.J. (1975). Regionalized variables and quantitative analysis of spatial data. In *Display and Analysis of Spatial Data*, edited by J.C. Davis and M.J. McCullagh. John Wiley and Sons, New York.

Hurvich, C.M. and Tsai, C.-L. (1989). Regression and time series model selection in small samples. *Biometrika*, **76**, 297-308.

Huynh, H. and Feldt, L.S. (1976). Estimation of the Box correction for degrees of freedom from sample data in randomized block and split-plot designs. *Journal of Educational Statistics*, **1**, 69-82.

Jeske, Daniel R. and Harville, David A. (1986). Mean squared error of estimation and prediction under a general linear model. Unpublished manuscript, Department of Statistics, Iowa State University.

Johnson, Richard A. and Wichern, Dean W. (1988). *Applied Multivariate Statistical Analysis*, Second Edition. Prentice-Hall, Englewood Cliffs, NJ.

Johnson, Wesley (1987). The detection of influential observations for allocation, separation and the determination of probabilities in a Bayesian framework. *Journal of Business and Economic Statistics*, **5**, 369-391.

Jolicoeur, P. and Mosimann, J.E. (1960). Size and shape variation on the painted turtle: A principal component analysis. *Growth*, **24**, 339-354.

Jolliffe, I.T. (1986). *Principal Component Analysis*. Springer-Verlag, New York.

Jones, R.H. (1980). Maximum likelihood fitting of ARMA models to time series with missing observations. *Technometrics*, **22**, 389-396.

Jöreskog, K.G. (1975). Factor analysis by least squares and maximum likelihood. In *Statistical Methods for Digital Computers*, edited by K. Enslein, A. Ralston, and H.S. Wilf. John Wiley and Sons, New York.

Journel, A.G. and Hüijbregts, Ch.J. (1978). *Mining Geostatistics*. Academic Press, New York.

Kackar, R.N. and Harville, D.A. (1981). Unbiasedness of two-stage estimation and prediction procedures for mixed linear models. *Communications in Statistics - Theory and Methods*, **A10**, 1249-1261.

Kackar, R.N. and Harville, D.A. (1984). Approximations for standard errors of estimators of fixed and random effects in mixed linear models. *Journal of the American Statistical Association*, **79**, 853-862.

Kalbfleisch, John D. and Prentice, Ross L. (1980). *The Statistical Analysis of Failure Time Data*. John Wiley and Sons, New York.

Kalman, R.E. (1960). A new approach to linear filtering and prediction problems. *Journal of Basic Engineering*, **82**, 34-45.

Kalman, R.E. and Bucy, R.S. (1961). New results in linear filtering and prediction theory. *Journal of Basic Engineering*, **83**, 95-108.

Khatri, C.G. (1966). A note on a MANOVA model applied to problems in growth curves. *Annals of the Institute of Statistical Mathematics*, **18**, 75-86.

Kitanidis, Peter K. (1983). Statistical estimation of polynomial generalized covariance functions and hydrologic applications. *Water Resources Research*, **19**, 909-921.

Kitanidis, Peter K. (1985). Minimum-variance unbiased quadratic estimation of covariances of regionalized variables. *Journal of the International Association for Mathematical Geology*, **17**, 195-208.

Kitanidis, Peter K. (1986). Parameter uncertainty in estimation of spatial functions: Bayesian analysis. *Water Resources Research*, **22**, 499-507.

Kitanidis, Peter K. and Lane, Robert W. (1985). Maximum likelihood parameter estimation of hydrologic spatial processes by the Gauss-Newton method. *Journal of Hydrology*, **79**, 53-71.

Koopmans, Lambert H. (1974). *The Spectral Analysis of Time Series*. Academic Press, New York.

Kres, Heinz (1983). *Statistical Tables for Multivariate Analysis*. Springer-Verlag, New York.

Lachenbruch, P.A. (1975). *Discriminate Analysis*. Hafner Press, New York.

Lachenbruch, P.A., Sneeringeer, C., and Revo, L.T. (1973). Robustness of the linear and quadratic discriminant function to certain types of non-normality. *Communications in Statistics*, **1**, 39-57.

Lawley, D.N. and Maxwell, A.E. (1971). *Factor Analysis as a Statistical Methodology*, Second Edition. American Elsevier, New York.

Levy, Martin S. and Perng, S.K. (1986). An optimal prediction function for the normal linear model. *Journal of the American Statistical Association*, **81**, 196-198.

Li, Guoying and Chen, Zhouglian (1985). Projection pursuit approach to robust dispersion matrices and principal components: Primary theory and Monte Carlo. *Journal of the American Statistical Association*, **80**, 759-766.

Lubischew, Alexander A. (1962). On the use of discriminant functions in taxonomy. *Biometrics*, **18**, 455-477.

Lütkepohl, H. (1985). Comparison of criteria for estimating the order of a vector autoregressive process. *Journal of Time Series Analysis*, **65**, 297-303.

McCullagh, P. and Nelder, J.A. (1989). *Generalized Linear Models*, Second Edition. Chapman and Hall, London.

McKeon, James J. (1974). F approximations to the distribution of Hotelling's T_0^2. *Biometrika*, **61**, 381-383.

McLeod, A.I. (1977). Improved Box-Jenkins estimators. *Biometrika*, **64**, 531-534.

Mardia, K.V., Kent, J.T., and Bibby, J.M. (1979). *Multivariate Analysis*. Academic Press, New York.

Mardia, K.V. and Marshall, R.J. (1984). Maximum likelihood estimation of models for residual covariance in spatial regression. *Biometrika*, **71**, 135-146.

Mardia, K.V. and Watkins, A.J. (1989). On multimodality of the likelihood in the spatial linear model. *Biometrika*, **76**, 289-295.

Marquardt, Donald W. (1963). An algorithm for least squares estimation of non-linear parameters. *Journal of the Society for Industrial and Applied Mathematics*, **2**, 431-441.

Marshall, R.J. and Mardia, K.V. (1985). Minimum norm quadratic estimation of components of spatial covariance. *Journal of the International Association for Mathematical Geology*, **17**, 517-525.

Matern, Bertil (1986). *Spatial Variation*, Second Edition. Springer-Verlag, New York.

Matheron, G. (1965). Les variable regionalisees et leur estimation: Masson, Paris, xxxp.

Matheron, G. (1969). Le krigeage universal: Fascicule 1, Cahiers du CMMM., 82p.

Matheron, G. (1973). The intrinsic random functions and their applications. *Advances in Applied Probability*, **5**, 439-468.

Meinhold, Richard J. and Singpurwalla, Nozer D. (1983). Understanding the Kalman filter. *The American Statistician*, **37**, 123-127.

Morrison, Donald F. (1976). *Multivariate Statistical Methods*, Second Edition. McGraw-Hill, New York.

Mosteller, Frederick and Tukey, John W. (1977). *Data Analysis and Regression*. Addison-Wesley, Reading, MA.

Muirhead, Robb J. (1982). *Aspects of Multivariate Statistical Theory*. John Wiley and Sons, New York.

Murray, G.D. (1977). A note on the estimation of probability density functions. *Biometrika*, **64**, 150-152.

Nelder, J.A., and Wedderburn, R.W.M. (1972). Generalized linear models. *Journal of the Royal Statistical Society, Series A*, **135**, 370-384.

Okamoto, M. (1973). Distinctness of the eigenvalues of a quadratic form in a multivariate sample. *Annals of Statistics*, **1**, 763-765.

Okamoto, M. and Kanazawa, M. (1968). Minimization of eigenvalues of a matrix and optimality of principal components. *Annals of Mathematical Statistics*, **39**, 859-863.

Pandit, S.M. and Wu, S.M. (1983). *Time Series and System Analysis with Applications*. John Wiley and Sons, New York.

Panel on Discriminant Analysis, Classification, and Clustering (1989). Discriminant analysis and clustering. *Statistical Science*, **4**, 34-69.

Patterson, H.D. and Thompson, R. (1974). Maximum likelihood estimation of variance components. *Proceedings of the 8th International Biometric Conference*, 197-207.

Pearson, Karl (1901). On lines and planes of closest fit to systems of points in space. *Philosophical Magazine*, **6(2)**, 559-572.

Phadke, M.S. (1981). Quality audit using adaptive Kalman filtering. *ASQC Quality Congress Transactions* - San Francisco, 1045-1052.

Potthoff, R.F. and Roy, S.N. (1964). A generalized multivariate analysis of variance model useful especially for growth curve problems. *Biometrika*, **51**, 313-326.

Press, S. James (1982). *Applied Multivariate Analysis*, Second Edition. R.E. Krieger, Malabar, FL.

Press, S. James and Wilson, S. (1978). Choosing between logistic regression and discriminant analysis. *Journal of the American Statistical Association*, **73**, 699-705.

Quenille, M.H. (1949). Approximate tests of correlation in time-series. *Journal of the Royal Statistical Society, Series B*, **11**, 68-84.

Rao, C.R. (1948). The utilization of multiple measurements in problems of biological classification. *Journal of the Royal Statistical Society, Series B*, **10**, 159-203.

Rao, C.R. (1951). An asymptotic expansion of the distribution of Wilks' criterion. *Bull. Inst. Int. Stat.*, **33**, 177-180.

Rao, C.R. (1952). *Advanced Statistical Methods in Biometric Research.* John Wiley and Sons, New York.

Rao, C.R. (1965). The theory of least squares when the parameters are stochastic and its application to the analysis of growth curves. *Biometrika*, **52**, 447-458.

Rao, C.R. (1966). Covariance adjustment and related problems in multivariate analysis. In *Multivariate Analysis - II*, edited by P.R. Krishnaiah. Academic Press, New York.

Rao, C.R. (1967). Least squares theory using an estimated dispersion matrix and its application to measurement signals. *Proceedings of the Fifth Berkeley Symposium on Mathematical Statistics and Probability*, **1**, 355-372.

Rao, C. Radhakrishna (1973). *Linear Statistical Inference and Its Applications*, Second Edition. John Wiley and Sons, New York.

Ripley, Brian D. (1981). *Spatial Statistics.* John Wiley and Sons, New York.

Roy, S.N. (1953). On a heuristic method of test construction and its use in multivariate analysis. *Annals of Mathematical Statistics*, **24**, 220-238.

Roy, S.N. and Bose, R.C. (1953). Simultaneous confidence interval estimation. *Annals of Mathematical Statistics*, **24**, 513-536.

Scheffé, Henry (1959). *The Analysis of Variance.* John Wiley and Sons, New York.

Schervish, Mark J. (1986). A predictive derivation of principal components. Technical Report 378, Department of Statistics, Carnegie-Mellon University, Pittsburgh, PA.

Schwarz, Gideon (1978). Estimating the dimension of a model. *Annals of Statistics*, **16**, 461-464.

Seber, G.A.F. (1984). *Multivariate Observations.* John Wiley and Sons, New York.

Seber, G.A.F. and Wild, C.J. (1989). *Nonlinear Regression.* John Wiley and Sons, New York.

Shumway, Robert H. (1988). *Applied Statistical Times Series Analysis.* Prentice Hall, Englewood Cliffs, NJ.

Shumway, R.H. and Stoffer, D.S. (1982). An approach to time-series smoothing and forecasting using the EM algorithm. *Journal of Time Series Analysis,* **3**, 253-264.

Smith, H., Gnanadesikan, R., and Hughes, J.B. (1962). Multivariate analysis of variance (MANOVA). *Biometrics,* **18**, 22-41.

Stein, Michael L. (1987). Minimum norm quadratic estimation of spatial variograms. *Journal of the American Statistical Association,* **82**, 765-772.

Stein, Michael L. (1988). Asymptotically efficient prediction of a random field with misspecified covariance function. *Annals of Statistics,* **16**, 55-64.

Thompson, G.H. (1934). Hotelling's method modified to give Spearman's g. *Journal of Educational Psychology,* **25**, 366-374.

Thurstone, L.L. (1931). Multiple factor analysis. *Psychological Review,* **38**, 406-427.

Tukey, John W. (1977). *Exploratory Data Analysis.* Addison-Wesley, Reading, MA.

van Houwelingen, J.C. (1988). Use and abuse of variance models in regression. *Biometrics,* **44**, 1073-1081.

Waldmeier, M. (1960-1978). *Monthly Sunspot Bulletin.* Swiss Federal Observatory, Zurich.

Waldmeier, M. (1961). *The Sunspot Activity in the Years 1610-1960.* Swiss Federal Observatory, Zurich.

Warnes, J.J. and Ripley, B.D. (1987). Problems with likelihood estimation of covariance functions of spatial Gaussian processes. *Biometrika,* **74**, 640-642.

Wegman, Edward J. (1982). Kalman filtering. In *Encyclopedia of Statistics,* edited by N. Johnson and S. Kotz. John Wiley and Sons, New York.

Weisberg, S. (1974). An empirical comparison of the cumulative distributions of W and W'. *Biometrika,* **61**, 644-646.

Whittle, P. (1954). On stationary processes in the plane. *Biometrika,* **41**, 434-449.

Whittle, P. (1963) Stochastic processes in several dimensions. *Bulletin of the International Statistics Institute,* **40(1)**, 974-994.

Williams, J.S. (1979). A synthetic basis for comprehensive factor-analysis theory. *Biometrics*, **35**, 719-733.

Zimmerman, Dale L. (1989). Computationally exploitable structure of covariance matrices and generalized covariance matrices in spatial models. *Journal of Statistical Computation and Simulation*, **32**, 1-15.

Zimmerman, Dale L. and Cressie, Noel (1989). Improved estimation of the kriging variance. Technical Report #161, Dept. of Statistics, Univ. of Iowa, Iowa City, IA.

Zimmerman, Dale L. and Harville, David A. (1990). A random field approach to the analysis of field-plot experiments and other spatial experiments. *Biometrics*, **47**, 223–239.

Zimmerman, Dale L. and Zimmerman, M. Bridget (1990). A Monte Carlo comparison of spatial variogram estimators and kriging predictors. *Technometrics*, **33**, 77–91.

Author Index

Subject Index

Principal Components

Spatial Data and Kriging

TIME SERIES